*Series on*
# Reproduction and Development in Aquatic Invertebrates
Volume 5

# Reproduction and Development in Platyhelminthes

**T. J. Pandian**

Valli Nivas, 9 Old Natham Road
Madurai-625014, TN, India

CRC Press is an imprint of the
Taylor & Francis Group, an **informa** business

A SCIENCE PUBLISHERS BOOK

Cover page: Representative examples of platyhelminth species. For more details, see Figure 1.1

CRC Press
Taylor & Francis Group
6000 Broken Sound Parkway NW, Suite 300
Boca Raton, FL 33487-2742

© 2020 by Taylor & Francis Group, LLC
CRC Press is an imprint of Taylor & Francis Group, an Informa business

No claim to original U.S. Government works

Version Date: 20190924

International Standard Book Number-13: 978-0-367-34805-2 (Hardback)

This book contains information obtained from authentic and highly regarded sources. Reasonable efforts have been made to publish reliable data and information, but the author and publisher cannot assume responsibility for the validity of all materials or the consequences of their use. The authors and publishers have attempted to trace the copyright holders of all material reproduced in this publication and apologize to copyright holders if permission to publish in this form has not been obtained. If any copyright material has not been acknowledged please write and let us know so we may rectify in any future reprint.

Except as permitted under U.S. Copyright Law, no part of this book may be reprinted, reproduced, transmitted, or utilized in any form by any electronic, mechanical, or other means, now known or hereafter invented, including photocopying, microfilming, and recording, or in any information storage or retrieval system, without written permission from the publishers.

For permission to photocopy or use material electronically from this work, please access www.copyright.com (http://www.copyright.com/) or contact the Copyright Clearance Center, Inc. (CCC), 222 Rosewood Drive, Danvers, MA 01923, 978-750-8400. CCC is a not-for-profit organization that provides licenses and registration for a variety of users. For organizations that have been granted a photocopy license by the CCC, a separate system of payment has been arranged.

**Trademark Notice:** Product or corporate names may be trademarks or registered trademarks, and are used only for identification and explanation without intent to infringe.

---

Library of Congress Cataloging-in-Publication Data

Names: Pandian, T. J., author.
Title: Reproduction and development in platyhelminthes / T.J. Pandian.
Description: Boca Raton : CRC Press, [2020] | Series: Reproduction and
    development in aquatic invertebrates ; volume 5 | Includes
    bibliographical references and index.
Identifiers: LCCN 2019040753 | ISBN 9780367348052 (hardcover)
Subjects: LCSH: Platyhelminthes--Reproduction. |
    Platyhelminthes--Development.
Classification: LCC QL391.P7 P366 2020 | DDC 592/.4--dc23
LC record available at https://lccn.loc.gov/2019040753

---

Visit the Taylor & Francis Web site at
http://www.taylorandfrancis.com

and the CRC Press Web site at
http://www.crcpress.com

# *Preface to the Series*

Invertebrates surpass vertebrates not only in species number but also in diversity of sexuality, modes of reproduction and development. Yet, we know much less of them than we know of vertebrates. During the 1950s, the multi-volume series by L.E. Hyman accumulated some information on reproduction and development of aquatic invertebrates. Through a few volumes published during the 1960s, A.C. Giese and A.S. Pearse provided a shape to the subject of Aquatic Invertebrate Reproduction. During the 1990s K.G. Adiyodi and R.G. Adiyodi in their multi-volume series on Reproductive Biology of Invertebrates elevated the subject to a visible and recognizable status.

Reproduction is central to all biological events. The life cycle of most aquatic invertebrates involves one or more larval stage(s). Hence, an account on reproduction without considering development shall remain incomplete. With the passage of time, various publications are being produced in a large number of newly established journals on invertebrate reproduction and development. The time is ripe to update the subject. This treatise series proposes to (i) update and comprehensively elucidate the subject in the context of cytogenetics and molecular biology, (ii) view modes of reproduction in relation to Embryonic Stem Cells (ESCs) and Primordial Germ Cells (PGCs) and (iii) consider cysts and vectors as biological resources.

Hence, the first chapter on Reproduction and Development of Crustacea opens with a survey of sexuality and modes of reproduction in aquatic invertebrates and bridges the gaps between zoological and stem cell research. With capacity for no or slow motility, the aquatic invertebrates have opted for hermaphroditism or parthenogenesis/polyembryony. In many of them, asexual reproduction is interspersed within sexual reproductive cycle. Acoelomates and eucoelomates have retained ESCs and also reproduce asexually. However, pseudocoelomates and haemocoelomates seem not to have retained ESCs and are unable to reproduce asexually. This series provides a possible explanation for the exceptional pseudocoelomates and haemocoelomates that reproduce asexually. For posterity, this series intends to bring out six volumes.

**August, 2015**                                                      **T. J. Pandian**
**Madurai-625 014**

# *Preface*

Platyhelminthes are unique because of the presence of neoblasts, which serve as a model for studies on cancer and senescence. Of ~ 27,700 species, 77% of them are parasites; they are harmful to man, and cause heavy loss to his food basket from livestock and fish. The available books on them are limited to one or other taxonomic group or to a specific theme. This book is a comprehensive synthesis of biological aspects of reproduction and development in platyhelminths and covers from Acoela to taeniids.

The book is organized in seven Chapters followed by References, Author (776), Species (1,140) and Subject indices. In the opening chapter, an analysis has revealed that the driving force for the evolution and diversification of helminth lineages is motility rather than body size of vertebrate host. Parasitic flatworms are more fecund to compensate the risks involved in transmission. They are clothed with an immune-resistant tegument permeable to inward flow of the abstracted low molecular nutrients, especially in cestodes, bereft of a mouth and gut. With possession of pluripotent neoblasts, flatworms are unique to renew and turnover of the somatic cell types to maintain a dynamic steady state. Regarding proliferation in stem cells, microscopic and molecular evidence has revealed a basic difference between the turbellarians neoblasts and germinal/germinative cells of digeneans/cestodes; the former undertake symmetric mitosis, whereas the latter lacking *vasa* and *piwi* orthologs undergo asymmetric mitosis.

In three parts, Chapter 2 narrates the features of the 6,500 speciose Turbellaria and their potency for regeneration and clonal multiplication. Part 1 elaborates the unique ectolecithality, adaptive stages in transition from free-living to parasitism, contrasting strategies for production of subitaneous eggs during the favorable season and dormant eggs to tide over the unfavorable season, and energy budgets in semelpares and iteropares. Part 2 elucidates neoblasts in the context of cell types and mitosis, blastema formation and regeneration, and genes in regeneration. In clonal multiplication, notable is the ability to switch sexuality in either direction. Disappointingly, reports available on regeneration is limited to ~ 70 species and clonal multiplication to < 50 species.

A vast majority of 4,500 speciose monogeneans are ectoparasites on fish; their indirect life cycle includes a single larval stage, the oncomiracidium but

## vi   *Reproduction and Development in Platyhelminthes*

involves no intermediate host. They can coexist with intra- and inter-specific competitors; however, some dactylogyrid species competitively eliminate others. The 150 speciose Polystomatidae have radiated into amphibians. Monogenean fecundity, incubation, hatching and larval duration are all viewed from the angle of dermal, branchial and bladder types. Known for the oiogenecity, > 70% monogeneans are host-specific. With a short life span, direct life cycle and amenability to rearing, the gyrodactylids are academically interesting and as parasites of salmonids, they are economically important. Differences in stress hormone cortisol level in the hosts are shown as responsible for contradictory reports on development of resistance.

The fourth chapter deals with 12,012 speciose Digenea. The digenean life cycle is highly complicated with four larval forms: miracidium, sporocyst, redia and cercaria, and one to three intermediate hosts. Rarely, the sporocysts simultaneously produce rediae and miracidia; following serial transplantations to naïve snails, sporocyst and redia produce rediae for many generations lasting for a year; metacercaria undertakes two generations of propagatory multiplications. A synthesis of these findings has led to suggest that the potency for propagatory multiplication is retained in all the larval forms and either direction by sporocyst. This has also led to suggest that clonal multiplication may be more likely rather than the so called 'asexual' reproduction by polyembryony or parthenogenesis, for which 'more widespread and compelling evidences' are needed. The importance of higher level clonal selection in relatively more motile Second Intermediate Host (SIH) is highlighted; the selection in SIH may purge deleterious clones from the propagatory multiplications and enhance probability of the fittest clone for onward transmission. Interestingly, more than motility of SIH, it is the euryxenic flexibility of digeneans/the choice to select SIH species that is responsible for lineage diversification. It is in this context, that the need for estimation of the number of digenean species engaging SIH becomes obvious. For the first time, it is shown that of 12,012 digeneans, 88.3% or 9,496 species may engage as many as 33,014 SIH species. These digeneans have the choice to select one among the awaiting/available 3.5 SIH species. In parasitic flatworms, the short-lived oncomiracidium, miracidium and cercaria involve the riskiest transmission phases but are guided by chemical cues arising from the host. The transmission efficiency ranges from 0.0005 to 0.3%; in *Haematoloechus coloradensis* involving two penetrative larvae and two intermediate hosts, the efficiency is 0.03%. With the need to support propagatory multiplication in sporocyst and/or redia, the snails suffer heavy loss on survival, growth and reproduction. Within the shell, the loss of tissues incurred by snail provides space for the growing fluke larvae, i.e. 'substitution', or new space is added by faster shell growth, i.e. 'addition'. In this process, the snail may suffer 'stunting', or 'gigantism' following castration. The short-lived freshwater lymnaeids suffer greater loss than the long-living marine prosobranchs. Again, for the first time, the wide differences between individual based 'proximate' and population level 'ultimate' responses are brought to light.

The 4,671 speciose Cestoda are unique for the lack of a mouth and gut, and the presence of proglottids and trophic transmission. Processing the superabundant nutrients to a staggering number of eggs by repetitive proglottids each with dual reproductive system has neutralized the obstacles encountered during trophic transmission. In Chapter 5, the life cycle of cestodes is divided into aquatic and terrestrial patterns. The former includes (i) oncosphere type, involving a single intermediate host, in which procercoid and plerocercoid are developed and (ii) coracidium type, in which these larval stages are completed in two intermediate hosts. The terrestrial pattern includes (i) hexacanth-cysticercoid, (ii) hexacanth-tetrathyridium and (iii) hexacanth-cysticercus types. For the first time, the share for the oncosphere, coracidium and hexacanth types are estimated as 17.0, 29.5 and 46.5%, respectively. The highest fecundity and adoption of intermediate hosts in herbivorous/insectivorous food chain have enriched Taenioidea as the most (2,264) speciose order. Amenability to *in vitro* culture up to adult stage (< 10–15 days of adult life span) has provided a unique opportunity to study the mating system in the hermaphrodite *Schistocephalus solidus*. Worms reared in pairs have more sperm in their seminal receptacle, are equally fecund but with higher hatching success than those reared in isolation. Unusually, fitness of these offspring depends on other features rather than the product of selfing or outcrossing.

Plathelminthes are hermaphrodites. Sporadic incidence of gonochorism is strewn over all the major classes. In some of these gonochores, sex specific genes *Smed-dmd-1* and *macbol* have been identified. In the absence of endocrine glands and the circulatory system, sexes are differentiated by neuropeptides and dipeptides. The worms are also capable of endocrine disruption in the hosts.

Chapter 7 is devoted to compare and highlight neoblasts, clonal selection, progenesis and parasitic flatworm distribution. Identification of only a few turbellarians species with potency for both regeneration and clonal reproduction suggests the existence of neoblasts of different types. With unpredictable water levels, a few freshwater monogeneans and cestodes have abbreviated the life cycle. However, unpredictable water levels, temperature and salinity, and/or unavailability (in 41% of progenetic digenean species) of SIH and/or DH have triggered progenesis in > 30–40 digenean families. The majority of these flukes have chosen facultative progenesis in SIH, in which inbreeding depression is purged by clonal selection. The absence of operculum and hard 'skin' unsuitable for deployment of hooks are adduced as reasons for elimination of elasmobranchs as hosts. The inability of trematodes to suck the body fluid/blood containing high urea level (764–873 mM/l) has been brought to light for the first time. Some elasmobranch's that host monogeneans are visitors/residents in freshwaters, in which their urea level is as low as 260 mM/l. However, the cestodes are dominant in elasmobranchs intestine, viii *Reproduction and Development in Platyhelminthes*

in which urea level poses no problem. In ~ 27,700 speciose platyhelminths, the share is 23.5, 16.2, 43.4 and 16.9% for turbellarians, monogeneans, digeneans and cestodes, respectively. Within platyhelminths, the digeneans are more speciose than other taxa as (i) their higher fecundity is supplemented by propagatory multiplication in one or more larval stage(s). (ii) 88% digeneans engage more mobile SIH; they are highly flexible and euryxenic to select one among 3.5 species awaiting/available to serve as SIH and (iii) the higher level of clonal selection in SIH and the consequent introduction of genetic diversity in the larval flukes.

This book is a comprehensive synthesis of over 800 publications selected from widely scattered information from 227 journals and 45 other literature sources. The holistic and incisive analyses have led to harvest several new findings related to reproduction and development in platyhelminths and to project their uniqueness. Hopefully, this book serves as a launching pad to further advance our knowledge on reproduction and development in the flatworms.

**June, 2019**                                                                 **T. J. Pandian**
**Madurai-625 014**

# Acknowledgements

It is with pleasure that I thank Drs. R.D. Michael and E. Vivekanandan for critically reviewing parts of the manuscript of this book and for offering valuable suggestions. In fact, I must confess that I am only a visitor to the theme of this book. However, editorial service on energetic of Platyhelminthes (Pandian, 1987, *Animal Energetics*, Academic Press) has emboldened me to author this book. The manuscript of this book was prepared by Mr. T.S. Surya, B.Sc. and I wish to thank him profusely for his competence, patience and co-operation.

I wish to thank many authors/publishers, whose published figures are simplified/modified/compiled/redrawn for an easier understanding. To reproduce original figures from published domain, I gratefully appreciate the permission issued by American Naturalist and International Journal of Developmental Biology. I welcome and gratefully appreciate the open acces policy of BMC Developmental Biology, Experimental Parasitology, Folia Parasitologica, Journal of Oceanography and Marine Research and Korean Journal of Parasitology. For advancing our knowledge in this area by their rich contributions, I thank all my fellow scientists, whose publications are cited in this book.

**June, 2019**                                                    **T. J. Pandian**
**Madurai-625 014**

# Contents

| | |
|---|---|
| *Preface to the Series* | iii |
| *Preface* | v |
| *Acknowledgements* | ix |

**1. General Introduction** — 1

Introduction — 1
1.1 Taxonomy and Diversity — 5
1.2 Life Cycles — 14
1.3 Host-Parasite Distribution — 19
1.4 Fecundity — 26
1.5 Osmotrophism and Digestion — 34
1.6 Neoblasts — 40

**2. Turbellaria** — 47

Introduction — 47

**Part A: General Features** — 48

A 2.1 Taxonomy and Distribution — 48
A 2.2 Reproductive System — 50
A 2.3 Transition to Parasitism — 53
A 2.4 Contrasting Strategies — 55
A 2.5 Semelparity and Iteroparity — 57
A 2.6 Sexuality — 59
    A 2.6.1 Hermaphroditism — 59
    A 2.6.2 Gynogenesis—Pseudogamy — 64
    A 2.6.3 Gonochorism — 66

**Part B: Regeneration** — 68

Introduction — 68
B 2.7 Distribution and Potency — 69
B 2.8 Cell Counts and Mitosis — 72
B 2.9 Blastema and Regeneration — 75
B 2.10 Genes in Regeneration — 79

xii  *Reproduction and Development in Platyhelminthes*

**Part C: Clonal Reproduction**                                    **80**

Introduction                                                        80
C 2.11 Surgery and Potency                                          81
C 2.12 Clonal Types                                                 83
C 2.13 Switching Sexuality                                          85
C 2.14 Telomere and Senescence                                      90

3. **Monogenea**                                                   **93**

Introduction                                                        93
3.1 Taxonomy and Distribution                                       94
    3.1.1 Taxonomy                              94
    3.1.2 Distribution                          95
3.2 Life Cycle and Characteristics                                 113
    3.2.1 Life Cycle                           113
    3.2.2 Egg Assemblage and Fecundity         117
    3.2.3 Incubation and Hatching              121
    3.2.4 Oncomiracidium and Dispersal         123
    3.2.5 Growth and Maturation                124
3.3 Host Specificity                                               126
    3.3.1 Specificity and Range                126
    3.3.2 The Gyrodactylids                    127
    3.3.3 Susceptibility and Resistence        133

4. **Digenea**                                                    **138**

Introduction                                                       138
4.1 Taxonomy: Parasites and Hosts                                  138
4.2 Larval Forms                                                   140
4.3 Intermediate Hosts                                             146
4.4 Clonal Selection                                               153
4.5 Amphiparatenic Transmission                                    154
4.6 Polyembryony and/or Parthenogenesis                            155
4.7 Sequence and Potency                                           158
4.8 The Transmission                                               159
4.9 'Short is Sweet'                                               170
4.10 Prevalence and Intensity                                      173
4.11 Causes and Losses                                             177
    4.11.1 Activity and Survival               177
    4.11.2 Growth                              179
    4.11.3 Reproduction                        186

5. **Cestoda**                                                    **194**

Introduction                                                       194
5.1 Taxonomy and Diversity                                         195
5.2 Reproductive Systems                                           196

Contents xiii

| | | |
|---|---|---|
| 5.3 | Life Cycles | 198 |
| | 5.3.1 Larval Forms | 198 |
| | 5.3.2 Ontogeneses and Life Cycles | 200 |
| | 5.3.3 Intermediate Hosts | 209 |
| 5.4 | Neoteny—Progenesis | 212 |
| 5.5 | Prevalence and Intensity | 214 |
| 5.6 | *Schistosocephalus solidus* | 216 |
| 5.7 | Clonal Multiplication | 220 |
| 5.8 | The Transmission | 222 |
| 5.9 | Fishes and Losses | 224 |

## 6. Sexualization — 227

| | | |
|---|---|---|
| | Introduction | 227 |
| 6.1 | Chromosomes and Genes | 227 |
| 6.2 | Endocrine Differentiation | 229 |
| 6.3 | Endocrine Disruption | 230 |

## 7. Comparison and Highlights — 234

| | | |
|---|---|---|
| | Introduction | 234 |
| 7.1-A | Neoblast Types | 234 |
| 7.1-B | Clonal Selection | 236 |
| 7.1-C | Progenesis | 237 |
| 7.2 | Habitat Distribution | 242 |
| 7.3 | Diversification and Speciation | 246 |

## 8. References — 249

*Author Index* — 283

*Species Index* — 291

*Subject Index* — 302

*Author's Biography* — 305

# 1

# *General Introduction*

## Introduction

THE phylum Platyhelminthes comprises free-living flatworms (turbellarians), and parasitic flukes (ectoparasitic monogeneans and endoparasitic digeneans) and tapeworms (endoparasitic cestodes). The platyhelminths are bilaterally symmetrical, dorso-ventrally flattened, triploplastic, acoelomic, mostly hermaphroditic, soft-bodied worms with either branched 'blind gut' tasked with digestion and distribution of nutrients or no gut (Fig. 1.1). They lack respiratory and circulatory systems and instead rely on diffusion to obtain oxygen. The following hallmark traits elevate them as unique in animal kingdom: (i) hermaphroditism, (ii) ectolecithal egg production, by separation of the ovary and vitellogenic glands and (iii) renewal of all somatic cell types, which cannot undergo mitosis, once their differentiation is completed (see Section 2.8). Consequently, the flatworms exist in a dynamic steady state between the single proliferating cell type and multiple short-lived differentiated cell types (Rink, 2013).

Thanks to the presence of pluripotent neoblasts, the turbellarian planarians are known for the extraordinary potency for regeneration and clonal reproduction. The planarian neoblasts serve as a model system to understand the proliferative control of stem cell, which is relevant to human cancers. Firstly, studies on planarian ortholog of the human tumor suppressor PTEN have revealed that RNAi-mediated knockdown leads to neoblast hyper-proliferation of the TOR signaling pathway. These disease-relevant genetic pathways in planarians may provide many keys to the homeostatic control of cell proliferation. Secondly, senescence in animals has been traced to the progressive shortening of telomeres following each round of DNA replication. Neoblasts seem to have overcome the end of replication problem indefinitely to facilitate their immortality. Hence, planarians may serve as a potential model system to understand how the diseased, damaged and ageing tissues can be regenerated (see Aboobaker, 2011).

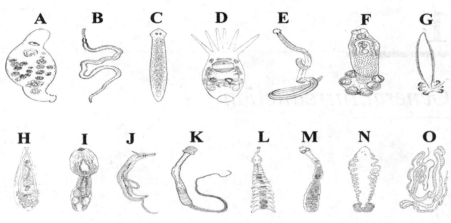

**FIGURE 1.1**

A. Ectocommensalic acoelid *Ectocotyle paguri* (after E.G. Richard), B. rhabdocoel *Microstomum*, C. *Dugesia tigrina* (from Hyman, 1951), D. *Temnocephala* (after Haswell, 1893), E. Land planarian *Bipalium kewense* and F. *Polystomoides* (after Stunkard, 1917). G. *Rajonchocotyle* (after Muller and Van Cleave, 1932), H. *Opisthorchis sinensis* (after A.E. Galigher), I. *Diplostomum* (after Dubois, 1938), J. *Schistosoma haematobium* (after Looss, 1990). K. *Lithobium aenigmaticum* (from Caira et al., 2014), L. *Taenia* (after Olsen, 1939), M. *Echinococcus granulosus* (after Southwell, 1930), N. *Gyrocotyle* and O. *T. saginata* (from Hyman, 1951) (All figures are free hand drawings).

Many vertebrate species serve as definitive or final host for the approximately 21,330 speciose flukes and tapeworms. The digenean flukes and cestodes engage a large number of intermediate host species. In these parasites, sexual reproduction occurs in vertebrate hosts but parthenogenic (or polyembryonic, see Galaktinov and Dobrovolskij, 2003, Littlewood et al., 2015, clonal, see p 158) multiplication occurs in invertebrate hosts, of which most of them are molluscs, and rarely annelids. For example, a single miracidium, the first larva to emerge from a fertilized egg of the digenean flukes after passing through propagatory multiplication in sporocyst and/or redia, is capable of producing more than a million cercariae (e.g. strigeids, see Hyman, 1951), the final or penultimate larva to infect the vertebrate host. The number of eggs and protoscoleces produced by a hydatid cyst of *Echinococcus granulosus* measuring 5 ml is an astonishing 25 million/day (Smyth, 1964).

The following examples may reveal the magnitudes of health problems emanating from these platyhelminth parasites: (i) Schistosomiasis, caused by the blood fluke *Schistosoma* spp, is the second most common disease in the world and is widespread in 70 subtropical countries. More than 200 million people are infected by the disease and an estimated 779 millions are at the risk of infection and ~ 200,000 die annually (see Pandian, 2017). Whereas

---

Name of most animal species are listed following Worms-World Register of Marine Species; however, some are named, according to author's citation.

General Introduction 3

not more than 0.7% die following malarial infection, ~ 10% of schistosome-infected patients succumb to death. Hence, the global expenditure on schistosomiasis amounts to US$ 67 million. Between 1980 and 2014 alone, 14,933 research articles were published on these parasites (Jurberg and Brindley, 2015). Available information shows the enormous cost and/or loss incurred by some countries and the world at large due to the platyhelminthic diseases to humans and their food basket from livestock and fish (Table 1.1). The loss suffered by livestock is assessed from the reduced yield of milk and meat, growth and fertility as well as increased morbidity and mortality. Global economic losses due to fasciolosis are estimated at ~ US$ 2.5 billion annually. A more recent study has estimated the loss of US$ 4.9 billion for India alone (see McCusker et al., 2016). Notably, the implementation of hygienic practices ranges from condemnation of fasciolosis-infected liver in a small developing African Ethiopia to others, where the meat/beef is not inspected and certified prior to marketing. For example, no information is available on implementation of hygienic practices on marketed meat in large developing Asian country like India. Sangunicoliasis is an important disease caused by *Sanguinicola inermis* in aquaculture farms. Unfortunately, the debilitated fish cannot be treated (see Pandian, 2017). An estimate suggests the possible presence of 25,000 monogenean parasitic species, i.e. the monogeneans are as speciose as teleostean fish are (Whittington, 1998). Despite the great aquaculture potency of China and India, Brazil seems to be the only country, which has made estimates on the loss due to ectoparasitic crustaceans and monogeneans in aquaculture farms. Briefly, the volume of loss due to health problems posed to human, livestock and fish is so huge and fluctuates widely between countries. Food and Agriculture Organization (FAO) has not been able to make a global estimate, as many countries either do not implement hygienic practices to check the health status of meat/beef or do not have valid data on the loss. Notably, aquaculture is perceived to have the greatest potential to produce more and good quality aquatic food (FAO, 2014). In aquaculture farms, efforts are made to produce more and more fish with less and less water. A consequence of this can increase the scope for infection by monogenean parasites. The foregone account may impress on the relevance of turbellarian stem cells for studies on cancer and senescence as well as that of parasitic trematodes and cestodes on the cost of human health and loss to his food basket from livestock and fish. Among invertebrates, Platyhelminthes ranks first, as they are harmful to man and cause heavy loss to his food basket.

Not surprisingly, many books on platyhelminths have been authored or edited. However, most of them are limited to one or other taxonomic group(s) or to a specific theme. For example, half a dozen books are available on regeneration and clonal reproduction in turbellarians (e.g. Brondsted, 1969); a few are also available on the biology and evolution of the flukes (e.g. Galaktinov and Dobrovolskij, 2003) and diseases and management of tapeworms (e.g. Muller, 2001). This book is a comprehensive synthesis of

4  *Reproduction and Development in Platyhelminthes*

## TABLE 1.1

Representative examples for the incidence and estimated cost or loss caused by parasitic platyhelminths. mil = million, bil = billion, y = year

| Disease | Host | Country | Reported observations | Reference |
|---|---|---|---|---|
| **Incidence** | | | | |
| Schistosomiasis | Human | | 200 mil infected, 0.2 mil die | see Pandian (2017) |
| Fasciolosis | Cattle | Global | 1.7 mil bovines infected | |
| | Human | Global | 35 mil people infected, especially in the Chinese countries | Lim (2011) |
| | Cattle | India | 30–80% incidence | Gupta and Singh (2002) |
| Helminths | Human | Global | 40 mil infected | A. Payne (Facebook) |
| | Poultry | Egypt | 14% (geese), 38% (ducks), 42% (turkeys), 46% (fowls) and 52% pigeons are infected | Nagwa et al. (2013) |
| **Cost and Loss** | | | | |
| Fasciolosis | Human | USA | Costs 100 mil US$/y | Roberts et al. (1994) |
| | Sheep | Ethiopia | Loss amounts to 114,678 $/y | Ayalneh et al. (2018) |
| | Sheep & Goats | Global Australia | Loss is 2.5 bil €/y Loss is 50–80 mil €/y | Love (2017) Boray (2007) |
| | Cattle | Ethiopia | Infected liver condemnation costs 8312 $/y | Abebe et al. (2010) |
| | Cattle | Switzerland | Loss @ 299 €/infected cattle/y costs 50 mil €/y | Dorchies (2007), Schweizer et al. (2005) |
| | Cattle | England | Loss is 23 mil £/y | Mazeri et al. (2016) |
| | Cattle | USA | Loss is 5 mil US$/y | Malone (1986) |
| | Cattle | Uttarkhand, India | 900 mil Rs/y on milk yield alone in a small state | Bardhan et al. (2014) |
| | Cattle | India | Estimated loss is 4.86 bil/y | McCusker et al. (2016) |
| Taeniasis | Human | USA | Cost 0.3 mil $/y with lowest (0.06%) Incidence. cf 12% incidence in Laos | Roberts et al. (1994) |
| Cysticercosis | Human | USA | 7% incidence costing 0.8 mil $/y | Roberts et al. (1994) |
| Parasites | Fishes | Global Brazil | Loss is 84 mil $/y Loss is 5 mil $/y | Tavares-Dias and Martins (2017), Shinn et al. (2015) |

relevant aspects of reproduction and development in platyhelminths covered from Acoela to Taenioidea. With availability of voluminous literature, it provides a 'snap-shot' of biological aspects of platyhelminths rather than an in depth or exhaustive account on diagnostics and management of platyhelminth diseases.

## 1.1 Taxonomy and Diversity

According to Hyman (1951), the phylum Platyhelminthes is classified into three classes namely Turbellaria, Trematoda and Cestoda; the trematodes are divided into two major orders: Monogenea and Digenea (Table 1.2). Further divisions of them have been subjected to many changes, which are described in the respective chapters.

*Phylogeny* traces the origin and evolution of taxonomic groups. Of theories proposing the origin of polyclad from ctenophores and acoeloid from planuloids, the second one is more acceptable to Hyman (1951). On the other hand, with the presence of simple pharynx, entolecithal eggs and absence of vitellaria, rhabditiphores are considered as the most basal taxon (Ehlers, 1985). There is unanimity in considering the monophyletic origin of the parasitic Monogenea, Digenea and Cestoda, collectively named as Neodermata and paraphyletic origin of the free-living Turbellaria. According to Laumer and Giribet (2014) and Ramm (2017), the neodermatans have originated from the turbellarian Bothrioplanida (Fig. 1.2B, C). With an indirect life cycle, polyclads are regarded as the ancestral condition for platyhelminths (Jagersten, 1972). In them, the Muller or Gotte larva is the dispersing stage. The secondary neodermatan larvae descended from the primary polyclad larva. The neodermatans have evolved elaborate life cycles with distinct larval stages. The first larval stage in the life cycle of neodermatans is the dispersing oncomiracidium in monogeneans, miracidium in digeneans and coracidium/oncosphere in cestodes. These larvae swim with a plate or bands of cilia. Based on these larval features, Rawlinson (2014) developed a phylogenetic cladogram (Fig. 1.2A). Ramm (2017) also made a similar one (Fig. 1.2B), based on free-living or parasitic mode of life. Notably, these phylogenetic cladograms exclude Acoela.

A peculiar feature of the phylum Platyhelminthes is the separation of the female gonad into two structures, the ovary proper and the yolk or vitelline glands. Whereas the ovary provides yolk and incorporates it into the egg in other animals, the eggs of vast majority of platyhelminths are devoid of the yolk, which is supplied by special yolk cells. Accommodated into the egg shell or capsule, these yolk cells accompany the egg and furnish nutrients to developing embryo (Hyman, 1951). Considering this unique feature, Laumer and Giribet (2014) subdivided the Platyhelminthes into the earliest branching

6  *Reproduction and Development in Platyhelminthes*

**TABLE 1.2**

Condensed version of systematic resume of Phylum Platyhelminthes (compiled from Hyman, 1951) 20,000 species (Chapman, 2009), 30,000 species (Caira and Littlewood, 2013), 100,000 species (Park et al., 2007)

---

Class: Turbellaria 3,000 species (Barnes, 1974), 3,416 species, after 100% checking of 6,376 nominal species including synonyms (Tyler et al., 2018), 5,500 species (Schockaert, 1996), 6,500 species (http://turbellaria.unimaine.edu)

    Order: Acoela, *Nemertoderma bathycola*
    Order: Rhabdocoela, *Macrostomum gigas*
    Order: Alloeocoela, *Baicalarctia gulo*
    Order: Tricladida, *Dugesia tigrina*
    Order: Polycladida, *Stylochus zebra*        800 species (Rawlinson, 2014)

Class: Trematoda 6,000 species (Barnes, 1974), 12,012 species (Littlewood et al., 2015)

    Order: Heterocotylea (Monogenea) ~ 4,500 species (Reed et al., 2012), 25,000 species (Whittington, 1998)
    Order: Aspidocotylea, *Stichocotyle nephrops* 25 species (see Littlewood et al., 2015)
    Order: Malacocotylea (Digenea), *Schistosoma haematobium* 10,000 species (Cribb et al., 2003), 12,012 species (Littlewood et al., 2015), 14,000 species (Gardner, 2002), 25,000 species (Esch et al., 2002)

Class: Cestoda, 4,000 species (Schmidt, 1986), 4,671 species (Littlewood et al., 2015)

  Subclass: Cestodaria
    Order: Amphilinidea, *Amphillina foliacea* 15 species (Littlewood et al., 2015)
    Order: Gyrocotylidea, *Gyrocotyle nybelini* 10 species (Littlewood et al., 2015)

  Subclass: Eucestoda (4,646 species, Littlewood et al., 2015)
    Order: Tetraphyllidea, *Phyllobothrium dohrnii*
    Order: Lecanicephalaoidea, *Tylocephalum*
    Order: Proteocephaloidea, *Proteocephalus*
    Order: Diphyllidea, *Echinobothrium benedeni*
    Order: Trypanorhyncha, *Haplobothrium globuliforme*
    Order: Pseudophyllidea, *Abothrium gadi, Ligula intestinalis*
    Order: Nippotaeniidea, *Nippotaenia*
    Order: Taenioidea, *Hymenolepis nana, Teania saginata*
    Order: Aporidea, *Nematoparataenia*

---

lineages constituting the paraphyletic Archoophora and the more divergent monophyletic Neoophora. Accordingly, Archoophora includes three orders characterized by endolecithal eggs namely the Catenulida, Polycladida and Macrostomorpha (see Egger et al., 2006) and Neoophora including all other taxonomic groups with ectolecithal eggs (Fig. 1.2C).

The advantages and constraints of parasitism involving exploitation of 'living environment' impose several effects on the evolution of parasite life history traits. For example, most parasites are more fecund than free-living species (p 23). Using selected life history traits namely (i) adult size, (ii) progeny volume, (iii) daily fecundity, (iv) prepatency, i.e. age/size at sexual maturity and (v) adult longevity (see also Table 1.5), Trouve et al. (1998) also

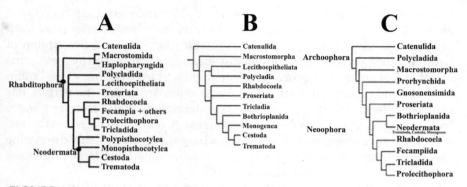

**FIGURE 1.2**
Phylogenetic cladograms within major taxonomic groups, especially as assembled by (A) Rawlinson (2014), (B) Ramm (2017) and (C) Laumer and Giribet (2014) (all are free hand drawings).

assembled a species specific cladogram. Clearly, phylogenetic cladograms assemblages based on structural and life history traits vary from each other. An unquestionable phylogeny may have to be assembled for Platyhelminthes using molecular characteristics. Based on a complete small subunit rDNA and partial (D1-D3) large subunit rDNA sequences, Cribb et al. (2003) assembled a less varied and more reliable phylogenetic cladogram for Digenea. This aspect of DNA based phylogeny is elaborated in the respective chapters.

*Species richness*: The phylum includes two diverse groups: (i) free-living flatworms and (ii) parasitic flukes and tapeworms. The differences in the reported species number are far wider for parasitic groups (4,000–25,000) than that for the free-living flatworms (3,000–6,500). Notably, > 77% of the described and to be described platyhelminths are parasites. The outer surface of free-living turbellarians is lined with a single layer of ciliated columnar epithelium, facilitating the worms to swim through the water column or to glide over the substratum. The inclusion of a syncytial integument in the place of a single layered epidermis is a key innovation of neodermatans (see Section 1.5). The neodermatans are known to infect every vertebrate species on earth and represent a single most successful transition to parasitism in the animal kingdom (Poulin and Morand, 2000). Covering the entire body, the syncytial integument protects the parasite against immune system, and extremes encountered in the gut and associated organs and blood of the host. Further, it serves also as a conduit for the worms to steal nutrients from the host, especially cestodes with no mouth and gut (Collins, 2017).

Parasites are ubiquitous. An estimate suggests that ~ 50% of known animal species are parasites (Windsor, 1998), implying that all free-living metazoans harbor at least one parasite species (Poulin and Morand, 2000). Numerous host species like deep water fish species are yet to be discovered (Poulin, 1996a). During the last decade alone, hundreds of new amphibian species

8 *Reproduction and Development in Platyhelminthes*

have been described and the increasing trend shows no sign of slowing down. New parasite species are recognized only after their host species have been described. Understandably, the trends for the discovery and description of new parasite species are also consistently increasing. Parasitism by metazoans on other metazoans has evolved independently at least 60 times (see Poulin and Morand, 2000). In contrast to nematodes, in which parasitism has evolved independently several times, neodermatans that infect vertebrates have arisen from a single evolutionary event (Collins, 2017). Nevertheless, several parasite lineages of neodermatans have diversified greatly during the checkered history of evolution and are now represented by a large number of distinct species (Poulin and Morand, 2000).

Since the 1850s, the trend for description and erection of new species has been consistently increasing in many aquatic invertebrate phyla (e.g. Echinodermata, see Pandian, 2018, Annelida, see Pandian, 2019). The Platyhelminthes are not an exception to this dictum. However, the number of described platyhelminth species from the 2010s differs widely from 20,000 to 100,000 (Table 1.2). This is also true for each taxonomic group within Platyhelminthes. The values differ from 3,000 to 6,500 for Turbellaria, from ~ 4,500 to 25,000 for Monogenea, 10,000 to 25,000 for Digenea and 4,000 to 4,671 for Cestoda. The reasons for these wide differences in the number may be traced to (i) description of more and more host species, (ii) erection of new parasite species without the description of the life cycle including intermediate host(s) and (iii) different approaches to characterize a species. By convention, helminth parasite species are erected following description based solely on adult worms. As platyhelminths are hermaphrodites, there is no need to describe female and male sexes. However, most of them have to complete an indirect life cycle, in which different stages must infect or accidentally be ingested in a specific ontogenetic sequence of intermediate hosts to complete a single generation. For example, the cestode *Schistocephalus solidus* is transmitted by accidental ingestion of its egg by the copepod *Cyclops* or *Macrocyclops*, which, in turn, is ingested by the stickleback *Gasterosteus aculeatus* and finally ingested by definitive host, the gray heron *Ardea cinerea* (see Blasco-Costa and Poulin, 2017). Similarly, the digenean trematode *Allocreadium fasciatusi* is transmitted by an infected snail *Amnicola travancorica* releasing a large number of miracidia, which infect the copepod *Mesocyclops leuckarti, Microcyclops varicans* or *Macrocyclops distructus*; in these second intermediate hosts, cercariae encyst as metacercariae, which develop into adults on being ingested by the definitive host *Aplocheilus melastigma* (Madhavi, 1978). An author like Dr. R. Madhavi has to be an expert in taxonomy of snails, cyclops and fish for describing the complete life cycle of a digenean species. Besides, the following points may be noted: (1) With vertebrates serving as a definitive host, the scope for dissemination of the parasite is enormously increased (see Shoop, 1988). (2) With the inclusion of propagatory multiplications in the first intermediate molluscan host, the number of infective cercariae/metacercariae is tremendously increased.

(3) However, the inclusion of one or more intermediate host(s) also poses the risks involved in transmission of the parasite to appropriate host(s) and time. Hence, the investigation and description of complete indirect life cycles of the parasite species throw a great challenge to the taxonomists. Not surprisingly, of ~ 1,000 described cestode species parasitizing elasmobranchs worldwide, complete life cycle is established only for 5 species (Caira and Jensen, 2014). Likewise, of 326 trematode species and another 1,000 to be discovered from fishes of the Great Barrier Reef (Cribb et al., 2014, 2016), complete life cycles have been resolved only for 4 species (Downie and Cribb, 2011). Hence, the cycle is known for only 0.3–0.4% of the described digenean and cestode species. Incidentally, the indirect life cycle is resolved only for < 3% of the described free-living aquatic invertebrates species (e.g. Annelida, see Pandian, 2019).

According to Blasco-Costa and Poulin (2017), different approaches are made to resolve the life cycles of parasites. The widely used first and simplest approach involves the morphological matching of the larval and adult forms that are found in a specific geographic location. For example, Hassanine (2006) described the complete life cycle of the lepocreadiid digenean *Diploproctodaeum arothroni* through (i) oyster *Crassostrea cucullata* and (ii) tetradontid fish *Arothron hispidus*—all the two hosts of the parasite species are resident of a lagoon within the mangrove swamp on the Egyptian coast of the Gulf of Aqaba. However, the matching may encounter the following difficulties: (1) With increasing mobility of the definitive host, for example, the aquatic birds (e.g. the gray heron 38 km) and humans, the area of geographic location is considerably enlarged. (2) A single parasite species may engage more than one intermediate host and a single species may serve as an intermediate host for many parasitic species. From molluscan database, Faltynkova et al. (2016) indicated that *Lymnaea stagnalis*, *Planorbis planorbis*, *Radix peregra* and *R. ovata* can serve as intermediate hosts for as many as 41, 39, 33 and 31 digenean species, respectively. A single digenean species has been recorded from 14 host species. (3) With the increasing complexity of the life cycle, diversity amidst the cercarial clones of 'propagatory' multiplication may induce morphological variations (Lagrue and Poulin, 2009).

The second approach is to elucidate the life cycle through experimental infection of one or more hosts. Typically, larvae collected from an intermediate host are fed to the suspected definitive host species, in which the adult form has been found or putative hosts are exposed to free-living infective stages (Blasco-Costa and Poulin, 2017). When natural host species cannot be used, it can be substituted by a laboratory model (e.g. rat, chick or guppy). Through this method, only a part of the life cycle can be resolved. Where there are logistical or ethical difficulties, culture media can be used to *in vitro* rearing of the parasite. Culture media that re-create conditions in the gut of birds have been used to grow adults of cestodes (Presswell et al., 2012) and trematodes (Presswell et al., 2014) to trace a part of the life cycle.

10  *Reproduction and Development in Platyhelminthes*

The third approach involves comparison of gene sequences such as mitochondria cytochrome *c* oxidase subunit I (COI) or the Internal Transcribed Spacers (ITS1 and ITS2) from the life stages. The basis is that different life stages of the parasite are different manifestations of the same genome expressed in an ontogenetic sequence. Using ITS2 rDNA sequences generated for the larval trematode *Gorgocephalus yaaji* from the snail *Echinolittorina austrotrochoides*, Huston et al. (2016) showed for the first time that the sequences are identical in both the encysted metacercariae on algae and adults of *G. yaaji* in the definitive host fish *Kyphosus cineraseans*. Using the sequence from 25 species of cestodes collected from elasmobranchs and 27 larval cestode species recovered from teleosts (46 species), molluscs (24 species) and crustaceans (5 species) collected from the same area in the Gulf of Mexico, Jensen and Bullards (2010) were able to assign some of them only up to the genera. It is not clear whether the cestodes are not amenable to the DNA techniques or being at the nascent stage, the technique is not precise enough to assign adults and larval stages to the species level (see also Hansen et al., 2007).

*Species number and publications*: The first dataset of Poulin and Presswell (2016) provides information on the number of species discovered and also described the life cycle. However, it is limited to (i) Digenea and Cestoda, (ii) publications in *Journal of Parasitology* and *Systematic Parasitology* and (iii) for 35 years from the 1980s onwards. For trematodes, the number of new species described increased from 15/year during the 1980s to the peak of 47/year between the 1990s and 2000s, and is then leveled at 40/year (Fig. 1.3A). For cestodes, it increased from 8/year during the 1980s to 40/year during the 2000s and subsequently leveled around 30/year (Fig. 1.3A). Lefebvre et al. (2009) established an electronic database for the cyclophyllidean family Hymenolepididae to describe life cycles of 230 valid species. There are > 90 publications describing the life cycle of one species, about a dozen for 2–10 species and < 1 publication for a few others. Hence, the parasite taxonomists endeavored to increase the number of publications on species description but not for elucidation of life cycle. As a consequence, the life cycle described for trematodes has remained ~ 5/year but that of cestodes lamentably at 2–3 species/year and that too in 4 'island years' (see dark spots at the bottom of Fig. 1.3A, B). Information on cumulative increase in the number of species description is also available, *albeit* limited to trematodes (Cribb et al., 2014). The Great Barrier Reef, the world's largest coral reef system stretches over 1,800 km and holds the richest array of marine life. Though the presence of the fish–Digenea combinations are known for 100 years, a systematic study has been made only during the last 25 years. The cumulative number of described trematode species increased from ~ 94 to 330 in 2010 and is predicted to increase to 2,270 species.

With regard to publications on monogeneans within Trematoda, comparable information is not available. However, an attempt has been

General Introduction    11

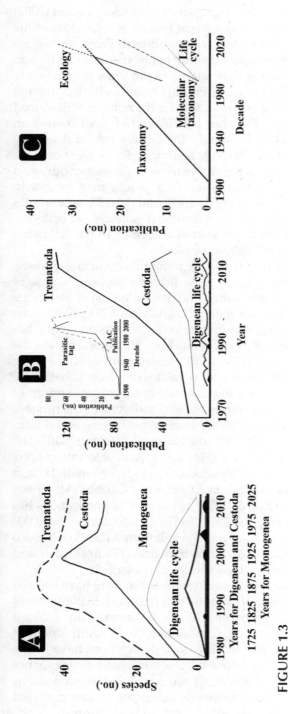

### FIGURE 1.3

(A) Number of species described for Monogenea, Trematoda and Cestoda. For monogenean species, values are drawn from Poulin (2002, 2005), trends for cestode and trematode species are compiled from Blasco-Costa and Poulin (2017). (B) Number of publications on Trematoda and Cestoda (compiled from Blasco-Costa and Poulin, 2017) and (in window) using parasitic tag (redrawn from Timi and Mackenzie, 2015). (C) Projected trends in publication numbers from taxonomic, molecular taxonomic, life cyclic and ecological approaches on helminth investigations in LAC countries (modified and redrawn from Aguirre-Macedo et al., 2016).

12  *Reproduction and Development in Platyhelminthes*

made to draw a trend for them. In his impressive publication, Poulin (2002) described the body length of 1,131 monogenean species as a function of the year, in which the relevant information was published. From the values shown in his Fig. 1, it is possible to count the number of species as a function of the year of publication. However, it is related to the years from as early as 1775 to 1980. In a previous publication, Poulin (1996a) indicated that the described trematodes species as of 1995 is > 1,000, i.e. the number of described species increased at the rate of ~ 50/year. It is likely to be not more than 10 for the monogenean species. Interestingly, the current rate of discovery is 2–3 species/year (Hansen et al., 2007). Accordingly, Fig. 1.3A shows the rapidly increasing trends during the early years and subsequent gradual decrease and leveling at ~ 3 species/year. Two points may be noted: Historically, it has been relatively easier to discover and describe new species in ectoparasitic monogeneans than in endoparasitic digeneans and cestodes. The number of species discovered and erected decreases in the following descending order: Monogenea > Digenea > Cestoda.

Ever since Harrington et al. (1939) used a naturally occurring parasite to investigate the stock structure of a marine fish, the parasite tags were regarded as a valuable tool for identification and determination of the degree of relation between populations. A chronological analysis of 294 publications on the use of parasites as biological tags for marine fishes indicates that it has steadily increased in the number of publications from 10/decade during the 1960s to 75/decade during the 2000s (Fig. 1.3B). However, the overwhelming majority of publications are concerned with parasites of teleost fishes; there are only a few on elasmobranchs and intermediate hosts such as shrimp of commercial interest. The described trends are also reflected in the number of publications. The publication number for trematodes has increased from ~ 15/year during the 1970s to ~ 120/year during the 2010s (Fig. 1.3B). The values are 3 and 48/year for cestodes (Fig. 1.3B). Available information on helminth parasites including Trematoda, Cestoda, Nematoda and Acanthocephala from 18 countries of South America and Caribbean has been examined by Aguirre-Macedo et al. (2016). The number of publications has increased from ~ 2/decade during the 1910s to 75/decade during the 2000s (Fig. 1.3B). Of these publications, 33, 20 and 8% hail from Argentina, Mexico and Brazil, respectively. As of 2010–2015, 136 trematode (76 freshwater and 60 marine) species and 14 marine cestode species are described.

Suffering from schistosomiasis, the South Americans seem to have evinced much interest in helminths. It is also possible to trace the trends in publications dealing with four different approaches of (i) taxonomy, (ii) ecology, (iii) molecular biology and (iv) life cycle. Considering the 18 South American and Caribbean countries, molecular and life cycle descriptions have begun to appear only from the 1990s, but those on taxonomy and ecology since the 1900s and 1980s, respectively (Fig. 1.3C). Encouragingly, the publications covering all the four approaches are consistently increasing and are projected to increase further in some of these developing subtropical countries.

*Structural diversity*: This description is based on Hyman (1951) and is limited to (i) external characters and (ii) internal blind gut. In general, freshwater turbellarians inhabit still waters. Strong currents elicit a clamping reaction and increased adhesion to the substratum. The adhesion is aided by (i) marginally distributed glandulo-epidermal adhesive organs (e.g. *Bdelloura candida*, Fig. 1.4M), (ii) glandulo–muscular adhesive organ present in the families of Dendrocoelidae and Kenkidae (e.g. *Kenkia rhynchida*, Fig. 1.4D) and (iii) muscular adhesive organ called acetabulum or sucker (present only in some triclads of the Lake Baikal (e.g. *Polycotylus validus*, Fig. 1.4N).

The true sucker is characteristic of the trematodes. Located at the posterior end, it aids the monogeneans to attach and/or cling to the gills of fishes (Fig. 1.4F). Besides adhesive glands, it is also armed with hooks and claws (Fig. 1.4E). To resist the peristaltic propulsion, the need for muscular suckers in the endoparasitic digeneans and cestodes is obvious. In digeneans, two suckers are present, the anterior oral sucker encircling the mouth and posterior acetabulum (Fig. 1.4G). These suckers lack hooks and spines. Bucephalids lack an acetabulum (Fig. 1.4O). In eucestodes, the knob-like or elevated head, the scolex bears the organ of attachment. The sucking scolex

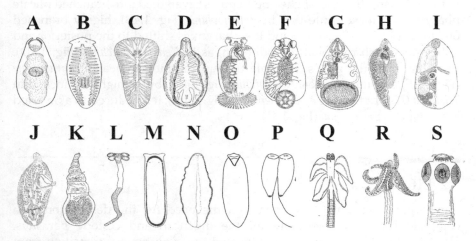

**FIGURE 1.4**

The 'blind gut' of platyhelminths. A. rhabdocoel with rosulate pharynx B. *Bothrioplana* with branched plicate pharynx and forked intestine, C. Highly branched gut of a polyclad and D. cave planarian *Kenkia rhynchida* (from Hyman, 1951), E. Monogenean *Acanthocotyle* (from Hyman, 1951) and F. *Tristoma* (after Goto, 1894), G. Digenean *Cladorchis* with pharyngeal sacs (after Fischoedor, 1903), H. *Fasciola hepatica* with highly branched gut (after A.E. Galigher), I. *Spelotrema* (after Rankin, 1940). J. Cestode *Phyllobothrium* (from Hyman, 1951), K. *Devainea proglottina* (after Jayeaux and Baer, 1936) and L. *Phyllobothrium dohrnii* (after Curtis, 1906), M. *Bdelloura candida*, N. *Polycotylus validus*: note the marginal suckers (from Hyman, 1951), O. *Prosorhynchus* (after Ozaki, 1928), P. Scoleces of *Bothridium pythonis* (after Southwell, 1931), Q. *Myzophyllobothrium* (after Shipley and Hornell, 1906), R. Scolex of *Echinobothrium* (after Linton, 1887) and S. *Taenia solium* (after Southwell, 1930). (All are free hand drawings).

14  *Reproduction and Development in Platyhelminthes*

is of three types: (i) Bothria, typical of Pseudophyllidae, are shallow in shape and weak in musculature (e.g. *Bothridium pythonis*, Fig. 1.4P), (ii) Phyllidia, characteristic of Tetraphyllidae, are four in number symmetrically placed around the anterior part of the elongated scolex (e.g. *Myzophyllobothrium*, Fig. 1.4Q) and (iii) Acetabula or the true suckers are present in Taenoidae, Lecanocephalidae and Proteocephalidae. They are four symmetrically placed hemispherical depressions sunk around the scolex. The apex may be provided with additional attachment organs like a glandular area or a single protrusible sucker or mass called myzorhynchus that may bear additional suckers (Fig. 1.4R). In taenoids, it forms a highly mobile core, the rostellum, usually armed with hooks withdrawable into a sac-like cavity in the scolex (Fig. 1.4S). Notably, the cestodarians lack the scolex but its anterior tip bears a protrusible proboscis (Fig. 1.4J).

In platyhelminths, the dorso-ventrally flattened body ensures the diffusion of oxygen and nutrients to their tissues. The absence of exoskeleton or shell has allowed the soft-bodied worms to undergo a dizzying array of diversity in its shapes, body plans and sizes (Collins, 2017). Among the internal systems, the blind gut and reproductive system have undergone a range of adaptive diversifications. The latter is elaborated in the respective chapters. In turbellarians, the unbranched blind gut is diversified into branched plicate pharynx and forked intestine in *Bothrioplana* (Fig. 1.4B), highly branched one in a polyclad (Fig. 1.4C) and the pharynx is shifted to the posterior end with implication to clonal reproduction, as in *Kenkia rhynchida* (Fig. 1.4D). In trematodes, the gut is a simple tube in the monogenean *Acanthocotyle* (Fig. 1.4E) and the digenean *Cladorchis* (Fig. 1.4G), but is highly branched in *Tristoma* (Fig. 1.4F) and *Fasciola hepatica* (Fig. 1.4H); it is reduced to a reverted Y in shape in *Spelotrema* (Fig. 1.4I).

## 1.2 Life Cycles

The forgone account has emphasized the need for the description of a complete life cycle, especially for the digenean and cestode parasites. Uniquely, platyhelminths have witnessed the evolutionary transition from a simple life cycle in free-living turbellarians to a complex one in ecto- and endo-parasitic flukes and tapeworms. The endoparasites rely on multiple hosts usually in an ontogenetic sequence to complete a single generation. To minimize the risks involved in transmission to appropriate host and time, some have chosen to reduce the number of intermediate hosts (Sections 4.9 and 5.4), while others have opted to have alternate intermediate host (see Table 4.3). Briefly, the life cycle of playthelminth is unique and fascinating (see Auld and Tinsely, 2015).

*General Introduction* 15

*Turbellarians*: In many turbellarian taxa, the cycle is simple and direct, and involves neither a larval stage nor an intermediate host species (Fig. 1.5.1a). The turbellarians, especially triclads are known for the unparalleled potency for clonal reproduction (Fig. 1.5.1b). This potency relies on a population of stem cells, the neoblasts. It ranges from the minimal requirement of the anterior 1/3rd of the body to a fragment as small as 1/279th size of an intact worm (Morgan, 1901). Interestingly, clonal reproduction is usually succeeded by sexual reproduction but can be intermittent and dependent on environmental cues including the bizarre example, where the extract from the sexually reproducing worm can induce sexualization in clonally reproducing worm (e.g. *Dugesia ryukyuensis*, Kobayashi et al., 2002a). In contrast, the cycle of polyclads is indirect and passes through the dispersive Muller (e.g. *Stylochus ellipticus*) or Gotte (e.g. *S. pilidium*) larval stage (Rawlinson, 2014) (Fig. 1.5.1c). Even amidst the overwhelmingly free-living turbellarians, 300 species belonging to 35 families are ectosymbionts (e.g. Rhabdocoela) or endosymbionts (e.g. Umagillidae) or true parasites (Jennings, 1997); for example, the neorhabdocoelid *Kronborgia amphipodicola* parasitizes the amphipods *Ampelisca macrocephala* and *Hoploops tubicola* (see Part 2.6.3).

*Monogeneans* are aquatic and so are their hosts, mostly marine and freshwater teleosts, and some amphibians. They are mostly oviparous. However, some of the 462 speciose dactylogyrids are viviparous (Poulin, 2005) and some polystomatids infecting amphibians can be ovoviviparous (Kearn, 1986b). The monogenean life cycle is indirect and involves a single larval stage, the oncomiracidium (Fig. 1.5.2, Table 1.3). They spawn ~ 100 eggs/fluke/day. Their tanned eggs, the cocoons are physically strong and chemically resistant. A detachable lid permits the release of the non-feeding, free swimming infective oncomiracidium. With limited energy reserve and life time (4–8 hours, *Pseudorhabdosynochus lantanensis*, Erazo-Pagador and Cruz-Lacierda, 2010, *Microcotyle sebastis*, 12–24 hours, Thoney, 1986b), the oncomiracidium, swimming at the speed of 4.6 mm/second (Thoney, 1986b), has to actively search, find and infect a specific host (Whittington and Kearn, 2011). The transmission is guided by chemical cues arising from the odor and mucus of an appropriate host (Kearn, 1986a). The oncomiracidium may then sink into the epidermis of the skin or attach to the gills of fish or reach the urinary bladder of amphibians. From these sites, they suck blood or feed on surrounding tissues. About 71% of monogeneans are strictly host specific (Yamaguti, 1963), i.e. oioxenic, the parasite species can infect only one host species (e.g. *Entobbdella soleae* on *Solea solea*, Kearn, 1967); the others are either stenoxenic (~ 8%, Yamaguti, 1963), i.e. the parasite species can infect a few closely related host species (e.g. *Dactylogyrus lenkoranoides* on *Barbus haasi*, *B. graellis*, *B. guiraonis*, Lambert and Gharbi, 1995) or euryxenic, i.e. the parasites can infect many unrelated host species. For example, *Benedenia hawaiiensis* can infect as many as 24 host species belonging to the genera *Priacanthus*, *Mulloidichthys*, *Parupeneus*, *Dascyllus*, *Amanses*, *Acanthurus*,

### 1a. Turbellaria: oviparous sexual reproduction

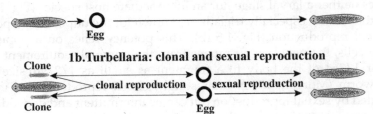

### 1b. Turbellaria: clonal and sexual reproduction

### 1c. Turbellaria: indirect life cycle

### 2. Monogenea: indirect life cycle involving no intermediate host

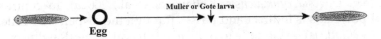

### 3. Digenea: indirect life cycle involving intermediate host, in which parthenogenic reproductive cycle occurs

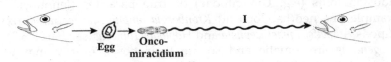

### 4a. Cestoda: Sexual reproduction involving two intermediate hosts in *Diphyllobothrium*

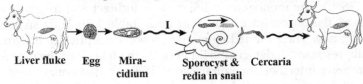

### 4b. Cestoda: sexual reproduction followed by asexual reproduction in intermediate host in *Echinococcus*

O = Oncosphere, H = Hydaid cyst, S = Scolex

**FIGURE 1.5**

Basic patterns of life cycle: 1a, b and c in free living turbellarians and 2, 3, 4a and 4b parasitic monogeneans, digeneans and cestodes. T indicates trophic transmission. I indicates penetrative transmission.

# TABLE 1.3

Life history traits that impose considerable effects on fecundity in major taxons of platyhelminths. Int = Intermediate, M = Miracidium, S = Sporocyst, R = Redia, C = Cercaria, MC = Metacercaria

| Traits | Turbellaria | Monogenea | Digenea | Cestoda |
|---|---|---|---|---|
| Life style | Free-living; a few commensals & parasites | Ectoparasites but a few are endoparasites | Endoparasites with mouth & gut. Ectoparasites: Didymozoonidae | Endoparasites with no mouth and gut |
| Reproductive modes | Sexual, clonal, parthenogenesis | Sexual only | Sexual adults + parthenogenic M, S, R + C larval stages | Sexual adults; 1% of cestodes asexual in larva |
| Life cycle | Mostly direct; a few indirect involving Muller/Gotte larva | Indirect involving oncomiracidium only; no int host | Indirect involving obligate molluscan host, M, S, R + C larval stages on 1 to 3 int hosts | Indirect involving coracidium, procercoid and pleurocercoid or hexaconth and cysticercus larvae in a wide range of hosts |
| Adult niches | Creep on substratum or swim in water | Skin and gills of fish, a few in urinary bladder of amphibians | Gut and associated organs, blood, kidney | In the absence of gut, entirely relegated to intestine only |
| Entry and exit | *Anoplodium hymanae* parasitic in *Sticopus californicus*, egg mass is relased through water vascular system (Shinn, 1985) | Active infection on fish; followed by migration to urinary bladder in frogs. Eggs continuously released into waters or through urine. Active search and infection | Eggs and larvae are continuously voided via feces, urine or saliva (e.g. Clinostomatidae). Ingestion of encysted C or MC, or penetration by M + ingestion of C or MC, or penetration by C and M. | Entry through ingestion alone, exit of eggs voided with feces or encysted muscle. |
| Transmission strategy & efficiency | 3.5% Gotte larvae are recruited (Allen et al., 2017) | Active | Active + passive, or active + active, passive + passive, or passive + active. Transmission efficiency is 0.03% in *Haematoloechus coloradensis* (Dronen, 1978) | Passive ingestion only. In *Bothriocephalus rarus*, Transmission efficiency is 0.08% (Jarroll, 1980) |

18   *Reproduction and Development in Platyhelminthes*

*Synodus, Abudefduf, Chromis, Chaetodon, Alutera, Pervagor, Naso, Helocentrus, Scarus* and *Xanthichthys*. They all belong to 11 families and are associated with coral reefs (see Rohde, 1979).

*Digeneans* are endoparasites of vertebrates in marine, freshwater and terrestrial habitats. As adults, they occur primarily in the digestive tract and associated organs such as the liver, lung, air-sacs in birds, gall-bladder and bile passages; the other sites are blood, kidneys, ureters, coelom, eye and head cavities. Their indirect life cycle includes four major stages, miracidium, mother sporocyst, redia and cercaria (Fig. 1.5.3). With no mouth, the miracidium is a non-feeding, free-swimming larva with a maximum life time of not more than 48 hours (Shoop, 1988). Guided by chemical cue(s), it must infect by active penetration a suitable molluscan host, in which it migrates to species specific location, i.e. hepatopancreas or gonad or mantle (see Esch et al., 2002), where it develops into a mother sporocyst. Whereas the ectoparasitic monogeneans reproduce sexually alone, the digeneans reproduce sexually in definitive host and propagatory multiplication in the molluscan host (see Table 1.3). The germinal cells within the osmotrophic mother sporocyst are bereft of a mouth and gut and undergo repeated multiplications into daughter sporocysts or rediae or cercariae. To meet the required cost of the multiplications, nutrients are acquired osmotrophically from the molluscan host. Redia is a quite distinct form from sporocyst in possessing a mouth, pharynx and intestine (see Hyman, 1951). Escaped from sporocyst, it moves within the host's hemocoelom and can ingest the host's tissues directly by tearing and swallowing. It may give rise to a new generation of rediae (e.g. Donges, 1971) or cercariae. Duration from the entry of miracidium to the escape of cercaria from the molluscan host lasts for some weeks or months. Cercaria is indeed the miniature of the immature adult. Guided by chemical cue(s), it has to infect the definitive vertebrate host. Alternatively, it may multiply (see Galaktionov et al., 2006) and/or develop into metacercaria, which may encyst on plants to gain access to herbivorous vertebrate definitive host or a second vertebrate intermediate host. As a consequence, the digenean life cycle may be more and more complicated involving propagatory multiplication in a second intermediate host (e.g. *Parvatrema margaritense*, Galaktionov et al., 2006). Conversely, the cycle may also be abbreviated (Poulin and Cribb, 2002, Largue and Poulin, 2009). This aspect is elaborated in Chapter 4.

*Cestodes*: Their life cycle is characterized by the following features: (1) Unlike trematodes, that are tied to the molluscan host, the cestodes exploit a large variety of invertebrates and vertebrates (teaniids) as intermediate host(s), (2) Also unlike trematodes, that involve active and/or passive transmission strategies, the transmission strategy of cestodes is always a passive trophic ingestion of one or more infective stage(s), (3) In trematodes, the larval stages undergo clonal multiplication. But not more than 20 cestode species

General Introduction 19

undergo larval clonal multiplications (Table 5.7), (4) In the absence of the mouth and gut, cestodes are relegated to the host's intestine as the sole habitat, (5) In the entire animal kingdom, the discovery of proglottization is unique to the eucestodes. In each of these 'pseudosegment' or proglottid, one set of the reproductive system is present. Irrespective of selfing or crossing, sexual reproduction results in continuous release of eggs or sometimes the proglottid itself and (6) Superabundance of almost fully digested semi-fluid food and reduced immune responsiveness has a profound effect leading to r-selection and evolution of diverse transmission strategies.

The life cycle of aquatic cestodes like *Diphyllobothrium* is indirect with three larval stages involving two intermediate hosts. The coracidium, emerging from the egg, is ingested by *Cyclops*, in which it develops into procercoid. When a fish eats the infected cyclops, the procercoid develops into plerocercoid, which develops into an adult, when the infected fish is eaten by a larger one (Fig. 1.5.4a). In the terrestrial cestodes, the egg voided with feces, on being eaten by an herbivorous vertebrate host(s), develops into the oncosphere, which subsequently undergoes clonal multiplication in the hydatid cyst. When the infected host is ingested by a dog, the protoscolex develops into an adult worm (Fig. 1.5.4b).

## 1.3 Host-Parasite Distribution

*Colonization and diversification*: Remarkably, vertebrates provide an array of niches for the adult ecto- and endo-parasitic flukes and tapeworms. For parasites, the host is the main habitat. Hence, the ability of the parasite species to infect and colonize more and more host species may facilitate the evolution and diversification of parasite lineages. At this point, a few terms have to be defined: *Incidence* refers to the number of host species infected by a parasite species. *Prevalence* is the proportion of a sample in a host population or species infected by a parasite species. *Density* can be assessed with reference to either the host or parasite; the host density refers to the number of a parasite species per host species. On the other hand, the parasite density indicates the number of a host species per parasite species (cf pp 9, 130, 150). *Intensity* is the number of helminth individuals of a species infecting a host individual.

Considering the close relation between parasite and host species, two alternate views have emerged. According to Paterson et al. (1993), parasites tend to be very host specific or oioxenic, infecting one or two closely related host species, as in 71% of monogeneans (Yamaguti, 1963). Hence, it is likely that phylogenies of parasite and host species are mirror images of one another. The second view is that the colonization of a new host by the parasite has played a major role in the diversification of parasitic lineages

(Hoberg et al., 1997). An analysis of monogenean (98 species), digenean (17 species), cestode (26 species), copepod (17 species), acanthocephalan (12 species) and nematode (6 species) had led Poulin (1992) to draw the following conclusions immediately relevant to this aspect: (i) The monogeneans are highly host specific (see also Lo et al., 1998); their specificity is limited to one, two and three host(s) in > 65, 24% and ~ 8% monogeneans, respectively (Fig. 1.6). (ii) They have more site restricted attachment than copepods and their adhesive organs are highly specialized to specific microhabitats on the host's gills, (iii) Surprisingly, the ectoparasitic copepods, that have a similar indirect life cycle involving no intermediate host (see Pandian, 2016), as in monogeneans, are the least specific and have colonized up to 20 host species, (iv) So are the digeneans with a most complicated life cycle. (v) At least 5% of digeneans and cestodes can successfully colonize a dozen host species and (vi) The same holds true for acanthocephalans and nematodes; 5 to 15% of them can colonize ~ 20 host species. Barring some gyrodactylids (Harris, 1993), colonization success of the ectoparasitic monogeneans depends on host-parasite phylogenies, represents a mirror image of one another and thereby confirms the view proposed by Paterson et al. (1993). Conversely, colonization of a new host has played a major role in diversification in ectoparasites, the copepods and endoparasites, the digeneans. However, this may not hold true for gyrodactylids (see Section 3.3.2).

Considering trematodes as representative of endoparasites, an analysis of their distribution in the four vertebrate taxa has been made by Poulin and Morand (2000). Being younger taxa than herptiles *in sensu* of evolutionary history, birds and mammals account for smaller shares (birds: 15.2%, mammals: 8.4%, see Table 1.4A) of vertebrate species. But they host relatively larger proportions (birds: ~ 25.5%, mammals: 22.5%) of known trematode species. However, reptiles with a larger share (26.4%) of vertebrate species serve as host for only 9.5% of trematode species (Fig. 1.7D). A reason adduced by Gibson and Bray (1994) that the trematodes exploiting herptiles as host is

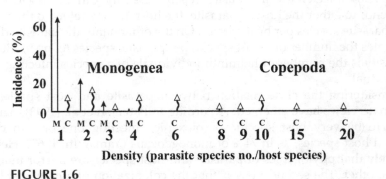

**FIGURE 1.6**

Incidence of parasite species (%) as a function of parasite species density in Monogena (M) and Copepoda (C) (compiled and redrawn from Poulin, 1992).

*General Introduction* 21

## TABLE 1.4

Distribution of intestinal helminth parasites in vertebrate taxons of aquatic and terrestrial habitats († Gaston and Blackburn, 1995, †† Shine et al., 2003)

### A. Distribution of vertebrate hosts

| Taxon | Vertebrates | | Sampled by Bush et al. (1990) | | |
|---|---|---|---|---|---|
| | (no.) | (%) | (no.) | (%)* | % of respective taxon |
| Fish | 32,900 | 50.0 | 245 | 0.372 | 0.745 |
| Herptiles | 17,340 | 26.4 | 113 | 0.170 | 0.646 |
| Birds | 10,038 | 15.2 | 83 | 0.128 | 0.847 |
| Mammals | 5,513 | 8.4 | 141 | 0.214 | 2.558 |
| Total | 65,791* | | 582 | | |

### B. Motility and size of host taxons

| Taxon | Motility (km/h) | | Body size | |
|---|---|---|---|---|
| | Aquatic | Terrestrial | Aquatic | Terrestrial |
| Fishes | 93 | – | 1 g–1100 kg | – |
| Amphibians†† | 16–45 | 3 | 12.5 g–23.5 kg | |
| Reptiles†† | 0.72 | 31 | 28.5 g | |
| | 4.7 | | | |
| Herptiles | 3.15 | 31 | 0.5 g–22 kg | 255–680 kg |
| Birds† | 19 | 77 | 875 g | 220 g |
| Mammals | 40.2 | 93 | 20 kg–50 ton | 100 g–20 ton |

### C. Host-helminth parasite distribution

| Taxon | Aquatic parasites | | | | Terrestrial parasites | | | |
|---|---|---|---|---|---|---|---|---|
| | Host species (no.) | Helminth | | | Host species (no.) | Helminth | | |
| | | Density (no./host) | Sub total (no.) | (%) | | Density (no./host) | Sub total (no.) | (%) |
| Fish | 245 x | 3.0 | = 735 | 63.4 | – | – | – | – |
| Herptiles | 66 x | 2.0 | = 132 | 11.4 | 47 x | 1.0 | = 47 | 9.5 |
| Birds | 44 x | 5.0 | = 220 | 19.0 | 39 x | 2.0 | = 78 | 15.7 |
| Mammals | 18 x | 4.0 | = 72 | 6.2 | 123 x | 3.0 | = 369 | 74.7 |
| Total | 373 | | 1,159 | | 209 | | 494 | |

### D. Host-helminth parasite distribution

| Total | | | | |
|---|---|---|---|---|
| Taxon | Host species (no.) | Helminth species | | |
| | | (no.) | (%) | Parasite/ host |
| Fish | 245 | 735 | 44.4 | 3.0 |
| Herptiles | 113 | 179 | 10.8 | 1.6 |
| Birds | 83 | 298 | 18.0 | 3.6 |
| Mammals | 141 | 441 | 26.7 | 3.1 |
| Total | 582 | 1653 | | |

22  *Reproduction and Development in Platyhelminthes*

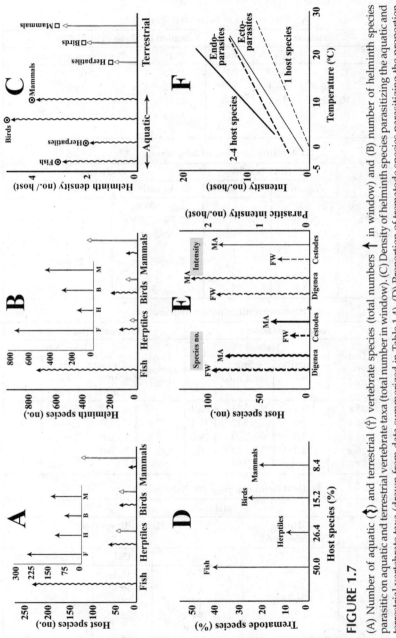

### FIGURE 1.7

(A) Number of aquatic (🐟) and terrestrial (🦌) vertebrate species (total numbers ← in window) and (B) number of helminth species parasitic on aquatic and terrestrial vertebrate taxa (total number in window). (C) Density of helminth species parasitizing the aquatic and terrestrial vertebrate taxa (drawn from data summerized in Table 1.4). (D) Proportion of trematode species parasitizing the proportion of vertebrate taxa (drawn from data reported by Gibson and Bray, 1994 and Table 1.4 A). (E) Number of mollusc and arthropod host species and helminth intensity of Digenea and Cestoda in freshwater (FW) and marine (MA) habitats in LAC countries (drawn from data reported by Aguirre-Macedo et al., 2016). (F) Intensity of ecto- and endo-parasites in fish as a function of temperature (modified and redrawn from Rohde and Heap, 1998). In Fig. 1.7A and B, F, H, B and M indicate Fish, Herptiles, Birds and Mammals, respectively.

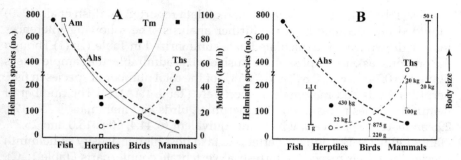

**FIGURE 1.8**

(A) Motility and (B) body size as function of number of intestinal helminth species parasitizing aquatic and terrestrial vertebrate taxa. Body sizes of vertebrate taxa are just indicated but not to the vertical cycle (data are drawn from Table 1.4). Am = Aquatic motility, Tm = Terrestrial motility, Ahs = Number of aquatic helminth species and Ths = Number of terrestrial helminth species.

more host specific than the other taxa. For example, if the density of trematode species parasitizing herptiles is 1.0, it ranges from 1.13 for fish to 1.53/host for mammals and birds. It is likely that the greater host specificity of digeneans in reptiles is more related to their motility (see Fig. 1.8B, amphibians sharing two habitats) rather than a phylogenetic phenomenon. Briefly, colonization has played a major role in diversification of trematode lineages and thereby confirms the view of Hoberg et al. (1997).

A preamble is required prior to the description and analysis of Bush et al. (1990). They examined the host-parasite relationship, considering the intestinal helminths mostly from north temperate regions. The intestinal helminths considered by them included digeneans, cestodes, nematodes and acanthocephalans. For one or other reason, they overlooked the publications from 17 Latin American Caribbean (LAC) countries. A recent analysis of Aguirre-Macedo et al. (2016) also included intestinal helminths from 17 LAC countries from 1990 to 2015. Of 479 helminth species, 356 (74%), 76 (16%), 38 (8%) and 9 (2%) were digeneans, cestodes, nematodes and acanthocephalans, respectively. Hence, some 90% of the intestinal helminth species analyzed by Bush et al. could have also been digeneans and cestodes. Further, the digeneans are distributed not only in the intestine but also in other organs (p 18, 141). Hence, the analysis of Bush et al. may be more of a digenean account. The number of host species sampled by Bush et al. (1990) is 245, 113, 83, and 141 for fish, herptiles, birds and mammals, respectively (Table 1.4A). The total number of helminth species parasitizing these vertebrate host taxa has been assessed by multiplying the average number of helminth species/host by the number of each of the vertebrate taxon. For example, 245 fish species are infected by three helminths/host and thereby make up 735 helminth species (Table 1.4C). Briefly, 1,653 helminth species parasitize 582 vertebrate

24   *Reproduction and Development in Platyhelminthes*

species (Table 1.4C, D). Reorganization of data reported by Bush et al. (1990) provided an adequate base for further analysis. The following inferences may be drawn from the calculated data summarized in Table 1.4: (1) Though the onerous task accomplished by Bush et al. is admirable, the sample size is limited to 0.745, 0.646, 0.847 and 2.558% of the total number of species in fish, herptiles, birds and mammals, respectively (Table 1.4A). (2) The number of helminth species parasitizing fish, herptiles, birds and mammals is 735, 179, 298 and 441, respectively, which are equivalent to 44.4, 10.8, 18.0 and 26.7% (Table 1.4D). (3) The aquatic vertebrate taxa host higher density of helminth species than their respective terrestrial vertebrate counterparts (Table 1.4C). (4) Remarkably, herptiles, either aquatic or terrestrial, are unable to serve as hosts, in comparison to birds and mammals (Fig. 1.7A, C, D). This conclusion is confirmed by both proportion (limited to trematodes, Fig. 1.7D) and number (Fig. 1.7B) of helminth species parasitizing the four vertebrate taxa. (5) Aquatic vertebrate taxa also host a higher parasitic density/host species (1,159 species, Table 1.4C) than their respective terrestrial counterparts (494 species, Table 1.4C, Fig. 1.7C). (6) Considering density parameter, aquatic bird serves the host for the highest number of helminth species/bird species. From this high density of 5 helminth species/aquatic bird host species, the density decreases to 4, 3 and 2 helminth species/host species in aquatic mammals, fish and reptiles, respectively (Table 1.4C, Fig. 1.7C). However, it is 3, 2 and 1 for helminth species/terrestrial species of mammals, birds and reptiles (Table 1.4C, Fig. 1.7C). Briefly, the fact that birds and mammals host more number of helminth or trematode species/host species than herptiles raises the question whether motility or body size of the host is a driving force in evolution and diversification of intestinal helminth lineages?

Within each of the vertebrate taxon, motility ranges vary widely from species to species, size to size, walking to running or flying, swimming (horizontally) or diving (vertically) and so on. For example, the swimming speed ranges from 56 to 129 km/hour within the sunfish family (sciencefocus.com). A few newt species belonging to the genus *Tyturus* walk at the speed of 8.5–24 km/hour but swim 24–60 km/hour (Wikipedia). Aquatic birds swim 3.6 km/hour (e.g. grebes, Johansson and Norberg, 2000) –19 km/hour but dives at the speed of 187–120 km/hour (e.g. penguins, nationalgeographic. com, Table 1.4B). Hence, it may be an onerous task to fix an average speed for a vertebrate group. So is the task for fixing the mean body size of a vertebrate group. However, the trends drawn for the relationships between motility of aquatic and terrestrial vertebrate taxa and proportion of vertebrate taxa colonized by helminths reveals more or less parallels between them (Fig. 1.8A). On the other hand, wide ranges of body size known for the vertebrate groups do not allow parallel trends to be drawn for the relationship between body size and number of helminth species (Fig. 1.8B).

In relatively denser medium of waters, the stream-line bodied fish swim (93 km/hour) as fast as terrestrial mammal can run (Table 1.4B). However,

General Introduction    25

the (horizontal) swimming speed of aquatic birds/mammals can be slower than that of their terrestrial counterparts, as their motility is intermittently interrupted by the obligate need for aerial respiration by vertical diving followed by emergence. Understandably, the aerial motility of flying by terrestrial birds is high (77 km/hour), despite their smallest average size of just 220 g. But even with a heavier body (3–31 kg), herptiles are unable to swim and run faster than 3 and 31 km/hour, respectively. Briefly, this analysis has brought to light for the first time that of the two, i.e. motility and body size, motility of vertebrates host has been the prime driving force for the evolution and diversification of helminth lineages. Of course, it is the inclusion of vertebrates as definitive hosts by the trematodes and cestodes that has facilitated the dissemination of them far and wide (e.g. Shoop, 1988). Again, why is the motility of herptiles so slow? For herptiles, the switch over of the musculature pattern from swimming to walking (as tetrapods) on land may have been a challenging task. Having returned to water, their swimming ability is also far slower than that of their walking ability (Table 1.4B). Another fact that has remained unrecognized thus far is the poikilothermy in herptiles and homeothermy in birds and mammals. As a niche for trematodes and cestodes, host intestine ensures not only abundant semi-digested nutrients but also a constant temperature for functioning of sexual reproduction at a constant rate. For example, embryonic and subsequent development or procercoid (in copepod) and plerocercoid (in fish) of the tapeworm *Diphyllobothrium dentrictum* are passed through at 5–15°C but adults at a constant temperature of 37–38°C in the natural definitive host the gull or experimental host the hamster (see Reuter and Kreshchenko, 2004).

To complement the analysis of Bush et al., Aguirre-Macedo et al. (2016) analyzed the distribution and density of digeneans and cestodes on mollusc and arthropod host, respectively. In both freshwater and marine habitat, there are more numbers of molluscan host species available for digeneans than arthropod host species available for cestodes (Fig. 1.7E). This is also true for intensity of parasite species per host species.

Temperature is another important factor that may significantly affect host-parasite species lineages in biogeographic distribution. Relatively, more information is available for the distribution pattern of metazoan parasites of marine fishes. Firstly, the diversity of parasitic helminths is greater in the Indo-Pacific than in the Atlantic (see Poulin and Morand, 2000). Secondly, the diversity of monogeneans decreases with increasing latitude (Poulin, 1996b). This observation is confirmed by the increase in parasite intensity with increasing temperature in ecto- (Fig. 1.7F) and endo- (Fig. 1.7F) parasites. The trends for the relation between parasite intensity vs temperature show that at a given temperature, the colonization is greater for endoparasites than for ectoparasites. For example, it is up to 10 for endoparasitic species/host fish but only 5 for ectoparasitic species/host fish at 10°C. Incidentally, it must be indicated that Rohde and Heap (1998) have included sample fish harboring endoparasites from marine habitats up to 24°C only (see Fig. 1.7F).

## 1.4 Fecundity

In view of platyhelminths composing both free-living and ecto- and endo-parasitic species, a comparative account on fecundity at class (and subclass) level is included here. However, fecundity is elaborated at intra- and inter-specific levels in the respective chapters.

The total number of oocytes contributing to fecundity is assured by waves of oogonial proliferation and subsequent oocyte recruitment (see Pandian, 2013). Fecundity is decisively an important factor in recruitment at population level. With inclusion of propagatory multiplication(s) in sporocyst, redia and/or cercaria of digeneans and clonal reproduction in 0.5% of cestodes, the terminologies related to fecundity, as used by fishery biologists, cannot directly be applied to platyhelminths. However, at least the terms like batch fecundity, cumulative and relative fecundity may be introduced: (1) Batch Fecundity (clutch) (BF) is the number of eggs released per spawning. Because BF is related to the volume of space available to accommodate the ripe ovaries ($F = aL^b$), geometry suggests that length exponent 'b' would be 3.0. The 'b' value is at 3, when growth in volume or weight is isometric, but without change in its shape. In bilaterally symmetrical vertebrates like fishes (e.g. Pandian, 2011), and invertebrates like crustaceans (e.g. Pandian, 2016), molluscs (e.g. Pandian, 2017) and in radially symmetrical echinoderms (e.g. Pandian, 2018), the 'b' values are mostly < 3 or rarely > 3 (due to change in body shape with growth) resulting in allometry. The cylindrical body shape of polychaetes and oligochaetes provides relatively less surface area/volume or weight for production and accommodation of eggs than the dorso-ventrally flattened hirudineans (see Pandian, 2019). Hence, it may be interesting to know how the dorso-ventrally flattened body and dizzying array of body shapes in the absence of exoskeleton in flatworms (see Figs. 1.1, 1.4) alter the expected exponent. Lifetime fecundity or cumulative fecundity is the number of eggs/spawning multiplied by the number of spawnings during the life time of a worm species. Relevant information is available for Batch Fecundity (BF), Relative Fecundity (RF) and cumulative or Lifetime Fecundity (LF) (= Reproductive capacity) of a few flatworms. The following are some examples: (1) In the free-living turbellarian *Stylochus ellipticus*, the BF ranges from 295 to 39,330 and LF is 1,56,000 eggs distributed in 14 batches (Chintala and Kennedy, 1993). In another turbellarian *Pleioplana atomata* with parental care, the BF increases with increasing body size from 8 eggs in a small worm (6 mm body length) to 12 eggs in the largest worm of 20 mm. The LF of *P. atomata* is ~ 400 eggs (Rawlinson et al., 2008). In the cestode *Schistocephalus solidus*, Wedekind et al. (1998) reported the BF in egg volume ($mm^3$/clutch) as a function of body size in single and paired worms during a 3-day experiment. BF increases from 60 to 75 $mm^3$/clutch in singles weighing 100 to 500 mg but decreases from 80 to 70 $mm^3$/clutch in paired worms in the weight range of 100 to 600 mg. In ectoparasitic monogeneans,

General Introduction 27

eggs are released continuously at the rate of an egg for every 5–10 minutes. However, considering the daily spawning in *Neobenedenia* sp from the 10th post-infection day (pid) to 17th pid, the BF is steadily increased from ~ 50 eggs/day on the 10th pid to 500/day on the 15th pid and subsequently progressively decreases to ~ 20/day on the 17th pid. In all, the LF was 3,229 ± 37 eggs (Hoai and Hutson, 2014). Relative fecundity refers to the number of eggs released per unit weight/volume of the worm.

Regarding temporal distribution of releasing eggs and emitting larvae, there are also vast differences between free-living turbellarians and parasitic flatworms. The free-living turbellarians spawn at 13 and 14 specific events in *P. atomata* and *S. ellipticus*, respectively. But the release of eggs and emittance of infective larvae are a continuous process for the parasite from the definitive adult host and from the intermediate host, respectively. For example, *Fasciola hepatica* releases eggs at an average rate of 16,900/day (see Whitfield and Evans, 1983) during its reproductive life span of 3,917 days totaling to the LF of 66.2 million eggs. An impressive example for the cercarial emittance by *Littorina littorea*, the intermediate host of another digenea *Cryptocotyle lingua*, is 2.1 million cercariae emitted at the rate of 830 cercariae/day for 7 years (see Hyman, 1951).

*Size*: In flatworms, the size, usually measured in length, ranges from as small as 0.3 mm to as large as 20 m (Fig. 1.9). It ranges from 0.3 mm in some rhabdocoels to 0.6 m in the terrestrial triclad *Bipalium kewense* (see Hyman, 1951) for turbellarians, from 1.3 mm in the viviparous *Dactylogyrus vastator* (see Trouve et al. 1998) to 7.6 cm in a gill parasite from a Mediterranean fish (Sasal et al., 1999) for monogeneans, from 0.9 mm in *Transversotrema patialense* (see Trouve et al. 1998) to 165 mm in *Gigantobilharzia ocotyla* (see Hyman, 1951) for digeneans and from 1 mm in *Protogynella* (Mackiewicz, 1988) to 20 m in *Diphyllobothrium latum* for cestodes (see Hyman, 1951). Interestingly, some values reported for the largest monogenean parasites are 36.2 mm in a chimaericolid (Poulin, 1996b) and 2.0 cm for *Entobdella hippoglossi* parasitic on *Hippoglossus hippoglossus* (Kearn, 2014). Briefly, the smallest are the monogeneans and the largest are the cestodes. The number of proglottids, a unique feature of eucestodes, ranges from 3 in *Echinococcus* to 4,500 in *Tetragonoporus* (see Mackiewicz, 1988).

*Life span* (LS) or longevity is another important life history trait that may have a 'telling effect' on fecundity. To their credit, Trouve et al. (1998) assembled some available information on longevity of flatworms. The LS values ranged from 54 days in *Phaenocora typhlops* to 3.01 years in *Dugesia derotocephala* for turbellarians, from 4.2 days in *Gyrodactylus bullatarudis* to 5.1 years in *Polystoma integerrimum* for monogeneans, from 13 days in *Apatemom gracilis* to 28.6 years in *Schistosoma japonica* for digeneans and from 16 days in *Schistocephalus solidus* to 25.2 years in *Taenia solium* and 35 years in *T. saginata* (Fig. 1.10). Briefly, LS increases in the following order: free-living turbellarians > ectoparasitic monogeneans > endoparasitic digeneans

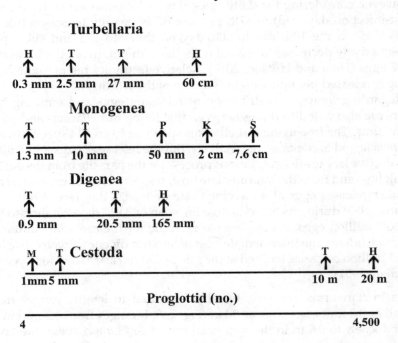
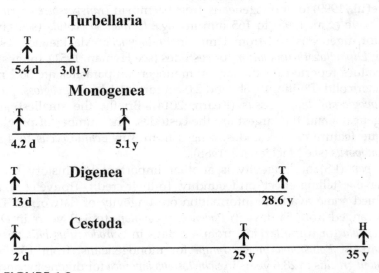

**FIGURE 1.9**

Ranges of body length and life span of platyhelminth taxa. The alphabets H, K, M, P, S and T indicate Hyman (1951), Kearn (2014), Mackiewicz (1988), Poulin (2002), Sasal et al. (1999), Trouve et al. (1998), respectively.

> cestodes. Interestingly, experimental elimination of reproduction extends the LS in the rhabdocoel *Stenostomum tenuicauda* to 11 years (see Hyman, 1951) and serial transplantations of the cestode *Hymenolepis diminuta* from an old to young rat also extends the parasite life span from 2 to 14 years (Collins, 2017). Notably, the shortest life span (16 days) and amenability to experimental cultivation of the cestode *S. solidus* has facilitated a large number of studies (e.g. Smyth, 1946, McCaig and Hopkins, 1963, Benesh and Hafer, 2012, Heins, 2012).

Estimates of Generation Time (GT) (egg to egg stage) and reproductive life span (life span minus post-reproductive or menopause period) are important for the estimation of the length of the reproductive life span. Information available for flatworms is, however, limited to age at sexual maturity and longevity. Considering these values, the approximate equivalent for Reproductive Life Span (RLS) can be assessed by deducting GT from Life Span (LS). The GT values are 47.7, 26.3, 12.3 and 17.6% for free-living turbellarians, ectoparasitic monogeneans, endoparasitic digeneans and cestodes, respectively (see Table 1.5); hence, the arrived RLS values are 52, 74, 88 and 82% for these worms. The RLS is the longest (88% of LS) for digeneans and shortest (52%) for turbellarians. Clearly, parasitism has extended both the LS and RLS. Notably, with the longest (99.5%) RLS and highest fecundity (2,579 eggs/day, see Appendix, Trouve et al., 1998), *Schistosoma japonicum* has higher probability of infecting a snail host than all other *Schistosoma* species.

Adoption of free-living or parasitic mode of life history may considerably influence reproductive strategy and fecundity (Whittington, 1997). Two hypotheses have been proposed to explain the high fecundity of parasites; (i) the increased fecundity may also increase the probability of transmission (e.g. Price, 1974) or (ii) it may be the result of ensured rich and super-abundant supply of nutrients (Jennings and Calow, 1975). To test these hypotheses, Trouve et al. (1998) subjected relatively smaller sample values (8 turbellarians, 10 monogeneans, 16 digeneans and 20 cestodes, Table 1.5) for fecundity–body size relationship to the complex independent contrasts method and processed them with the CAIC program. From their analysis, they concluded that irrespective of free-living or ecto- or endo-parasite life history, fecundity is determined by the adult body size and the idea that parasite species are more fecund than free-living species is largely exaggerated (Trouve and Morand, 1998). Further, the total reproductive capacity (= cumulative fecundity) and adult body size are significantly correlated in both free-living and parasitic platyhelminths, i.e. most values cited in their Fig. 2a are significantly greater than unity. In other words, the larger the adult body length, the greater is the 'reproductive capacity'. Hence, the modes of life cycle do not impose any effect on the said relationship.

Firstly, the size range, for which data are assembled by Trouve et al., is fairly small (see Fig. 1.10). Secondly, it must be indicated that Trouve et al. have estimated total reproductive capacity for CF (= cumulative fecundity)

## TABLE 1.5

Life history traits of selected platyhelminthic species (condensed from Trouve et al., 1998, for species wise details see Appendix of Trouve et al., 1998, mean values are given in bold letters; values for the range are also given below)

| Species | Size (mm) | Progeny (no x 10³ mm³ or no./mm³)* | Fecundity (no/d) | GT (d) | LS (d) | GT/LS (%) |
|---------|-----------|------------------------------------|------------------|--------|--------|-----------|
| Turbellaria | **16.4** 2.5–32.5 | **30.6** 0.4–166.4 | **0.123** 0.04–0.41 | **230** 25–300 | **564** 54–1095 | **47.7** 13.7–72.0 |
| Monogenea | **4.1** 0.7–10.1 | **41,718** 365–43,987 | **9.64** 0.32–30.0 | **190** 1–1,095 | **631** 4–1,825 | **26.3** 5.5–60.0 |
| Digenea | **11.6** 0.9–25.0 | **2,201** 24–5,359 | **4,088** 1.9–25,000 | **31.6** 3–98 | **2,102** 13–10,220 | **12.3** 0.2–40.9 |
| Cestoda | **3,053** 5–25,000 | **204** 8–913 | **223,124** 230–720,000 | **43.3** 2–255 | **1,540** 16–9,000 | **17.6** 0.4–85.0 |

* For turbellarians, progeny volume (in no/mm³) is indicated. GT = Generation time, LS = Life span

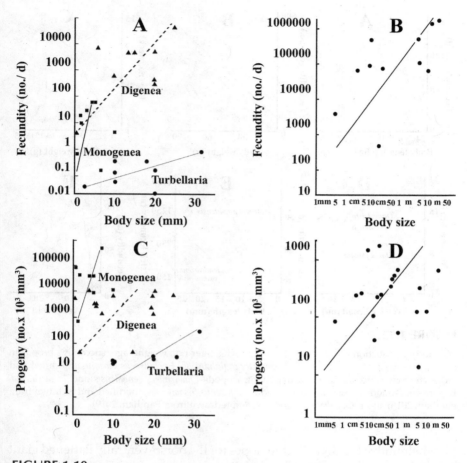

**FIGURE 1.10**

Fecundity (A, B) and progeny volume (C, D) as a function of body size in Turbellaria, Monogenea, Digenea (A, C) and Cestoda (B, D) (Values are drawn from Appendix of Troue et al., 1998. Linear trends are drawn for each taxonomic group considering the simple average values between fecundity/progeny volume and body size.

and daily fecundity. When the values assembled by Trouve et al. are plotted in single logarithmic graph for turbellarians, monogeneans and digeneans and on double logarithmic graph for cestodes, the apparent trends fall at different levels covering different ranges of body size. Clearly, parasitic flatworms are more fecund than free-living flatworm at any given body size (Fig. 1.10). Thirdly, the slope for the fecundity–body size relationship in spherical body shaped polychaetes and oligochaetes is less than unity, as the surface area available/volume or weight to accommodate the ripening eggs is relatively smaller (see Pandian, 2019). However, the area to accommodate both eggs and yolk cells is relatively larger in the dorso-ventrally flattened

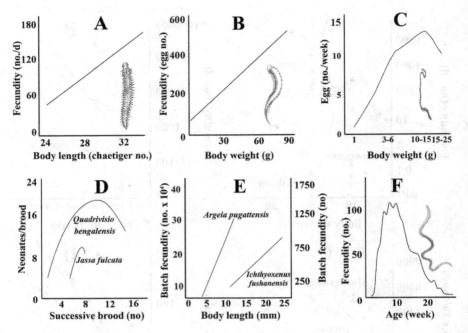

**FIGURE 1.11**

Fecundity as function of body size (in length, chaetiger no., weight, age, successive brood) in (A) *Ophryotrocha puerilis puerilis* B. *Haementeria ghilianii* C. *Enchytraeus variatus* (modified and redrawn from Pandian, 2019) D. free-living isopods *Quadrivisio bengalensis* and *Jassa fulcata*, E. parasitic isopods *Argeia pugattensis* and *Ichthyoxenus fushanensis* (modified and redrawn from Pandian, 2016) and F. *O. adherens* (modified and redrawn from Pandian, 2019).

platyhelminths. Incidentally, the slopes for the dorso-ventrally flattened giant leech *Haementeria ghilianii* (Fig. 1.11B) and polychaete *Ophryotrocha puerilis puerilis* (Fig. 1.11A) are also more than unity. But the trends for the BF–body size relationship for the cylindrical bodied oligochaete *Enchytraeus variatus* (Fig. 1.11C) and polychaete *O. adherens* (Fig. 1.11F) are different, though initially more than unity. Remarkably, the slopes of the bopyrid isopod parasites are always more than unity (Fig. 1.11E). Hence, it is not surprising that the slope for either the fecundity–body size (Fig. 1.10A, B) or progeny volume-body size (Fig. 1.10C, D) relation is more than unity for the dorso-ventrally flattened free-living turbellarians and ecto- and endo-parasitic flatworms. But they are all located at different levels covering different ranges of body size. Most importantly, a single trend with a single slope covering all the four groups of platyhelminths has not emerged. The slope is at the highest level for the smallest size range of ectoparasitic monogeneans involving neither intermediate host nor propagatory multiplication. It is at the next lower level for the endoparasitic digeneans, whose life cycle involves 1–4 host(s) and propagatory multiplications in sporocyst and/or redia.

The third is for cestodes with life cycle involving 1–3 host (s) and < 0.5% of them undergoing clonal reproduction. The lowest slope is for the free-living turbellarians. For example, the CF of turbellarian *Stylochus elipticus* is 1,56,000 eggs, whereas that of the endoparasitic digenean *Fasciola hepatica* is 6.2 mil eggs. The rich and superabundant nutrient resource of the vertebrate hosts allows the platyhelminth parasites to allocate the greater fraction of their resources for reproduction and may compensate the huge loss incurred during transmission event(s). Incidentally, it may also be noted that the parasitic flatworms store the surplus resource as glycogen (up to 30–40% dry weight) and adopt r-strategy (MacArthur and Wilson, 1967) to produce a large number of smaller eggs. But the free-living turbellarians store it as lipid (13–20% of dry weight) and adopt a K-strategy to produce fewer but larger eggs with more lipids, which are hatched at an advanced development stage (see Jennings, 1997).

Fourthly, it may also be indicated that the information available or selected by Trouve et al. is limited (35 species for > 27,700 speciose flatworms) and may also be biased. For example, (i) the smaller size range, for which the relevant data are assembled (see Fig. 1.10). Amidst the 8 assembled digenean species, the life cycle of 5 (*Schistosoma* spp, *Fasciola hepatica*) involves only one intermediate host, while the cycle involves two to three intermediate hosts in others. However, of 11 assembled cestode species, 3 (*Taenia*, *Echinococcus*, *Monieza*) of them reproduce clonally also in the intermediate host. Therefore, the parasitic flatworms are certainly more fecund than the free-living turbellarians. This analysis supports the hypothesis of Jennings and Calow (1975) and remains unable to support the conclusion of Trouve et al. (1998).

Regarding the hypothesis of Price (1974), limited information is available. The probability of recruitment of an individual from an egg in free-living turbellarians may be high. Indicated by hatching success in *Pleiaplana atomata*, the potential recruitment is 65.1 and 35.6% for the incubated eggs with and without parental care, respectively (Rawlinson et al., 2008). Experimental study has shown that 3.5% of Gotte larvae are metamorphosed and recruited in free-living polyclad *Stylochus ellipticus*. Another value available for recruitment is for the dragonfly *Brachthemis contaminata* from the Idumban Pond, Tamil Nadu, India (Mathavan and Pandian, 1974). In this terrestrial dragonfly, the indirect life cycle involves relatively large and highly mobile (to escape predation or predate) aquatic nymph. Of 7.27 million eggs oviposited by the dragonfly, only 0.33% (or 23,990) of them emerged as an adult dragonfly. Notably, the indirect life cycle, despite involving a single larval stage and no intermediate host, reduces the recruitment from 35.6% in incubated eggs without parental care in *P. atomata* to 0.33% in the dragonfly with an indirect life cycle. This may be relevant to monogeneans with an indirect life cycle involving no host but with obligate need to actively infect a specific host. However, a single publication reporting laboratory rearing indicates 3.5% recruitment in a turbellarian (Allen et al., 2017). But the

34  *Reproduction and Development in Platyhelminthes*

probability may be < 0.33%. For the endoparasitic flatworms, a few values are available for the recruitment, 0.03% for the digenean *Haematoloechus coloradensis* (Dronen 1978) and 0.08% for the tapeworm *Borthriocephalus rarus* (Jarroll, 1980). Interestingly, the indirect life cycle of these flatworms involves two and one intermediate host(s), respectively. Evidently, the high risk involved in the transmission requires corresponding high fecundity. Of 58, 320 eggs/m$^2$/y released by *H. coloradensis*, only 3.25% miracidia penetrate into the snail *Physa virgata*; the snail emits 2,100 cercariae/m$^2$/y and of them only 1.55% gain access into the dragonfly, the second intermediate host. Finally, 0.03% of the digenean egg is recruited into the population (see also Whitfield and Evans, 1983). Interestingly, the mean fecundity values estimated from the assembled data of Trouve et al. in Table 1.5 also indicate that the parasite flatworms are far more fecund (9.64 eggs/day for monogeneans, 4,088 eggs/day for digeneans and 223,124 eggs/day for cestodes) than 0.12 egg/day for free-living turbellarians. Clearly, the parasitic flatworms compensate the huge loss incurred during the transmission phases (see also Granovitch et al., 2009).

In support of this observation for flatworms, evidence is also available for other aquatic invertebrate taxa. For example, the free-living isopod *Quadrivisio bengalensis* produce ~ 16 neonates/batch and *Jassa fulcata* only ~ 8 neonates/batch (Fig. 1.11D). These values are less than those of the parasitic cymothoid isopod *Ichthyoxenus fushanensis* characterized by the direct life cycle involving no intermediary host. In *I. fushanensis*, the BF increases but from 16 to 26 manga with increasing body size from 107 to 820 mm. These BF values are again less than those of the bopyrid isopod *Argeia pugattensis*, a parasite inside the branchial chamber of the prawn *Crangon franciscorium* (Fig. 1.11E). The indirect life cycle of *A. pugattensis* involves three larval stages, namely, epicardium, microniscus and cryptoniscus as well as a copepod as an intermediate host. Its BF increases from 1,600 to 33,300 eggs/batch with increasing body size from 4 to 9 mm. Hence, parasite species not only in flatworms but also in Crustacea are more fecund than their respective free-living counterparts. Clearly, parasites are more fecund and compensate the risk involved in the indirect life cycle as well as transition from one host to the next and thereby validate the hypothesis of Price (1974).

## 1.5 Osmotrophism and Digestion

In Platyhelminthes, parasitic monogenean and digenean trematodes, and cestodes constitute ~ 77% of known flatworms. Some monogeneans feed on the soft and relatively easily digestible non-keratinized epidermal cells of fish (e.g. *Entobdella soleae* from the skin of *Solea solea*, Kearn, 1963), while others on blood from the richly vascularized tissues of fish gills (e.g. *Diclidophora merlangi* on *Merlangius merlangus*, Halton, 1975). The trematodes are essentially suctorial feeders. Irrespective of their location in the gall bladder of the wolf-fish *Anarhichas lupus* by *Fellodistomum fellis* (Halton, 1982) or the liver in mammals (e.g. *Fasciola hepatica*) or vascular system itself (e.g. *Schistosoma*), they suck body fluids drawn from vasculature of the gall bladder/liver or blood directly. Hence, blood is the major food of digenetic trematodes. For example, male and female adult *S. mansoni* ingest some 39,000 and 330,000 erythrocytes with a turnover time of 3 and 4 hours, respectively. For more details on abstraction of cells/blood from the hosts by these parasites, Halton (1997) may be consulted.

In a very good review, Halton (1997) provided a comprehensive description of the tegument, digestion and residue elimination in flukes. A striking histo-morphological feature of these parasitic flatworms is that they all share a tegument, which prompted their collection into a monophylum named Neodermata. Ultrastructural studies have revealed that the tegument of these parasites is organized into an anucleate surface layer connected at intervals to subsurface nucleated regions or cell bodies (Fig. 1.12A). It is a living membrane that facilitates selective transport of low molecular materials into the parasite from the host and regulates a 2-way flux of molecules and information between the host and parasite. Further, its confers a number of advantages: (1) Due to the absence of boundaries in the syncytium, the parasite is less vulnerable to attack and breakdown by host agents like digestive enzymes, detergent, bile acids and components of immune system. (2) Transport of substrates and ions is possible laterally without restriction. (3) The sunken nucleated region is located relatively far from the syncytial surface and reduces any adverse influence of the host. (4) The tegument supplements a bulk of parasite diet through extra-gut uptake of low molecular substances.

Significant concentrations of dissolved organic materials are present in sea water; for example, 1 mg dissolved organic carbon/l including neutral amino acids like glycine and leucine. Indeed, sea water is an organic 'soup' (see Pandian, 1975). The tegumental surface of monogeneans is elevated into numerous villi numbering to 12 million/mm$^2$ and thereby increases the cumulative surface by a factor of three (Fig. 1.12B), as in *D. merlangi* (Halton, 1975). Experimental studies with labeled amino acids have revealed the entry of these amino acids solely across the integument. Hence, the tegumental uptake of low molecular substance across the marine monogenean parasites on the fish gills does supplement their diet. Living amidst the juicy habitats of vertebrate hosts, the endoparasitic digeneans also tegumentally acquire low molecular substances, despite being localized in a variety of habitats within the hosts. In schistosomes, for example, the tegument is ~ 4 µm thick and is limited externally by a heptalaminate membrane, which extends and interconnects to form a lattice and thereby increases the tegumental surface at least 10-fold. In others, the membrane invaginates (instead of pitting) and thereby amplifies the surface, as in *Fasciola hepatica*. Briefly, the digenean tegument is structurally adapted for transport, immune-evasion

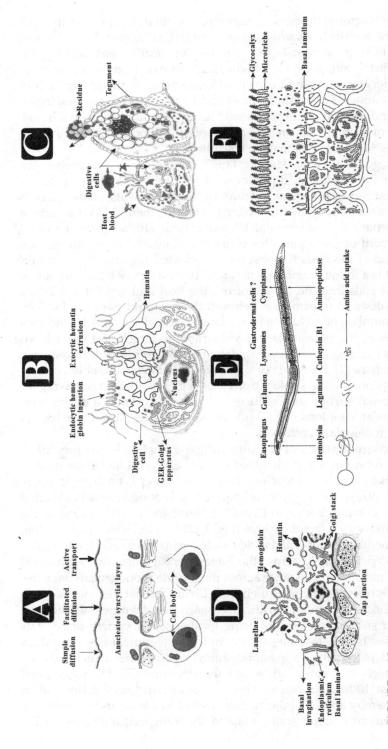

### FIGURE 1.12

The digestive cell and associated syncytial tegument of (A) typical parasitic flatworm, (B) monogenean *Dicliophora merlangi*. (C) digeneans *Fellodistomum fellis*, (D) *Schistosoma mansoni* (free hand drawings from Halton, 1982, Smyth and Halton, 1983, Threadgold, 1984) and (E) Hemoglobin digestion process in blood fluke *Schistosoma mansoni* (free hand drawings from Dzik, 2006). (F) Illustration to show the cross section through tegument and underlying tissues in cestode *Hymenolepis diminuta* (free hand drawing from Threadgold, 1984).

*General Introduction* 37

and communication with the neuromuscular system via the gap junction. In them, the inwardly directed $Na^+$ gradient maintains glucose flux by an active $Na^+$ dependent carrier-mediated process. Adult schistosomes depend solely on host plasma glucose for energy. However, amino acid uptake relies exclusively on diffusion. Trematodes do not synthesize cholesterol *de nova* but tegumentally acquires it from host (Thompson and Geary, 1995).

The highly specialized cestodes, bereft of any vestige of a mouth, pharynx and gut, have to rely solely on the tegumental surface to acquire the entire spectrum of nutrients. Their tegumental external surface is evaginated into finger-like structures called 'microtriches', reminiscent of the apical microvilli of the vertebrate intestinal epithelial cells (Fig. 1.12F). The microtriches amplify the functional tegumental surface area by 2–12 folds. For example, the amplification is 2.05, 2.61 and 2.94 fold in *Hymenolepis microstoma, H. diminuta* and *H. nana*, respectively (Berger and Mettrick, 1971). Indeed, the cestode body plan has been conceptualized as an inside-out intestine with tegumental surface performing the intestinal mucosa. In them, uptake of nutrients across the tegument is summarized hereunder: (1) Cestodes derive most of their energy from glucose, which is actively absorbed across the tegument by a single $Na^+$ mediated transport system. Galactose and glycerol can also be absorbed by both carrier mediated uptake or diffusion. (2) Their amino acid transport system is characterized by high affinity for D-amino acids, as against L-amino acids in mammals. In them, six transport systems have been identified for amino acids, of which four transport neutral amino acids and one each for acidic and basic amino acids. They can synthesize at least 17 proteins in the microtriches (e.g. *H. diminuta,* Halton, 1997, see also Pappas and Read, 1975). A gradient of DNA and PNA (pentosenucleic acid) along the strobila and a corresponding protein gradient are reported in *H. nana*. The highest rate of protein synthesis is detected in mature and gravid proglottids of *H. diminuta*. A high level of protein synthesis occurs in the testis and ovarian primordial of *H. nana* (see Henderson and Hanna, 1988). (3) Cestodes cannot synthesize lipids and have to depend entirely on their hosts. With relatively a larger surface, the poultry tapeworm *Raillietina echinobothrida* is able to absorb short and long chains of fatty acids through a mixture of diffusion and mediated transport much faster than nematode parasites (see Mondal et al., 2016). Cholesterol, cardiolipids and phosphatidylethanolamine are all tegumentally absorbed. Recently, Navarette-Perea et al. (2014) found 17 host proteins in the vesicular fluids of *Teania solium* cysticercus; these proteins are assumed to have been absorbed through a non-specific mechanism of fluid pinocytosis. (4) Cestodes can synthesize pyrimidines *de nova* but purines must be absorbed from the host (Thompson and Geary, 1995).

*Digestion*: In parasitic flatworms, intracellular digestion is the rule except in schistosomes. Dzik (2006) listed among others, the known proteinases of digeneans and cestode (Table 1.6). Proteinases hydrolyze peptide bond. On

## 38 Reproduction and Development in Platyhelminthes

## TABLE 1.6

Proteinases of platyhelminths known as important for host-parasite relationships (modified from Dzik, 2006)

| Proteinase families | | | Platyhelminthes |
|---|---|---|---|
| Cysteine proteinases | Family C1 Papain-like | Cathepsin B-like | *Schistosoma, Fasicola* |
| | | Cathepsin L-like | *Paragonimus, Schistosoma, Fasciola, Spirometra* |
| | | Cathepsin S | ***Spirometra mansoni* plerocercoid** |
| | | Cathepsin C | *Schistosoma japonicum* |
| | Family C2 Calpain-like | | *S. mansoni, S. japonicum* |
| | Family C13 Legumain-like | Asparaginyl endopeptidase | *S. mansoni, S. japonicum, F. hepatica* |
| Serine proteinases | Family S1 Chymotrypsin | Elastase | *Schistosoma* |
| | | Chymotrypsin-like | *Schistosoma* cercaria, ***Schistocephalus solidus* procercoid, *Spirometra mansoni* plerocercoid** |
| | | Trypsin-like | ***Spirometra mansoni* plerocercoid,** *Schistosoma* cercaria |
| | | Kallikrein-like | *S. mansoni,* ***Hymenolepis diminuta*** |
| Aspartic proteinases | Family A1 Pepsin | Cathepsin D | *Schistosoma japonicum* |
| | | Aspartic endopeptidase | *S. mansoni* |
| Metallo-Proteinases | Other helminth metaloproteinases | Collagenase | *S. mansoni* |
| | | Dipeptidyl peptidase III | *S. mansoni* |
| | | Metallominopeptidase | *S. mansoni* |

the basis of the hydrolytic site, they are grouped into four classes namely (i) cysteine- (ii) serine- (iii) aspartic acid- and (iv) metallo-proteinases. Owing to their economic importance, the tested species are limited to *Schistosoma* spp, *Fasciola hepatica* and their larvae among digeneans as well as *Schistocephalus solidus* procercoid, *Spirometra mansoni* plerocercoid, and *Hymenolepis diminuta* among cestodes. As the tegumental uptake in cestodes is limited to amino acids and peptides, incidence of proteinases is also limited to chymotrypsin and cathepsin but cathepsin L-like cysteine proteinase is added to *Spirometra*.

As blood is the major constituent of food for digeneans and branchial monogeneans, its digestion has received some attention. By the time the blood meal reaches the gut, it is already hemolyzed and is rapidly abstracted from the gut lumen for intracellular digestion (e.g. *D. merlangi*). In the digenean *Fellodistomum fellis* that feeds on the blood of *Anarhichas lupus*, it is neither intracellular nor extracellular (in the gut lumen) but occurs inside a pocket

General Introduction  39

of vascularized lamellar folds of the surface membrane in the apical cavity of the gut cell (Fig. 1.12C). In *Schistosoma*, it is extracellular, i.e. occurs in the gut lumen. Digestion of hemoglobin involves cleaving of the globin moiety to its constituent amino acids and peptides by digestive enzymes. But it leaves the residual haem accumulated as insoluble hematin. To eliminate the hematin, hematophagous flukes have invented different routes. In *D. merlangi*, the digestion involves two distinct components: (1) Digestive cells sunken below the syncytium are bulged into the lumen and a syncytial tissue that serves to separate and support individual digestive cells. The syncytium is distended, with blood meal or irregularly folded within the empty gut. Into the digestive cells, digestion occurs with enzymes secreted by ribosomes, packed by Golgi apparatus and released through lyzosomal channels. Transitory outlets are provided for the exclusion of hematin into the gut lumen, from where it is regurgitated (Fig. 1.12B). In *F. fellis*, the hemolyzed blood is entrapped and digested within the just described pocket that is permanently open to the gut lumen via a small opening. The indigestible residue amassed and periodically released from the apical cavity of the digestive cell into gut lumen through an overlying syncytium (Fig. 1.12C). In schistosomes, the proteinases include all types of cystine (e.g. cathepsins), serine (chymotrypsin, trypsin), aspartic (e.g. aspartic endopeptidases) and metallo (e.g. collagenase) proteinases (Table 1.6). As a consequence, the digestion is a rapid process and occurs in the following sequence: erythrocyte lysis in esophagus, cleaving hemoglobin into polypeptides and then to oligopeptides in the anterior and posterior gut, respectively followed by absorption of amino acids by the gastrodermal cells (Fig. 1.12D, E).

*Metabolism*: Despite a superabundant supply of nutrients, the cestodes have to encounter the challenge posed by an anoxic climate prevailing in the intestinal lumen. The challenge is more acute and intense, as they have neither a circulatory system nor respiratory pigments (Alvite and Esteves, 2011). Not surprisingly, their metabolism is predominantly anaerobic. More surprisingly, they continue anaerobiosis, even in the presence of oxygen (Komuniecki and Harris, 1995). Bathed by blood or body fluids, digeneans excrete primarily lactate as an end product of anaerobic glucose metabolism. In contrast, the lumen-dwelling cestodes rely on stored glycogen for energy and possess unique pathways to increase the number of ATPs production in the absence of oxygen. For example, glycogen content ranges from 10–15% for trematodes but 20–50% for cestodes (see Halton, 1997). While maintaining redox balance by efficient reduction of fumerate to succinate, cestodes produce succinate and acetate as end products of their anaerobic dissimilation (e.g. *H. diminuta*). But digenetic trematodes decarboxylate, further the succinate to propionate and produce primarily propionate and acetate as end products (e.g. *F. hepatica*) (see Komuniecki and Harris, 1995). Hence, digeneans and cestodes have devised unique pathways to use unsaturated organic acids as terminal electron acceptors instead of oxygen and extract more ATPs.

## 1.6 Neoblasts

In flatworms, the differentiated somatic cells do not divide (Peter et al., 2004, Cebria et al., 2015). Despite high diversity in terms of morphology, embryology, life cycle complexity and potency for regeneration and clonal reproduction, Platyhelminthes are unique in possessing a population of undifferentiated stem cells called neoblasts with mitotic power (see Baguna, 1974); the neoblasts are the only source of new cells for normal tissue renewal and turnover (Koziol et al., 2014). They are small in size (~ 10 µm), round to ovoid in shape, possess a high nucleus/cytoplasm ratio and its scanty cytoplasm have a few mitochondria, many free ribosomes and a few other organelles (Hori, 1992). Classical methods of staining neoblasts involved the labeling of S-phase cell through incorporation of thymidine analog 5'-bromo-desoxy-uridine (BrdU) or utilization of anti-phospho-histone H3 antibodies in immunohistochemistry (Brehm, 2010). In flatworms, different nomenclatures for stem cells have been used: German authors have named them Buildungzellen, Wanderzellen, Stofftager, Stamenzellen until Randolph (1897) called them neoblasts. In English, they are named as regenerative cells, replacement cells or reserve cells in the family Planuridae, neoblasts or regenerative cells in the order Acoela, neoblasts in the family Macrostomidae, germinal cells in the class Trematoda and germinative cells or stem cells in the class Cestoda (Reuter and Kreshchenko, 2004). For reasons described below (see also p 235), it may be correct to retain the terms neoblasts for turbellarians, germinal cells for digeneans and germinative cells for cestodes.

The asymmetrically dividing precursor stem cells in cestodes and digeneans are distinguished from symmetrically mitotically dividing proliferative stem cells in turbellarians. Ultrastructural studies have revealed the co-occurrence of so called 'light' stained and 'dark' stained undifferentiated cells in developing metacestode tissue (Sakamoto, 1982, see also Fig. 8B in Reuter and Kreshchenko, 2004). The light-stained undifferentiated cells may be precursors of the dark-stained type. They may represent true self sustained stem cell lineage and their asymmetrical cell divisions give rise to dark-stained and light-stained daughter cells during larval development (Brehm, 2010). In the germinal cells of *Schistosoma mansoni*, this morphological heterogeneity is also apparent. In them, the molecular heterogeneity indicates that *nano*-2[+] cells displaying slower cell cycle may be the precursor pluripotent stem cells and the *nanos*-2[−] cells are more primed toward somatic fate (Wang et al., 2013). In the acoel *Isodiametra pulchra*, BrdU pulse-chase tracking has indicated that after a 10 day chase, only ~ 7% of labeled cells display stem cell morphology and the remaining 93% differentiated somatic cells (De Mulder et al., 2009).

The classic publications of Wolff and Dubois (1947, 1948) and Dubois (1949) have revealed that (i) the planarian neoblasts are susceptible to irradiation, (ii) with irradiation, they are destroyed and the planarian loses its potency to regenerate and renew its cells, (iii) shielding different portions of the body from irradiation and subsequent estimation of regenerative potency of the irradiated planarian, the time required for regeneration has been found to be proportional to the length of irradiation and (iv) this discovery has also confirmed the existence of neoblasts and their expansion toward the wound. Using nuclear and cytoplasmic markers in conjunction with the grafting technique, Salo and Baguna (1985a) conclusively showed that neoblast migration was not an active one toward the blastema but rather cell spreading due to proliferation (Newmark and Alvarado, 2001). The neoblasts are identified by their proliferation and other cytological features like the presence of electron 'dense lump', 'nuclear satellite material' and/ or chromatoid body. To identify and trace their distribution more precisely, Orii et al. (2005) cloned the Proliferating Cell Nuclear Antigen (PCNA) from *Dugesia japonica*. The identified PCNAs, on being treated with X-ray irradiation, are rapidly lost. Hence, the PCNA identified cells are indeed the neoblasts.

The turbellarian neoblast contains a unique cytoplasmic organelle called the chromatoid body. During cytodifferentiation, the chromatoid bodies decrease in number and size, and subsequently disappear completely in differentiated cells during regeneration (Hori, 1982). Thereby, the differentiated cells in the flatworms have lost the potency to divide further. This seems to be a more appropriate mechanism for manifestation of homeostasis in regenerating proliferating neoblasts. But it is likely that the suggested autocrine(s) is secreted by neoblasts at their high density (Salo and Baguna, 1985b). The chromatoid body structurally resembles that of germline granules of germline cells of other animals. The products of *vas* or *vas*-related genes are associated with germline granules. Hence, one of the chromatoid bodies may be the product of *vas*-related gene. In *Dugesia japonica*, Shibata et al. (1999) succeeded in isolating two *vas*-related genes *DjvlgA* and *DjvlgB* from the testicular and ovarian regions. Their sequence homology was closer to the mouse *PL10* and *An3* than to *vas*-homologs in other animals. *PL10* is expressed only in testis (Gururajan et al., 1991). But both *DjvlgA* and *DjvlgB* are expressed in testis as well as in the ovary of the hermaphroditic planarians, although that of *DjvlgB* is more restricted to testis. In sexual and clonal planarians, the transcripts of these genes are detectable at the same level. However, expression of *DjvlgA* is also distributed in the mesenchyme from the head to tail. Hence, *vas*-related genes are expressed not only germ cells but also in putative neoblasts. This suggests that *DjvlgA* and *DjvlgB* may play a master role in differentiation and maintenance of homeostasis of neoblasts. In the acoel, *Isodiametra pulchra*, De Mulder et al. (2009) isolated two *piwi*-like genes *ipiwi1* and *ipiwi2*, characterized by germline expression. In fact, several genes with conserved roles in germ cell development are expressed in neoblasts including the homologs of *vasa, piwi, pumilo* and *bruno* (see Newmark et al., 2008). Interestingly, the expression of *ipiwi* 1 is also extended to a subpopulation of somatic neoblasts, a situation only known from rhabditophoran flatworms. In fact, the border between somatic and

42  *Reproduction and Development in Platyhelminthes*

germ cell lineages is not clearly demarcated in acoels, rhabditophorans and triclads (see also Baguna et al., 1999), as it is in sponges and cnidarains. In all these primitive aquatic invertebrates, germ cells can be formed *de novo* from somatic stem cells.

In *I. pulchra*, the number of germline specific *ipiwi* positive cells is diminished following prolonged starvation. In this acoel, the body is dramatically reduced in size and is devoid of reproductive organs at morphological level. After refeeding, the worm regrows to a reproductive adult stage within a month (De Mulder et al., 2009). A neuroendocrine factor called *npy-8* is essential for the maintenance of reproductive system in adult planarians, as it is expressed exclusively in the nervous system of sexually reproducing planarians (Collins et al., 2010, see also p 230). The abrogation of *npy-8* function leads to regression of the reproductive organs in adult planarians (e.g. *Planaria maculata*). Since an ortholog of *npy-8* is found in the genome of *Schistosoma mansoni*, it is likely that planarians and schistosomes may engage a similar neuroendocrine mechanism to control the development of their reproductive organs. As if to support this view, the reproductive organs of mature female *S. mansoni*, following separation from its male partner, are resorbed and are able to regenerate them, only when the male-female pairing is reestablished (Collins and Newmark, 2013).

The inhibition of DNA by hydroxyurea (HU) results in the arrest of proliferation of cells in the S-phase of cell cycle and a pause of cell cycle progression is inhibited by ribonucleotide reductase. An interruption of cell proliferation is ultimately reflected in germ cell lineage. Following a 10-day treatment of *I. pulchra* with HU, *ipiwi1* expression persists only in mature eggs but not in testes. This indicates a faster turnover of germ cells in testis than in the ovary. Ultimately, egg production is dramatically decreased from 1.1 eggs/worm/day in an untreated worm to 0.8, 0.3 and 0.2 egg/worm/day in those treated for 5, 10 and 15 days, respectively (De Mulder et al., 2009).

*Neoblast distribution*: In the turbellarian body, three major patterns of neoblasts distribution can be recognized. Pattern 1 includes the acoel *Isodiametra pulchra* (De Mulder et al., 2009), macrostomorphid *Macrostomum hystricinum* (see Peter et al., 2004) and proseriate *Monocelis* sp (Girstmair et al., 2014) with limited potency for regeneration and low potency for clonal reproduction. F-action staining in *Monocelis* sp, (Girstmair et al., 2014) and laser scanning micrograph of BrdU labeled S-phase neoblasts displaying bright fluorescence against the dark background in the macrostomorphid *Microstomum* sp have revealed the distribution of neoblasts in two broad lateral bands along the antero-posterior axis but its anterior extension is delimited by the brain. In them, no proliferating neoblasts are detectable in the brain and epidermis (Fig. 1.13A, B, C). In *Monocelis* sp, adhesive organs at the tip of the tail plate and genital organ can also be regenerated within 2–3 days and 8–9 days after amputation, respectively. Proliferation within blastema has been detected in anterior and posterior blastemas but notably missing in triclads (Girstmair et

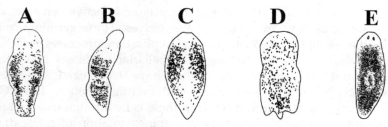

**FIGURE 1.13**

Distribution of proliferating neoblats in turbellarians. A. *Macrostomum hystricinum*, B. *Microstomum* sp, C. *Isodiametra pulchra*, D. *Convolutriloba longifissura* and E. *Dugesia japonica* (modified free hand drawings from Peter et al., 2004, De Mulder et al., 2009, Akesson et al., 2001, Newmark and Alvarado, 2002).

al., 2014). Pattern 2, polyclads like *Notoplana humilis* (Okano et al., 2015) and some macrostomorphid like *M. hystricinum* can be included (Rieger et al., 1999) and have the potency for posterior regeneration but not the anterior. In them, one population of neoblasts namely Pluripotent Stem Cells (PSCs) is located in lateral bands along the main longitudinal nerve cods and another namely Intestine Stem Cells (ISCs) in the gastrodermis. In animals, the intestinal epithelium is physically eroded and biochemically damaged during digestion and absorption. In *N. humilis*, the ISCs proliferate maximally within 5 hours. Clearly, the non-triclad turbellarians carry out morphallactic and epimorphic regeneration by ISCs. Pattern 3, which includes most clonally reproducing triclads like *Dugesia japonica* (Orii et al., 2005) and acoels like *Convolutriloba longifissura* (Akesson et al., 2001). In them, the BrdU labeled S-phase epifluorescence and PCNA marker, respectively have revealed the homogenous distribution of neoblasts throughout the body (Fig. 1.13D, E). However, the distribution in *D. japonica* is relatively denser especially, the neoblasts being present in clusters along the midline and bilateral bands in the dorsal mesenchyme. Evidently, homogenous distribution of neoblasts throughout the body of turbellarians is a pre-requisite to accomplishing clonal reproduction.

*Neoblasts in neodermatans*: In view of their economic importance, research on the equivalents of neoblasts has been limited to a few species of digeneans, for example, *Schistosoma mansoni* (Collins and Newmark, 2013, Wang et al., 2013) and *Fasicola hepatica* (McCusker et al., 2016) as well as a few cestodes, e.g. *Mesocestoides corti* (Koziol et al., 2010) and *Echinococcus multilocularis* (Koziol et al., 2014). Despite the monogenean *Gyrodactylus* is known to undergo polyembryonic clonal reproduction, research on their germinal cells remains to be undertaken (however, see Ohashi et al., 2007). Serial transplantations of sporocysts of *S. mansoni* into naïve snails have led to continuous clonal production of sporocysts and cercariae (Jourdane and Theron, 1980). Briefly, the sporocyst is essentially a sac filled with germinal cells that produce more daughter sporocysts or infective cercariae in the same manner, as they are generated themselves. These germinal cells are responsible for the transplantation and replication processes and have a molecular signature similar to those of turbellarian neoblasts. In *S. mansoni*, ultrastructural and histological studies have recognized the presence of stem cell-like morphology and rapid cycling kinetics. The POPO-1 staining technique is useful in tracking these cells through different stages of intramolluscan development. *In vivo* transplantation of miracidia into sporocyst triggers germinal cell proliferation. The *S. mansoni* genome has the argonant homologs, two *nanos* homologs and three *vasa*/PL10 (a paralog of *vasa*) are abundantly expressed in sporocysts. Notably, schistosomes do not have a true *vasa* ortholog; however, the *vasa*/PL10 homolog *vlg* 3 may assume a '*vasa*-like' role in them. In monogeneans too, the role of *vasa* is played by *vlg* 3 (Ohashi et al., 2007). In *Neobenedenia girellae*, an important pathogen of yellowtail and amberjack, *vasa*-like genes, *Ngvlg1*, *Ngvlg2* and *Ngvlg3* are reported to express only in germ cells. More importantly, that the commonality between the germ cells and adult stem cells identified in adult schistosomes (Collins et al., 2013) suggests that the germinal cells may persist throughout the entire life cycle of schistosomes (Wang et al., 2013).

A significant breakthrough in the cultivation of isolated parasitic cells of the cestode *Echinococcus multilocularis* (Brehm, 2010) and a combination of electron microscopy and uptake of radioactively labeled thymidine studies on the cellular dynamics of proliferating cells have revealed that germinative cells—the counterpart of turbellarian neoblasts—are the only mitotically active cells in cestodes that give rise to all other differentiated cells. Interestingly, isolated cells from the cysticerci of *Taenia crassipes*, on intraperitoneal injection into mice, develop into complete cysticerci, which give rise to sexually reproducing adults (Toledo et al., 1997). Brehm (2010) showed that it is possible to obtain the completely regenerated metacestode of *E. multilocularis* from cultured dispersed cells. However, these germinative cells lack Chromatid Bodies (CBs) and true *vasa* and *piwi* orthologs. Incidentally, it is not known whether or not the *vlg* 3 is present and assumes the role of *vasa* in cestodes as well as whether or not germinal cells of trematodes possess CBs. Clearly, these are morphological and molecular heterogeneties between turbellarian neoblasts, trematode germinal cells and cestode germinative cells.

Evidence is accumulating for the existence of a neoblasts pool as a heterogenic pluripotent system. Different staining and quantitative video microscopic DNA and RNA studies have revealed the heterogeneity. In planarians, the existence of four subtypes of neoblasts has been confirmed (Schurmann et al., 1998, Behensky et al., 2001). This account proposes the existence of at least three types of neoblasts: (i) In all flatworms, stem cells responsible for renewal and turnover of somatic cells may exist. (ii) Stem cells responsible for regeneration of missing body parts are present in some

*General Introduction*   45

## TABLE 1.7

Factors controlling morphogenesis in flatworms (condensed from Reuter and Kreshchenko, 2004, Cebria et al., 2015)

---

### (i) Neuronal substances

1 a. Tissue extract isolated from *Dugesia japonica* inhibits head regeneration and growth by 40%

b. Neuronal substance influences fission and regeneration by I. Inducing mitogenic effect, II. Playing a role in pattern formation via peptidergic immune-reactivity in the peripheral nerve net. III. Growth factor in free-living and parasitic flatworms

### (ii) Serotonin (S-HT)

2 a. Regenerative process is regulated by a signaling pathway through serotonin receptors. Regeneration is inhibited by dihydroergocriptine and methiothepin, i.e. antagonists of serotonin receptors (most turbellarians, Strigeidida, Pseudophyllidae)

b. Melatonin, displaying a circadian time-keeping mechanism, inhibits head and tail regeneration

### (iii) Neuropeptides (NPs)

3 a. Six neuropeptides (FMRF amide-related peptides) and a NPF have been isolated

b. NPF precursor gene (npf) has been identified and characterized in the cyclophyllid cestode *Moniezia expansa* and turbellarian *Artioposthia triangulata* (Dougan et al., 2002).

c. NPF stimulates regeneration of the cephalic ganglia and pharynx in *Girardia tigrina*

d. NP substance P and K stimulate cell proliferation during planarian regeneration. NPY also stimulates the same in planarians and acoels

e. $Ca^{2+}$-transporting peptide, vasopressin, dalargin somatosatin and hydra head activator stimulate and release luteinizing hormone releasing hormone (LH-RH), which inhibits regeneration of cephalic ganglia in *Dugesia tigrina* and cestode *Diphyllobothrium dendriticum*

### (iv) Growth factors

4. Correlation between mitotic activity and Epidermal Growth Factor (EGF) in *D. dendriticum* suggests a role for EGF

### (v) Weak electromagnetic fields (EMF)

5. Weak EMF (DC, 42 mT) and extra weak alternated current (AC, 100 nT, 1–60 Hz) magnetic fields increase fission intensity in *D. tigrina*. Weak EMF increases mitotic index by ~ 30% in the blastema of planrians

---

turbellarians (Table 2.7) and a few trematodes and cestodes (see Cebria et al., 2015). (iii) Stem cells, responsible for clonal multiplication, are also present in a few turbellarians, monogeneans (e.g. *Gyrodactylus*), all digeneans (intramolluscan clonal multiplication) and cestodes (clonal reproduction during larval stage of eucestodes).

*Nervous system and its role*: In the absence of endocrine glands and circulatory system, humoral signaling by neuropeptides (NPs) may play a key role in regeneration and clonal reproduction. For example, the inability of postpharyngeally amputated *Macrostomum* can be correlated with a lack of serotogenic neurons in the posterior regions. The presence of serotogenic

46  *Reproduction and Development in Platyhelminthes*

neurons along the main nerve cords has also been related with clonal reproduction in *Microstomum lineare* (see Cebria et al., 2015). Hence, a brief description of the nervous system and its secretory neurons is a necessary prelude. In flatworms, the presence of multiple numbers of longitudinal nerve cords seems to be a characteristic feature. The cords can be placed at different positions, dorsal, lateral, marginal, ventro-lateral or ventral. The evolutionary tendency seems to reduce the number of cords by thickening and possessing more neuron positive for serotonin and catecholamine markers. In turbellarians, the cord number can be three to five. The main cords are lateral in Catenulida, Macrostomida and Proseriata, but ventral in Tricladida. In trematodes, three pairs of the cords (dorsal, lateral, ventral) are present; of them, the ventrals are the largest and are interconnected at the caudal tip. A variable (one to 38 or 60) number of cords are present in cestodes but the lateral ones are the main cords (Bullock and Horridge, 1965, Joffe and Reuter, 1993). In contrast to other invertebrates, flatworms have a large variety of neural cells including uni-, bi- and multi-polar neurons. A characteristic of their neurons is the high content of secretory vesicles. Some of these vesicles are small and cholinergic, others are aminergic and still others are peptidergic.

Research in recent years has revealed the importance of biologically active morphogene molecules in regeneration of flatworms. Table 1.7 briefly summarizes the neuronal factors controlling morphogenesis in flatworms. These factors include (i) neuronal substances, (ii) serotonin, (iii) neuropeptides (NPs), (iv) growth factors and (v) electromagnetic field. They seem to control the processes of proliferation, 'migration' and differentiation of neoblasts. Notably, some of them like the gene *npf* (see p 230) and neuroendocrines like serotonin are common to turbellarians, trematodes and cestodes.

# 2

# Turbellaria

## Introduction

The 3,000–6,500 speciose (Table 1.2) Turbellaria compose the commonly known planarians and other free-living flatworms. They are clothed with ciliated cellular or syncytial epidermis and are free-living hermaphrodites characterized by sexual reproduction with a simple direct life cycle. However, each of these characteristics includes exceptions. (1) Of them, ~ 300 species are ecto- (e.g. *Bdelloura candida* on *Limulus polyphemus*), endo-symbionts (e.g. *Syndesimus antillarum* in the gut of echinoderms (Jennings, 1997) and parasites (e.g. *Ichthyophaga* sp in the parrotfish *Scarus rivulatus*, Cannon and Lester, 1988). The voracious predator *Stylochus frontalis*, known as 'oyster leech', is deleterious to the oyster industry. The ovary of the alleocoel *Plagiostomum* is degenerated by the parasitic rhabdocoel *Oekiocolax plagiostomorum* (Hyman, 1951). Several daleyellioids, some typhloplanids (see Schockaert et al., 2008) and acoels like *Convoluta convoluta* and *C. roscoffensis* ingest, acquire and harbor autotrophs like *Platymonas convoluta* intracellularly in epidermis and draw nutrients; they consume less food than the acoels holding no symbionts (see Pandian, 1975). (2) A few of them are gonochorics (e.g. *Kronborgia amphipodicola*, Christensen and Kanneworff, 1964); some others can be parthenogens (e.g. *Schmidtea polychroa*, Navarro and Jokela, 2013). (3) Some rhabdocoels like *Typhloplana* produce self-fertilized subitaneous eggs and cross fertilized dormant eggs (see Hyman, 1951). (4) The life cycle of some triclads includes clonal reproduction initially (or seasonally) and subsequently sexual reproduction (e.g. *Dugesia ryukyuensis*, Kobayshi et al., 2002b). (5) In turbellarians, the indirect life cycle is described in two distantly related groups. The cycle involves the free-swimming Muller or Gotte larva in polyclads (Rawlinson, 2014) and Luther larva in catenulids (*Rhynchoscolex simplex*, see Norena et al., 2015).

Moved by ciliary action, some turbellarians swim in waters (500 μm/s, Wikipedia); others glide, creep or crawl on the substratum; still others float

48  *Reproduction and Development in Platyhelminthes*

on weeds like *Sargassum* (e.g. *Amphiscolops sargassi*) and yet others drift as plankton in pelagic waters (e.g. *Alaurina composita*). The habitat distribution of turbellarians ranges from the exclusive marine acoelids, extends from marine (Maricola) to freshwater (Paludicola) and to *terra firma* (Terricola in the moist forest floors of tropics and subtropics) in polyclads. In rhabdocoel, it is limited to marine and freshwater habitats. The triclad family, Kenkiidae includes exclusively cave-dwellers. Reynoldson et al. (1965) reported that the upper limit of vertical distribution ranges from 122 m altitude for *Dugesia lugubris* to 183 m for *Polycelis tenuis*. However, the distribution ranges from 1,000 m depth in mostly sediment-inhabiting polyclads to 3,180 m altitude (e.g. *Mesostomum arctica* in Canadian tundra) and to 4,545 m in *Mesostomum* sp in Tibet, i.e. their vertical distribution covers over 5,500 m. For them, hot soda water is not an uncommon habitat, for example, 45°C for *Microstomum thermale* and 40–47°C for *M. linae*. Briefly, their distributions, modes of motility and reproduction are fascinating and have attracted the attention of zoologists for more than a century.

# Part A: General Features

## A 2.1 Taxonomy and Distribution

Turbellarians have the reputation of a group 'difficult to identify' (Schockaert et al., 2008). Not surprisingly, the number of taxonomists capable of identifying these flatworms to species level is fewer than 10 worldwide (Dumont et al., 2014). Nevertheless, the identified species ranges from 3,000 to 6,500 (Table 1.2). A cladogram assembled by Schockaert (1996) indicates the presence of 5,500 species distributed in 10–11 orders (Fig. 2.1A). Of them, Polycladida (2,000 species) and Tricladida (1,000 species) are speciose. Attempts have also been made to build cladograms for polyclads and triclads. Based on life history traits, Rawlinson (2014) assembled a cladogram to describe the phylogeny among the acotylea characterized by the direct life cycle or indirect development with Gotte larva, and cotylea with Muller larva (Fig. 2.1B). In triclad planarians, the use of molecular markers to resolve phylogenesis has uncovered more diversity and taxonomic richness of the group. However, the resolving power of the markers is limited (Alvarez-Presas and Riutort, 2014). Hence, a combination of cladograms has been assembled by several authors, who have used various methods for tree-building and to assess all the clades of turbellarians (Fig. 2.1C). The taxa marked by light letters at the top row in Fig. 2.1C are monophyletic and exclusively continental in habitat. Briefly, the number of turbellarian species may not exceed 7,000. Due to paraphyletic

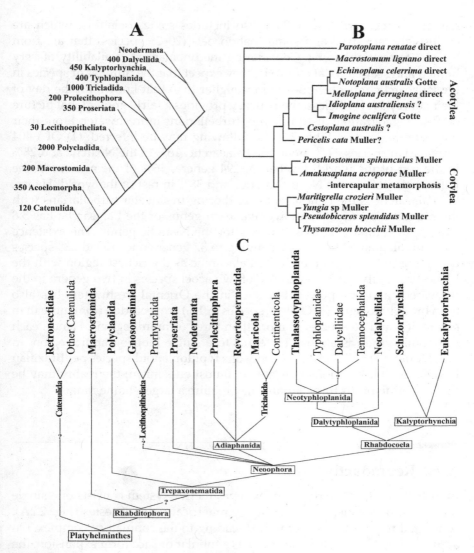

**FIGURE 2.1**

(A) Cladogram of Turbellaria with approximate number of known species, (B) Cladogram of Polycladida (free hand drawings from Schockaert, 1996, Rawlinson, 2014) and (C) Combined cladogram of turbellarians based on the molecular marker 18S rDNA. Polyphyletic taxa are indicated by bold letters (modified and redrawn from Schockaert et al., 2008).

origin, the assemblage of phylogenetic cladograms using taxonomic or life history traits or molecular markers is expected to vary widely.

Acoels and polyclads are exclusively marine habitants. However, not much is known about their geographic distribution. To describe the distribution of freshwater turbellarians (1,404 species) in the eight recognized biogeographic

50    *Reproduction and Development in Platyhelminthes*

regions, Schockaert et al. (2008) also includes some tricladids, which are exclusively terrestrial and some rhabdocoels (20–25 species) that are from the wet terrestrial habitat. Considering the unbalanced availability of very limited and scattered data for Asia, they expect that the number of species in the Palaearctic must be ~ 5–10 times higher than that known on the day of their survey. Nevertheless, their survey provides a fairly good global picture on the biogeographic distribution of turbellarians in freshwater. From their data summarized in Table 2.1, the following may be inferred: (1) Of 1,404 turbellarian species, 56% are in the Palaearctic, 16% in the Nearctic and 28% in the rest of the world. Further, of 294 genera, 46% of them occur in the Palaearctic, 16% in the North America and 38% in rest of the world. Hence, the Palaearctic is the richest region harboring freshwater turbellarians with the highest diversity. On an average, each genus of the Palaearctic has 5.8 species, in comparison to 4.7 species for the Nearctic genus. The existence of typhloplanoid 233 species belonging to 37 genera and triclad 238 species belonging to 33 genera enrich the Palaearctic as the richest region with the highest diversity. (2) The presence of two acoel species in two genera in the Palaearctic and one polyclad species in the Oriental freshwaters has also been brought to light. (3) Only 76 species (6.1%) are recorded from more than one region and 16 species in more than three or more regions. Hence, each taxonomic group is almost 'endemic' to a single biogeographic region. In fact, Dumont et al. (2014) found only one typhloplanoid species per Brazilian Lake. These observations indicate an almost endemism; the reality may be that the carnivorous turbellarians may require a larger 'home range'.

## A 2.2  Reproductive System

In turbellarians, the fairly complex reproductive system consists of a single (Fig. 2.2B) or a series of bilaterally symmetrical pairs of testes (Fig. 2.2A), connected to a common sperm duct leading to the copulatory complex. On approaching the complex, the duct has a tubular or sacciform expansion, the seminal vesicle serving to store mature sperm. The terminal duct opening into the copulatory complex is the ejaculatory duct. The complex consists of (i) the tubular or pyriform seminal vesicle, which helps to propel the sperm onward during the ejaculation, (ii) the penis and (iii) also a gland complex named the prostate apparatus. The seminal vesicle may open directly into the penis or through the apparatus. From its simple to complex form, the penis may develop into papilla, stylet, bulb or cirrus. Multiplication of male copulatory organ or its part is very common and the extra parts may stimulate copulation. The penis is projected into the gonopore through an antrum prior to penetration into that of the mating partner. Individual testis possesses an outer layer of spermatogonia that undergo three rounds

## TABLE 2.1

Number of species and genera in freshwater turbellarians of different geographical regions (compiled from Schockaert et al., 2008). In each item, the first number represents the number of species and the second, the number of genera

| Taxon | PA | NA | NT | AT | OL | AU | PAc | ANT | TOT |
|---|---|---|---|---|---|---|---|---|---|
| Acoela | 2/2 | –/– | –/– | –/– | –/– | –/– | –/– | –/– | 2/2 |
| Catenulida | 36/9 | 36/6 | 45/8 | 10/5 | 1/1 | 1/1 | –/– | –/– | 90/10 |
| Macrostomida | 43/4 | 26/2 | 3/2 | 14/1 | 2/2 | 1/1 | –/– | –/– | 84/4 |
| Polycladida | –/– | –/– | –/– | –/– | 1/1 | –/– | –/– | –/– | 1/1 |
| Lecithoepitheliata | 20/3 | 4/2 | 4/3 | 3/2 | –/– | 3/1 | –/– | 1/1 | 31/8 |
| Proseriata | 6/5 | 1/1 | 3/3 | 1/1 | –/– | –/– | –/– | 1/1 | 11/3 |
| Prolecithophora | 12/5 | 2/1 | –/– | –/– | 1/– | 1/1 | –/– | –/– | 12/9 |
| Dalyellioida | 98/14 | 28/6 | 25/3 | 13/4 | 1/– | 3/3 | –/– | –/– | 159/16 |
| Typhloplanoida | 233/37 | 56/15 | 13/6 | 19/10 | 4/3 | 10/2 | –/– | 1/1 | 307/42 |
| Temnocephalida | 18/5 | –/– | 20/1 | 1/1 | 3/1 | 56/10 | –/– | –/ | 98/15 |
| Kalyptorhynchia | 82/20 | 2/2 | 1/1 | 1/1 | 1//1 | 1/1 | –/– | –/ | 82/20 |
| Tricladida | 238/33 | 66/12 | 36/6 | 23/3 | 23/3 | 40/10 | 2/2 | 3/3 | 426/51 |
| Total | 788/137 | 221/47 | 150/33 | 85/28 | 36/12 | 116/30 | 2/2 | 6/5 | 1,404/294 |
| % obs. of total obs. | 56.2/46.4 | 15.8/15.9 | 10.7/11.2 | 6.1/9.5 | 2.6/4.1 | 8.3/10.2 | 0.1/0.7 | 0.4/2.0 | –/– |
| Species/genus | 5.8 | 4.7 | 4.5 | 3.0 | 3.0 | 3.9 | 1.0 | 0.8 | 7.2 |

PA = Palaearctic, NA = Nearctic, NT = Neotropical, AT = Afrotropical, OL = Oriental, AU = Australasian, PAc = Pacific and Oceanic Islands, ANT = Antarctic, TOT = total number of species

52  *Reproduction and Development in Platyhelminthes*

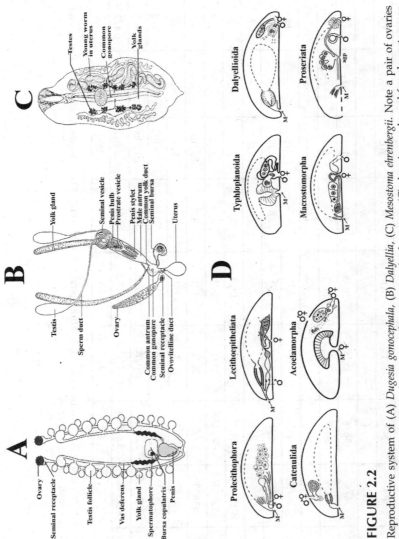

### FIGURE 2.2

Reproductive system of (A) *Dugesia gonocephala*, (B) *Dalyellia*, (C) *Mesostoma ehrenbergii*. Note a pair of ovaries in (A), a single ovary in (B) and presence of young ones in the uterus (C) showing male and female systems on the left and right sides (modified and redrawn from Vreys and Michiels, 1998, Hyman, 1951). (D) Diagrammatic representation of reproductive system in turbellarians. Note the exits for the male and female gametes. M = mouth, Agp = accessory genital pore. (free hand drawings from Noreña et al., 2015).

of divisions with incomplete cytokinensis. These divisions generate eight primary spermatocysts that progress through meiosis to generate 32 spermatids. Spermiogenesis packages the DNA of the elongated sperm with flagella (see Newmark et al., 2008).

The female reproductive system is even more complicated and varied. The system consists of a single ovary with a single oviduct (Fig. 2.2B) or a pair with a corresponding pair of oviducts (Fig. 2.2A). As mentioned earlier (p 5), most flatworms are unique among animals in that the yolk is supplied to eggs from the special yolk cells. Following the entrance of the yolk cells, the oviduct becomes the ovovitelline duct, which proceeds toward the female copulatory complex. The complex consists of (i) seminal bursa for temporary sperm storage, (ii) seminal receptacle for long term sperm storage and (iii) the bursal canal. The bursa opens into the exterior through the bursal pore. The canal leading to the pore is the copulation canal or ductus vaginalis. In many turbellarians, the terminal canal serves as storage for accumulation of ripe eggs. This part of the canal is the uterus. But in others, in which eggs are laid in singles or a few at a time (Batch fecundity), the uterus is not present. In some species like *Mesostoma ehrenbergii* (Fig. 2.2C), the eggs develop into young ones in the uteri, and the young ones escape through the mouth, gonopore or by rupture and death of the parent. In Acoela and Polycladidae, the fertilized eggs, enclosed in delicate capsules, are laid embedded in gelatinous material. In others, they are enclosed singly or severally in a hard shell capsule, which is usually formed in the male antrum (Hyman, 1951).

Regarding the germinal and nutritional (yolk cells) tissues, it is not known whether they have a common origin. However, the gametes of some turbellarians use the mouth as an exit. In hermaphroditic turbellarians, the presence or absence of exit for the gametes, and the presence of separate exits for female and male gametes have implications for self- or cross-fertilization. For example, the gametes released by rupturing of the body of a worm in the absence of an exit may provide greater chances for selfing among the released gametes. Not surprisingly, the exit for the gametes has been used as a key for classification of turbellarians. Accordingly, the mouth is used as an exit by prolecithophores to release the female and male gametes but only to the male gametes in lecithoepitheliates (Fig. 2.2D). Rupturing the body is used as exits for the gametes in acoelomorphs and catenulids. A single common exit is available in typhloplanoids and dalyellioids. However, separate male and female exits are used by the gametes of macrostomorphs and an additional one for accessory genital pore in proseriates.

## A 2.3 Transition to Parasitism

As indicated elsewhere, some 300 turbellarian species from 35 families live in permanent association with other aquatic invertebrates: echinoderms,

54    *Reproduction and Development in Platyhelminthes*

crustaceans, molluscs, annelids (Jennings, 1968, 1997) and coelenterates (e.g. *Prosthiostomum* sp, a coral parasite in Hawaii, Jokiel and Townsley, 1974). Notably, the turbellarians are unable to engage sponges as 'a feeding platform', while a large number of crustaceans, molluscs and annelids can do it (see Pandian, 2016, 2017, 2019). The turbellarian cysts are found on the skin/gills of theraponid *Pelates quadrilineatus* and platycephalid *Platycephalus fuscus* and *Ichthyophaga* sp on parrotfish *Scarus rivulatus* (Cannon and Lester, 1988). The gonochoric fecampiids are endoparasites in crustaceans. Among the rhabdocoel families, Umagillidae, Graffillidae and Fecampiidae have received much attention. In some umagillids and graffillids, evolution from free-living predators to parasitism via ecto- and endo-symbiosis has occurred through at least four recognizable transitional stages. For example, the umagillid ectosymbiont *Syndesmis antillarum* inhabiting the gut of echinoids and feeding on the co-symbionts represents stage (1). At stage (2), *S. franciscana* ingests co-symbionts along with host's gut wall tissue. The endosymbiont *S. dendrostomum*, at stage (3) feeds on the host's gut tissue and partially digested food as well as host's enzymes. At stage (4) *S. echinorum* totally depends on feeding the host's tissues alone and thereby demonstrate the complete transition from ectocommensalism to endoparasitism within a single genus of the Umagillidae. In graffillids too, the first three stages are reported among unrelated genera namely *Pseudograffilla arenicola*, *Paravortex scrubiculariae* and *Graffilla buccinicola* (see Jennings, 1997). In the ecto-, endo-commensals and parasites, modes of food acquisition, digestion, absorption and respiration are briefly summarized in Table 2.2. Whereas some endosymbionts have

## TABLE 2.2

Selected traits of ecto-, endo-commensalic and parasitic turbellarians (compiled from information reported by Jennings, 1997)

| Traits | Ectocommensals | Endocommensals | Parasites |
|---|---|---|---|
| Survival | Can survive without the host | Can survive without the host for a few days but not permanently | Cannot survive in the absence of host |
| Food supply | Highly variable | Variable | Assured |
| Feeding | Feed on co-symbionts | Feed on co-symbiont + host's tissue | Feed only on host's tissue |
| Digestion | Have their own enzyme | Partially depends on host's enzyme | Depends totally on host's enzyme |
| No. and size of epidermal microvilli | As in free-living turbellarians | As in free-living turbellarians | Longer and more number of microvilli |
| Reserve storage | More as glycogen than lipid | More as glycogen than lipid | More as lipid than glycogen |
| Respiratory adaptation | Not present | Switch to glycolysis or possess hemoglobin | Possess hemoglobin |

opted to glycolysis, others have developed respiratory pigment that provides them preferential access to oxygen or even allows them to abstract oxygen directly from the host's tissue. More importantly, the occurrence of more efficient hemoglobin in parasitic turbellarians (with no circulatory system) is relatively common in parenchymatous tissues. It highlights the value of the hemoglobin to acquire oxygen from the echinoderm hosts possessing no respiratory pigments, and crustacean and molluscan hosts having less efficient hemocyanin.

## A 2.4 Contrasting Strategies

In freshwater system, especially in ephemeral ones, the freshwater inhabiting rhabdocoels and triclads enjoy an abundant food supply during the favorable spring and summer in temperate zones and rainy days in the tropics. But the ephemeral water bodies may freeze during winter in the temperate zone or may partly or totally be dried during summer in tropics (e.g. see Pandian, 2017). To effectively utilize the favorable season, the freshwater turbellarians have developed contrasting strategies namely the rhabdocoels for subitaneous eggs and the triclads for clonal reproduction. In the triclads, the maximum fecundity is 125 cocoons (e.g. *Dugesia benazzii*) of which, assuming 36% recruitment (see p 33), the number of progenies added to recruitment shall be ~ 44 progenies. On the other hand, the triclads produce as many as 50 clonal offspring within the favorable season of 20 weeks (e.g. *D. ryukyuensis*, see Kobayashi et al., 2012). As clonal reproduction is elaborated in Part C of this Chapter, an account is included here on rhabdocoelan strategy of producing subitaneous eggs followed by dormant eggs.

In *Mesostoma* spp, the same worm can produce subitaneous (S) eggs in its first clutch and dormant (D) eggs in the last one prior to death (see Dumont et al., 2014). In *M. ehrenbergii*, the viviparous S egg (100 μ diameter) is bound by a thin (~ 1,400 Å), porous (for uptake of nutrients from parent), translucent, proteinaceous cover (Fig. 2.3A). In contrast, the orange-red colored D egg is covered by a thick shell (Fig. 2.3B), presumably composed of sclerotin, synthesized together with yolk by the yolk cells (Domenici and Gremigni, 1977). The parturition of young ones from S eggs and emergence of young ones from D eggs are triggered and/or regulated by temperature and feeding level. In a D egg, the young one requires a duration of 34 days at 11°C or 8 days at 25°C to emerge from the egg and another 75 days at 11°C or 17 days at 26°C (Heitkamp, 1977) to attain the minimum body size of 12 mm$^2$, i.e. 10–12 mm length and 4–5 mm width to produce the first clutch of ~ 15 S eggs and the subsequent second clutch of ~ 45 S eggs (see Beisner et al., 1997). In all, S eggs are produced in European waters almost throughout the favorable spring and summer. The D eggs, deposited during late summer

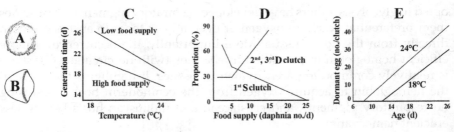

**FIGURE 2.3**

*Mesostomum ehrenbergii*: (A) Subitaneous egg, (B) Dormant egg (redrawn from Domenici and Gremigni, 1977), (C) Effect of temperature and food supply level on generation time, (D) Proportion of subitaneous egg produced at 1st and 2nd-3rd clutches as function of food supply at 24°C and (E) Batch fecundity of dormant eggs as function of age (modified and redrawn from Beisner et al., 1997).

and early autumn, hatch after an obligate dormancy during unfavorable winter, when the adults are no longer present. Hence, the D eggs ensure population survival during the unfavorable cold season. Thus, young ones, hatched from the D eggs in May in the temperate mid-Europe, is followed by production of successive 'clutches' (see Dumont et al., 2014) or generations up to October within a year. In other words, *M. ehrenbergii* can produce up to seven batches of S eggs within 7 months in a year. Similarly, *M. lingua* and *M. productum* can produce 8 and 8.5 clutches, respectively but within 5.5 months in a year (see Heitkamp, 1997). But the arctic, alpine and other northern European species like *M. nigrirostrum* and *M. rhynchotum* have only a single batch of D eggs involving duration of ~ 5.5 months (see Dumont et al., 2014).

Studies were undertaken to identify whether temperature, food supply and/or 'crowding effect' trigger the shift from S egg clutch to D egg clutch. Rearing experiments indicated that increase in temperature from 20 to 28°C altered neither fecundity nor longevity of the Brazilian *M. ehrenbergii*. Rietzler et al. (2018) have fed *M. ehrenbergii* with 2–10 daphnids at temperature from 20 to 32°C and reported consistent production of S egg clutch, irrespective of changes in food supply and temperature, and thereby confirmed the earlier observation of Dumont et al. (2014). Being stenothermic, *M. craci* produced an intermediate (between S and M) type of egg. Incidentally, survival of *M. ehrenbergii* D eggs was 100% for 2 months in moist sediment but > 6 months at low temperatures (Heitkamp, 1977). Investigating it in greater details, Beisner et al. (1997) reported reduced generation time with increasing food supply and temperature (Fig. 2.3C). With a proportion of *M. ehrenbergii* switching to produce D eggs from the second and third clutch, those producing S egg clutch decreased (Fig. 2.3D). Clearly, a high level of food supply and temperature at 18°C trigger more and more worms to switch and produce D egg clutch. At 18 to 24°C, fecundity in D egg clutch increased

**TABLE 2.3**

Crowding effect on fecundity of *Mesostoma ehrenbergii* reared at 24.5°C (compiled from Dumont et al., 2014)

| Crowding of worms (no./volume) | Subitaneous eggs (no.) | Dormant eggs (no.) | Longevity (d) |
|---|---|---|---|
| 1/100 ml | 18 | 40 | 32 |
| 1/250 ml | 24 | 24 | 32 |
| 4/400 ml | 9 | 18 | 25 |
| 4/1000 ml | 24 | 22 | 32 |

with advancing age (Fig. 2.3E), i.e. with increasing body size more and more dormant eggs were produced. Dumont et al. (2014) reported that changes in 'crowding' increased fecundity in D clutch from 18 eggs in four worms kept in 400 ml water to 40 eggs in one worm kept in 100 ml (Table 2.3). Fiore (1971) also reported that adult *M. ehrenbergii* chemically induced the young ones to switch from S to D egg production. However, it is not known whether a single factor or a combination of factors trigger(s) D egg production.

## A 2.5 Semelparity and Iteroparity

In animals, energy budget is assessed by estimation of $C = F + U + R + P$, where the C is the food energy consumed, F, U and R, the energy lost on feces, urine and metabolism, respectively, and the P, the energy gained and utilized for somatic growth (PS) and reproductive output (PR) (e.g. see Pandian, 1987). With a blind gut or no gut, indigestible residue is regurgitated by flatworms. Hence, C may be equivalent to the food energy absorption (A). However, most aquatic animals 'excrete' secreting mucus to chemically clean the body and to facilitate motility as well. Mucus secretion amounts to ~ 9% of assimilated energy by the snail *Hydrobia ventrosa* (see Pandian, 2017), 0.2–2.8 g (dry weight) by the holothurian *Cucumaria frondosa* (see Pandian, 2018) and constitutes 64% of U in earthworms (see Pandian, 2019). The values estimated for excretion of nitrogenous wastes together with mucus is 28% in *Dendrocoelum lacteum* and 43% in *Polycelis tenuis* (Woodhead and Calow, 1979).

In a rare study, Woodhead and Calow (1979) assessed the energy partitioning strategies during egg production in semelparous triclad *D. lacteum* (body size: 60 mm²) and perennial iteroparous triclad *P. tenuis* (body size: 30 mm²). These worms were offered decreasing rations at 1 *Asellus*/day, 1/2 day, 1/4 day, 1/8 day and 1/16 day, which are equivalent to food supply index (FSI) of 1.0, 0.5, 0.25, 0.125 and 0.063, respectively. From Fig. 2.4, the following inferences

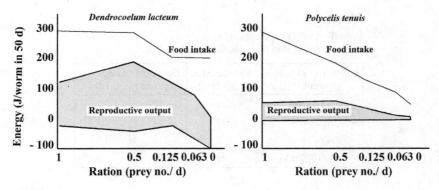

**FIGURE 2.4**

Energy partitioning between reproductive output and other bioenergetic components in semelparous *Dendrocoelum lacteum* (left panel) and iteroparous *Polycelis tenuis* (right panel) (modified and redrawn from Woodhead and Calow, 1979).

may be noted: (1) Though food intake is decreased with decreasing ration, it remains higher throughout the range of decreasing ration in semelparous *D. lacteum* than in iteroparous *P. tenuis*. Hence, *P. tenuis* is more sensitive to variations in food supply. Not surprisingly, the iteroparous flatworms, especially the relatively smaller sized trematodes enjoy an assured food supply (see Jennings, 1997), once they successfully infect a suitable host. (2) Despite decreasing ration, reproductive output is greater in *D. lacteum* at all FSI levels than in *P. tenuis*. (3) With decreasing ration, the output is decreased in both *P. tenuis* and *D. lacteum* but the decrease is profoundly more in *P. tenuis* than in *D. lacteum*. (4) To sustain the level of reproductive output, when ration is decreased or with variable food supply, the semelparous *D. lacteum* meets the cost by 'sacrificing' or degrowing its body size. Consequently, it succumbs to death at or after breeding. When the inferences two to four are applied to the iteroparous parasitic flatworms, the following shall become apparent. To them, food supply is not a variable, as they enjoy an assured supply from the host. In these iteroparous parasites, reproductive output is sustained at an almost constant level, i.e. egg release is a continuous process in them. At best, the cestodes 'sacrifice' a fraction of their body as proglottids, but the reproductive output is constantly maintained.

In triclads, Calow et al. (1979) also analyzed the relevant information on the duration required to regain the 'sacrificed' soma in semelpares and iteropares. Firstly, the hatchling size of semelpares is 2-times larger than that of iteropares. Secondly, semelpares require a longer duration of 7 days to produce a single cocoon, from which a larger hatchling arises. Thirdly, the semelpares also require a longer duration of 77 days to regain the biomass lost on degrowth, in comparison to 28 days in the iteropares (Table 2.4).

## TABLE 2.4

Selected reproductive traits of semelparous and iteroparous triclads (modified and compiled from Calow et al., 1979)

| Species | Adult size (mm²) | Hatchling size (mm²) | Days to produce one capsule | Days to grow equivalent soma |
|---|---|---|---|---|
| Semelpares | | | | |
| *Planaria torva* | 30 | 1.5 | 6 | 40 |
| *Dendrocoelum lacteum* | 50 | 2.0 | 7 | 92 |
| *Bdellocephala punctata* | 60 | 2.4 | 7 | 100 |
| **Average** | **47** | **2.0** | **7** | **77** |
| Iteropares | | | | |
| *Polycelis tenuis* | 35 | 0.9 | 5 | 16 |
| *Dugesia lugubris* | 50 | 1.0 | 7 | 40 |
| **Average** | **42** | **1.0** | **6** | **28** |

# A 2.6 Sexuality

Hermaphroditism is defined as the expression of both female and male sexes in a single individual either simultaneously or sequentially. A vast majority of platyhelminths are hermaphrodites. In turbellarians, sequential hermaphroditism is limited to Catenulida and a few rhabdocoels. For example, the protandric males of *Stenostomum* reproduce clonally but the females sexually alone (Hyman, 1951). However, manifestation and maintenance of dual reproductive systems within an individual is costlier, *albeit* providing the scope for selfing in patchily distributed populations and species that are low motile or sessile. In all aquatic invertebrate taxa, repeated incidences have occurred to reduce the dual cost by reduction/elimination of male component to evolve parthenogenesis or rarely gynogenesis, as in turbellarians.

## A 2.6.1 Hermaphroditism

*Mating and copulation*: In turbellarians, mating is a prolonged affair and includes the pre-copulatory circling and reeling, and post-copulatory sucking of the sperm and seminal contents to be used as rich nutrients after digestion (Scharer et al., 2004). Although the externally visible pre-copulatory behaviors have been the subject of many studies (e.g. *Microstomum* sp, Scharer et al., 2004), the study of post-copulatory behavior has remained a challenge, as the event occurs internally. However, the challenge has been overcome with the development of transgenic line that expresses Green Fluorescent Protein (*GFP*) gene in all the cell types of the transparent *Microstomum lignano*. This

60  *Reproduction and Development in Platyhelminthes*

development offers a unique opportunity to non-invasively visualize and quantify the sperm of a *GFP*-expressing donor inside the reproductive tract *in vivo*. Marie-Orleach et al. (2014) also found that *GFP* expressing worms do not differ from wild type in terms of morphology, mating frequency and reproductive success.

With amenability to rearing, many turbellarians provide an opportunity to study the copulation and insemination processes. In turbellarians, internal fertilization is the rule. Selfing may be possible in typhloplanoids and dalyelloids (see Fig. 2.2D), in which both female and male ducts open through a common exit. But it does not occur in freshwater triclads and polyclads (Hyman, 1951). Two methods of impregnation are recognized: (i) copulation and (ii) hypodermic injection. Michiels and Newman (1998) have narrated a vivid description of 'violent' mutual hypodermic injection in the marine polyclad *Pseudoceros bifurcatus*. Striking of one another is followed by 'penis fencing' to increase the benefit of sperm donation over sperm receipt. During 39 contests, 287 strikes have led to 46 inseminations in 12 pairs. The first to inseminate obtains a longer injection time (9.7 minutes) than the second (5.7 minutes).

*Dugesia polychroa* mate at the frequency of 0.6 times/pair/day. And the copulation lasts from a few minutes to 2.5 hours with pronounced maxima of short copulation type with < 35 minutes and long copulation type with > 35 minutes. Copulation may occur almost throughout the day but the long type occurs mostly between 8 pm and 2 am in the night and the short type between 10 am and 4 pm during the day. The former involves reciprocal inseminations but the latter may involve unilateral. Pairs, that have had longer copulation, produce more cocoons at an increased frequency. However, *D. polychroa* produces three or more cocoons/worm/week (Peters et al., 1996). Clearly, the mating partners firstly 'assess' each other during the initial few minutes of copulation. If they find compatibilty, copulation is continued for longer than 35 minutes; lest, the partners separate peacefully, i.e. the sexual selection is made after the penial insertion and assessment of compatibility between the partners.

In an interesting investigation, Vreys and Michaels (1998) compared the volume of sperm exchanged reciprocally between mating partners of *D. gonocephala*. In it, copulation occurs every 2.5 days, lasts for 4.7 hours and insemination is reciprocal. Sperm donation continues until the spherical, stalked spermatophores are reciprocally (85%) exchanged into the bursa of the partners up to 3.0–4.5 hours after the commencement of mutual penial insertion. Mating is induced by high quantum of autosperm accumulation in seminal vesicle, but not by the lack of allosperm in the sperm receptacle. Hence, the partners copulate mainly to donate rather than to receive sperm (see also Charnov, 1979). They may mate to receive other benefits like nutrients from digested sperm. With greater store of more autosperm, a larger worm can transfer a bigger sperm clump. However, individuals mate in an assortative manner by size. Hence, autosperm reserves are not correlated within mating pairs. Mating partners exchange an equal volume of sperm, i.e. each partner donates sperm only as much as it has received. This sperm trading by volume commences with the transfer of sperm from the partner with relatively less sperm store, and subsequent transfer of equal volume of sperm from the partner with relatively rich sperm store. The evolution of conditional gamete trading may stabilize hermaphroditism.

*Generation time*: Estimates on Generation Time (GT) and Life Span (LS) provide valuable information on Reproductive Life Span (RLS) of an animal. For example, the RLS of semelparous *Dendrocoelum lacteum* is shorter, in comparison to that of iteroparous *Dugesia lugubris*. In slow motile turbellarians (163–352 cm/day in *Polycelis* spp, see Calow et al., 1979), changes in temperature may greatly affect the GT and LS. Reynoldson et al. (1965) is perhaps the only publication on the influence of temperature in turbellarians. At 3.5°C, GT is the shortest (393 days) in semelparous *D. lacteum* and longest (632 days) in iteroparous *P. tenuis* (Fig. 2.5A). At 20°C, the corresponding values are 17 and 194 days in *D. lugubris* and *P. nigra*. Consequently, *D. lugubris* may pass through 3.5 generations within a year at 16°C but *P. tenuis* ~ 2 generations (Fig. 2.5B).

*Fecundity*: At temperatures ranging from 1.5°C to 23°C, batch fecundity remains equal but at a higher level in *Dugesia lugubris* (Fig. 2.6A1) than in *Polycelis nigra* (Fig. 2.6A2). But it peaks at 3.5°C and 10°C for semelparous *P. tenuis* and *Dendrocoelum lacteum*, respectively (Fig. 2.6A3, A4), prior to decrease with increasing temperature. The iteropares produce relatively more non-viable eggs than semelpares (Fig. 2.6A2, A4). In oviparous *Stylochus ellipticus*, the number of eggs per batch increases linearly with increasing body size but at higher levels for the fed worms than the starved and starved-fed worms. The increases are at higher levels at 30°C than at 21°C (Fig. 2.6B, C) but for the

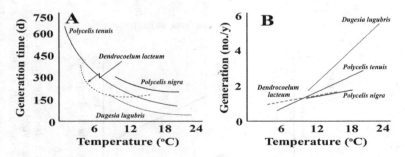

**FIGURE 2.5**

Effect of temperature on (A) generation time (drawn from data reported by Reynoldson et al., 1965) and (B) number of generations in four triclad species (modified and redrawn from Reynoldson et al., 1965).

62  *Reproduction and Development in Platyhelminthes*

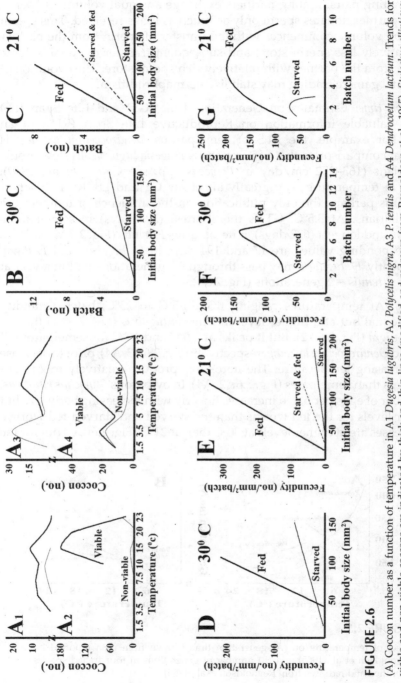

**FIGURE 2.6**

(A) Cocoon number as a function of temperature in A1 *Dugesia lugubris*, A2 *Polycelis nigra*, A3 *P. tenuis* and A4 *Dendrocoelum lacteum*. Trends for viable and non-viable coccons are indicated by thick and thin lines (modified and redrawn from Reynoldson et al., 1965). *Stylochus ellipticus*: number of batch fecundity as a function of body size (B) at 30°C and (C) at 21°C in fed, starved and starved-fed worm. Relative fecundity as a function of body size (D) in fed and starved worms at 30°C, and (E) fed, starved and starved-fed worms at 21°. Relative fecundity as a function of successive batches in fed and starved worms at 30°C (F) and 21°C (G) (modified and redrawn from Chintala and Kennedy, 1993).

relative fecundity, it was higher at 21°C than at 30°C (Fig. 2.6D, E). Relative fecundity remains at a higher levels in successive batches of fed worms but egg production is ceased at the 10th and 14th batch at 21°C and 30°C, respectively. Likewise, egg production is terminated beyond the 3rd and 4th batches in starved worms at 21°C and 30°C (Fig. 2.6F, G). Remarkably, the starved worms continue to produce eggs at the cost of degrowth.

Information on fecundity of turbellarians with parental care is available for the polyclads *Pleioplana atomata* and *Imogine zebra*. In them too, the number of batches and relative fecundity increase with increasing body length. Comparisons of the batch number and relative fecundity (no./mm²/batch) between oviparous, gynogenic and parental caring species show that (i) body size is a decisive factor on both the batch number (Fig. 2.7A) and relative fecundity (Fig. 2.7B), irrespective of the mode of reproduction, (ii) a large body size is attained by oviparous *S. ellipticus* and the size decreases in

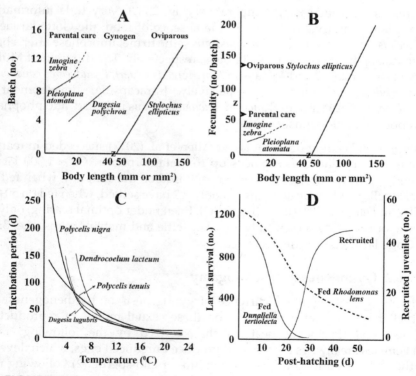

### FIGURE 2.7

Batch (A) and relative fecundity (B) as a function of body length in oviparous, gynogenic and parental caring turbellarians (compiled and redrawn from Reynoldson et al., 1965, Weinzierl et al., 1999, Rawlinson et al., 2008), (C) incubation period as a function of temperature in triclads (modifed and redrawn from Reylondson et al., 1965) and (D) number of larval survival and recruitment in *Stylochus ellipticus* fed on different diets (modifed and redrawn from Allen et al., 2017).

64    *Reproduction and Development in Platyhelminthes*

the following descending sequence: oviparous > gynogenic > parental caring species, (iii) the number of batches also decreases in the same sequence and (iv) as may be expected, fecundity of oviparous *S. ellipticus* is nearly 2-times higher than that of the polyclads with parental care (Fig. 2.7B).

Available information of egg size of some polyclad species is summarized by Rawlinson (2014) and Allen et al. (2017). The life cycle of 13 species belonging to the acotyleans Stylochidae and Notoplanidae involves the four-lobed Gotte larva and that of the acotyleans (Leptoplanidae, Planoceridae and the cotyleans Pseudocerotidae, Euryleptidae and Prosthiostomidae) involves 6–8 lobed Muller larva (Rawlinson, 2014). The egg size ranges from 85 to 105 μm in polyclad species involving Gotte larva and averages 101 μm. Most of them are planktotrophic (e.g. *S. ellipticus*, Allen et al., 2017). On the other hand, the lecithotrophic egg size of polyclad species with Muller larva ranges from 120 to 330 μm and averages 188 μm (Rawlinson, 2014).

*Incubation and larval duration*: In the four triclad species, incubation period decreases with increasing temperature (Fig. 2.7C). Very little information is available on larval duration. In the absence of food, the Gotte larvae of *Stylochus uniporus* and *Notoplana australis* settle to metamorphose after a brief planktotrophic phase of 5 and 14 days, respectively. The duration of Muller larva lasts for 25, 40 and 40 days in *Maritigrella crozieri*, *Pseudoceros canadensis* and *Stylostomum sanjuania*, respectively. Intracapsular development is followed by Kato larva in *Planocera reticulata*. In this polyclad, adelphophagy is reported (see Rawlinson, 2014).

*Recruitment*: Feeding different algae, Allen et al. (2017) succeeded in rearing Gotte larva of *Stylochus ellipticus* up to metamorphosis. Of ~ 1,300 larvae reared, ~ 8 have successfully metamorphosed on the 50th day, when fed on 50,000 cells/ml *Rhodomonas lens* but only < 2 have settled, when fed on 50,000 cells/ml *Dunaliella tertiolecta* (Fig. 2.7D). Even under optimal rearing, only 45 larvae out of 1,300 successfully survive, settle and metamorphose, i.e. 3.5% of Gotte larvae are finally recruited.

## A 2.6.2 Gynogenesis—Pseudogamy

Prior to describing the occurrence of gynogenesis or parthenogenesis in turbellarians, a brief explanation on these sexual modes of reproduction is necessary. Parthenogenesis is characterized by the following traits: (1) Females are always present and produce unreduced eggs, which develop without the need for a trigger by an auto- or allo-sperm. (2) Following rare and sporadic occurrence of males, sexual reproduction may be restored. Typically, three types of parthenogeneses are recognized: (1) Ameiotic or apomictic parthenogenesis, in which cytological events do not involve synapsis between homologous chromosomes; hence, segregation and crossing over are eliminated. Ameiotic oogenesis generates unreduced egg, requiring neither activation nor fertilization by a sperm. (2) Conversely, meiosis occurs in automictic parthenogenesis. However, development is commenced without activation or fertilization. Diploidy is restored in its egg by fusion of two haploid homologous nuclei at the first mitotic division or by fusion of the haploid egg nucleus with non-homologous nucleus from the polar body. (3) The telytochorous parthenogenesis involves pre-meiotic doubling of chromosomes at the last oogonial division resulting in endomitosis. This is followed by formation of chiasmatic bivalence and regular meiosis with extrusion of two polar bodies. The genetic consequences of this cytological mechanism are parallel to that of apomictics, as synapsis is limited between sister chromosomes that are exact molecular copies of one another. As a result, diploid eggs are produced without recombination leading to the production of genetically indistinguishable clones (e.g. *Tubifex*, Christensen, 1984). Notably, the incidence of parthenogenesis is limited to a few species in Annelida, Mollusca and Echinodermata but occurs in almost all crustacean major taxa and is elaborated into different types in Cladocera (Table 2.5).

In gynogenesis or pseudogamy too, the oogenetic process facilitates the exclusive matrilineal inheritance. In it, activation of the unreduced egg by an auto- or allosperm to trigger zygote development is obligatory. But the paternal set of chromosomes is subsequently eliminated (Beukeboom et al., 1998). In fishes, natural gynogenesis occurs in unisexual diploid *Poecilia formosa* and in triploid gonochoric *Carassius auratus gibelio* and *C. a. langsdorfi*. Employing different methods including activation by homo-specific or hetero-specific sperm either in intact or genome-inactivated status, diploid, triploid, tetraploid, pentaploid and hexaploid gynogens have been developed (see Pandian, 2011).

In turbellarians, the incidence of natural gynogenesis is limited to three triclad species: *Dugesia benazzii*, *D. lugubris* (Benazzi-Lentati, 1966, 1970) and *Schmidtea polychroa* (Beukeboom et al., 1996a, Weinzierl et al., 1999,

**TABLE 2.5**

Incidence of parthenogenesis in aquatic major invertebrate phyla/class (compiled from Pandian, 2016, 2017, 2018, 2019)

| Phylum/class | Species (no.) | Incidence |
|---|---|---|
| Annelida | 22,000 | 75 species, of which 56 are earthworms |
| Mollusca | 110,350 | Limited to 4 genera *Compeloma, Hydrobia, Melanoides, Potamopyrgus* |
| Echinodermata | 7,000 | *Ophidiaster granifer, Diadema antillarum, Asterina miniata†, Ophiomyxa brevirima†* |
| Crustacea | 52,000 | Anostraca, Cladocera (620 species), Ostrocoda, Copepoda, Isopoda, Amphipoda, Decapoda |

† bisexual and parthenogens co-occurs

66  *Reproduction and Development in Platyhelminthes*

Storhas et al., 2000, Pongratz et al., 2003, Navarro and Jokela, 2013). In *S. polychroa*, gynogens can be triploid (3n = 12), tetraploid and pentaploid (Baguna et al., 1999). Unfortunately, the authors have wrongly named them as parthenogens, although Benazzi-Lentati called it pseudogamics. It must also be noted that there is a succinct difference between the gynogens in gonochoric fishes and gynogens of hermaphrodites. In gonochorics, only female gynogens are generated in male heterogametic fishes (see Pandian, 2011). In hermaphroditic gynogens of turbellarians, hermaphrodites are generated with dominant matrilineal expression.

In parthenogens, elevation of ploidy occurs not infrequently and its advantages are described by Comai (2005). Among the three triclad gynogenic species, not only elevation of ploidy but also inclusion of B chromosomes considerably increases the complexity. A large number of publications are available on the distribution of B chromosomes in European natural populations of *S. polychroa* (e.g. Pongratz et al., 2003). Benazzi-Lentati (1970) described how some deviations in cytological events have generated new biotypes in *D. benazzii*. A climax seems to be that in hexaploid, oogenesis undergoes true meiosis to restore triploidy in *D. benazzii* (see Beukeboom et al., 1996b). The sperm from gynogen can fertilize eggs of hermaphrodites and thereby produce 'hybrid' zygotes, which may also generate new biotypes, as in *S. polychroa* (Weinzierl et al., 1998). For example, the cross between diploid hermaphrodite and triploid gynogen produces diploids and triploids in equal ratio (Storhas et al., 2000). In general, eggs of gynogenic triclads do not accept self (auto-) sperm to trigger zygote development.

As no paternal genomic contribution is required in gynogenesis, natural selection is expected to reduce male allocation in hermaphroditic gynogens. Using the Feulgen DNA staining to visualize testes filled with sperm and related techniques, Weinzierl et al. (1998) found that hermaphroditic *S. polychroa* has larger testes. But copulation frequency and duration of gynogens are as high as those of hermaphrodites. Consequently, gynogens produce more cocoons by number and volume and also more number of embryos per cocoon than hermaphrodites. But the number of fertile cocoons produced is 3.65 for hermaphrodites, in comparison to 2.8 in gynogens. Clearly, a gynogenetic triclads are unable to produce as many fertile sperm as hermaphrodites can do. Consequently, the number of embryos commencing to develop is also more in hermaphrodites than in gynogens.

### A 2.6.3 Gonochorism

Sporadic incidences of gonochores are reported for a few turbellarians, digeneans and cestodes (Table 6.1). In turbellarians, the incidence of gonochorism is reported from distantly related free-living triclad *Sabussowia diocia* (Delogu and Galletti, 2011) and parasitic fecampiid rhabdocoel *Kronborgia amphipodicola* (Christensen and Kanneworff, 1964, 1965, Kanneworff and Christensen, 1966). In the Fecampiidae and Acholadidae, all the traces of alimentary system are lost in the adult stage. In the former, all the described species are gonochoric and endoparasitic in the hemocoelom of crustaceans or in various tissues of myzostomid annelids (e.g. *Glanduloderma myzostomatis*, Jennings, 1997). For example, *Femcampia erythrocephala* is a parasite of the prawn *Palaemon serratus*. *K. amphipodicola*, is an endoparasite in the hemocoel of the amphipods *Ampelisca macrocephala* and *Haploops tubicola*. The data of Christensen and Kanneworff are recalculated and processed to draw the following inferences: (1) The life cycle of *K. amphipodicola* is one year. (2) Following sexual maturity, both female and male *K. amphipodicola* leave the host through its posterior end and the female commences to secrete a cocoon, in which the eggs are fertilized by the male. Following 48–59 days of incubation, the larvae are hatched and their life time is ~ 48 hours. (3) On successful infection, the dwarf male grows to 0.4 cm, while the female to ~ 30 cm. (4) Subsequent to infection, the differentiation process proceeds. As a result, the ratio of undifferentiated individuals decreases from 0.77 to 0.19 in *A. macrocephala* and from 0.37 to 0.0 in *Haploops* sp on the 69th day after infection (Table 2.6). (5) In both hosts, the male ratio increases to 0.64 in the former and to as high as 0.94 in the latter. The corresponding values for the female ratio are 0.17 and 0.06. Hence, for every female, there are 3.76 males in *A. macrocephala* and as many as 15.7 males in *Haploops* sp. (6) In *A. macrocephala*, the incidence of infection steadily decreases from ~ 60% in April–May to 4–0% in November–December (Fig. 2.8A) but remains at 22% level in *Haploops* sp (Fig. 2.8B). The parasitic intensity increases from 4.8 to 7.9–10.3/host in *A. macrocephala* and 1.3 to 1.4 in *Haploops* (Table 2.6). Hence, susceptibility to *K. amphipodicola* is higher for *A. macrocephala* than for *Haploops* sp. (7) The number of *A. macrocephala* hosting both female and male parasites increases to a peak of ~ 25 during June–July (Fig. 2.8C); a parallel trend is also observed for the isolated single females. Whereas the incidence of 2–4 females/host increases up to 10% in July–August (Fig. 2.8C), the incidence of isolated occurrence of male is rare (Fig. 2.8C). Briefly, the chances for sexual reproduction are greater during spring and summer. But it is not clear whether or not the isolated males (in June) and females during spring and summer can reproduce.

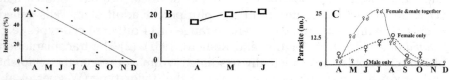

**FIGURE 2.8**

Incidence of *Kronborgia amphipodicola* in (A) *Ampelisca macrocephala* and in (B) *Haploops* sp during calendar months. (C) number of co-occuring females and males, females alone and males alone of *K. amphipodicola* during calendar months (drawn using data reported by Christensen and Kanneworff, 1965).

68   *Reproduction and Development in Platyhelminthes*

## TABLE 2.6

Increasing parasite intensity of *Kronborgia amphipodicola* and its sex ratio as function of day following infection in *Ampelisca macrocephala* or *Haploops* sp. O = undifferentiated (recalculated and processed from Christensen and Kanneworff, 1965)

| Day | Parasite intensity (no./ host) | | | | Sex ratio | | |
|---|---|---|---|---|---|---|---|
| | ♀ | ♂ | O | Subtotal | ♀ | ♂ | O |
| *A. macrocephala* | | | | | | | |
| 0 | 0.84 | 0.16 | 3.78 | 4.78 | 0.05 | 0.04 | 0.77 |
| 28 | 1.28 | 3.37 | 5.65 | 10.30 | 0.12 | 0.33 | 0.55 |
| 69 | 1.36 | 5.06 | 1.51 | 7.93 | 0.17 | 0.64 | 0.19 |
| *Haploops* sp | | | | | | | |
| 0 | 0.06 | 0.74 | 0.48 | 1.28 | 0.05 | 0.58 | 0.37 |
| 28 | 0.00 | 1.10 | 0.30 | 1.40 | 0.00 | 0.82 | 0.17 |
| 69 | 0.07 | 1.06 | 0.00 | 1.13 | 0.06 | 0.94 | 0.00 |

# Part B: Regeneration

## Introduction

Regeneration is a multicellular process of remolding tissues into an exact replica of damaged or missing tissues, organs and systems (Mouton et al., 2018). In planarians, it commences with epimorphosis leading to blastema formation and subsequent morphallaxis, and restitution pattern through a sequential antero-posterior series of inductions and inhibitions (see Salo and Baguna, 1984a). Ever since Pallas (1774) described it in *Dendrocoelum lacteum* and *Bdellocephala punctata*, researchers have inflicted partial or complete amputation longitudinally and meridionally along the axes of a number of planarian species (see Fig. 2.10). Their results led them to conclude that planarians are "almost immortal under the edge of the knife" (Newmark and Alvarado, 2001). Subsequently, a series of researches have shown that (1) Regeneration originates from the pluripotent adult stem cells called neoblasts, which give rise to the entire range of 14 different cell types (e.g. *Schmidtea mediterranae*, Baguna and Romero, 1981). (2) The transplantation of a single neoblast into a lethally irradiated worm can rescue the recipient and produce a perfectly healthy worm of the donor type (Wagner et al., 2011). For example, about one third of all cells are renewed within 2 weeks in *Macrostomum lignano* consisting of 25,000 cells (Nimeth et al., 2007). Interestingly, the dynamic architecture of planarians renders an astonishing regenerative potency to completely turnover the entire triploplastic animal within a few weeks from the remnant tissue(s). Hence, as an experimental model, planarians provide excellent opportunities to address a spectrum of problems in stem cell research like the (i) evolutionary conservation of pluripotency, (ii) dynamic organization of differentiation lineages, (iii) mechanisms underlying organismal stem cell homeostasis (Rink, 2013) and proliferative control of stem cells, relevant to human cancers and (iv) understanding of the ageing phenomenon (Aboobaker, 2011).

## B 2.7 Distribution and Potency

Incidence of regeneration is reported from almost all turbellarian orders (Table 2.7). The known turbellarian species, for which reports are available, are listed in Table 2.9, but the list is not exhaustive. In all, ~ 65 turbellarians species are reported to possess the regenerative potency. The vital organs like the brain, sensory organs, eyes, which enable the worm to locate a prey or predator, and the ventral mouth, if it is included, are all located in the anterior part. Hence, regeneration of anterior part can be costlier than posterior part. Notably, the potency is limited to the eyes and statocysts in the acoelids, gut, gonad and sensory organs in a proseraite and posterior fragment in macrostomorphids (Table 2.8). Triclads alone are capable of regenerating any amputated or missing body part from the remnant body. They are grouped into eight types on the basis of the amputation level in antero-posterior the axis required to ensure regeneration of the head: regeneration of a head is possible from any part of the body but with a varying probability of success in (type 1) *Phagocota velata*, (type 2) *Dugesia deratocephala*, (type 3) *Polycelis auriculata* and (type 4) *Dendroceolopsis lacteum* types. In *D. lacteum*, from the Kuroishi population, the head is not regenerated, when amputated below the male gonopore (type 5). In *Dendrocoelopsis ezensis* (type 6) and *D. lacteum* (type 7), the head is regenerated, only when amputated above the pharynx but *Bdelloura candida* (type 8) is not capable of regenerating a head (see Egger et al., 2006). These aspects are elaborated below but a comparison of the potency with other aquatic invertebrates has to be made.

Morgan (1898) classified regeneration into two types: (1) Morphallaxis involving remolding the existing tissues into missing ones without extensive cell proliferation and blastema formation. (2) Epimorphosis involving massive proliferation of undifferentiated stem cells and blastema formation. Epimorphosis can be more energy expensive than morphallaxis, as the latter involves remodeling of some pre-existing structure but the former requires *de nova* formation of missing parts after blastema formation. Remarkably, clonal reproduction is rare in hemocoelomic crustaceans and molluscs (see Pandian, 2016, 2017). In crustaceans, regeneration is limited to amputated fractions of appendages, which involve (i) epidermal and nerve cell types of ectodermal origin and (ii) muscle and connective tissue types of mesodermal

70 *Reproduction and Development in Platyhelminthes*

## TABLE 2.7

Regenerative and clonal potency of turbellarians (based on Tables 2.8, 2.10). – = no incidence, ++ = > 2.0% incidence for regenerative potency or 5.0% clonal potency, + = < 2% incidence of regenerative potency or < 5% clonal potency (see also Table 7.1)

| Taxon | Regenerative potency | Clonal potency |
|---|---|---|
| Tricladida | ++ | + |
| Acoela | ++ | + |
| Catenulida | + | ++ |
| Macrostomorpha | ++ | + |
| Polycladida | + | – |
| Proseriata | ++ | – |
| Lecithoepitheliata | ++ | – |
| Bothrioplanida | + | – |
| Prolecithophora | + | – |
| Rhabdocoela | + | – |

## TABLE 2.8

Reported regenerative potency of organs and systems in turbellarian taxa (compiled from Egger et al., 2006)

### Acoela

Regenerates eyes (*Amphiscolopos langerhansi*, *Convolutriloba longifissura*) and statocyst (*Hofstenia giselae*)

### Macrostomorpha

Regenerates posterior fragment including copulatory organs in the presence of a quarter of the gut but not anterior fragment (*Microstomum lignano M. hystricinum, M. marinum, M. pusillum*)

### Polycladida

Regenerates posterior fragment including gonads, copulatory organs but not the anterior fragment and pharynx (*Leptoplana alcioni, L. tremallaris, L. velutinus, Thysanozoon brocchii, Cryptocelis alba, Notoplana humilis*). Can regenerate brain (*L. littoralis, T. brocchi*)

### Proseriata

Regenerates the gut, gonads and sensory organs (*Otomesostoma auditivum*)

### Rhabdocoela

Regenerate anterior fragment in front of the brain (*Mesostoma lingua*). Regenerate the tail tip but not brain and pharynx.

origin. It is also true of clonal reproduction in colonial parasitic rhizocephalans. It involves morphallaxis and Oligopotent Stem Cells (OlSCs) but without blastema formation (see Pandian, 2016, 2017). In molluscs, respiratory siphon, tentacle, proboscis, nerve and brain can be regenerated. Some molluscs have

Turbellaria 71

## TABLE 2.9

Reported incidences of regenerative potency in turbellarian taxa (from Egger et al., 2006 and others). Species in rectangle boxes are capable of clonal multiplication also

**Acoela**: *Amphiscolops langerhansi,* Convolutriloba longifissura, *Hofstenia giselae, Paramecynostomum diversicolor, Polychoerus caudatus, Praesagittifera naikaiensis, Pseudohaplogonaria sutcliffei* (7)

**Catenulida**: *Catenula* (1)

**Macrostomorpha**: *Macrostomum grande, M. hystricinum, M. ligano, M. lineare, M. marinum, M. pusillum, M. tuba* (7)

**Polycladida**: *Cestoplana, Cryptocelis alba, Leptoplana alcioni, L. automata, L. littoralis, L. saxicola, L. tremellaris, L. velutinus, Notoplana humilis, Planocera californica, Pseudostylochus intermedius* (Sato et al., 2001), *Pucelis litoricola, Thysanozoon brocchii* **(13)**

**Lecithoepitheliata**: *Geocentrophora sphyrocephala* (1)

**Prolecithophora**: *Plagiostomum girardi* (1)

**Proseriata**: *Archilopsis, Bothriomolus, Coelogynopora, Itaspiella, Monocelis fusca, M. lineata, Otomesostoma auditivum* (7)

**Bothrioplanida**: *Bothrioplana semperi* (1)

**Rhabdocoela**: *Mesostoma lingua, M. productum, M. punctatum, Rhynchomesostoma rostratum* **(4)**

**Tricladida**: *Bdellocephala brunea, B. punctata* (Brondsted, 1969), *Bdelloura candida* (Egger et al., 2006), *Dendrocoelum lacteum* (Brondsted, 1969), *Dendrocoelopsis lactea, D. ezensis* (Egger et al., 2006), Dugesia gonocephala (DeVries, 1986), D. japonica (Malinowski et al., 2017), D. lugubris (Benazzi et al., 1970), *D. ryukyuensis* (Nakagawa et al., 2012a, b), *D. sicula* (Baguna et al., 1999), *D. subtentaculata* (Chandebois, 1980), *D. tigrina* (Baguna and Romero, 1981), *D. torva* (Brondsted, 1969), *Euplanaria polychroa* (Brondsted and Brondsted, 1961), *Planaria derotocephala* (Child, 1911), *P. maculata* (Curtis, 1902), *Phagocata gracilis* (Rose and Shostak, 1968), *P. vitta* (Brondsted, 1969), *Polycelis auriculata* (Egger et al., 2006), *P. nigra* (Lender, 1956), *P. tenuis* (Newmark et al., 2012), Schmidtea mediterranea (Baguna and Romero, 1981) **(23)**

retained the potency to regenerate the gonads, especially testis and penis until sexual maturity but even beyond it in aeolids. No author has clearly indicated the blastema formation in regeneration of some of these organs; hence, the molluscan regeneration may be limited to morphallaxis. In *Clio pyrimidata* too, the only mollusc claimed as capable of clonal reproduction, also involves morphallaxis. In this pteropod, Multipotent Stem Cells (MSCs) present in hepatopancreas gland, an endodermal derivative, regenerates the alimentary canal system. Similarly, the MSCs present in the columnar muscle, a mesodermal derivative, differentiate all the missing mesodermal and its derivative lineages. The Primoridal Germ Cells (PGCs) present in the septum, a mesodermal derivative, manifest sex and develop the reproductive system. Briefly, the MSCs present in the triploplastic layers are instrumental in regeneration of missing body parts (Pandian, 2017).

In eucoelomic Echinodermata, regeneration is more prevalent. In them, it is epimorphic with formation of a true blastema in crinoids, ophiuroids and holothurians but morphallaxis in asteroids and echinoids. It is catalyzed by

72   *Reproduction and Development in Platyhelminthes*

neurotransmitters, neuropeptides and other neural factors as well as specific genes *Anbmp, Afuni, EpenHg, ArHox1*, and *vasa* and *piwi* have been identified in crinoids, ophiuroids, holothuroids, asteroids and echinoids, respectively (Pandian, 2018). In all these aquatic invertebrates, the presence of neoblasts has not been demonstrated so far. Randolph (1891) was the first to discover the neoblast from the posterior surface of the septa adjacent to the ventral nerve cord of *Lumbriculus variegatus*. Subsequently, the presence of neoblast has been confirmed in a large number of annelids and platyhelminths. Yet, the annelidan multipotent neoblast is responsible only for epimorphic regeneration of mesoderm and its derivatives. All other missing body parts are regenerated by respective ectodermal and endodermal MSCs (Pandian, 2019). Though discovered first in an annelid, the annelidan neoblast is not considered as pluripotent. In platyhelminths, the neoblast is capable of regenerating all other tissue type lineages arising from the ectoderm, endoderm and mesoderm. Hence, it is pluripotent and unique to platyhelminths in the animal kingdom. Further, regeneration is accomplished mostly by epimorphosis. Still the distribution of such pluripotent neoblasts may be limited to triclads. Even assuming ~ 100 platyhelminth species (Table 2.9 accounts for 65 species only) and considering the species number as 27,559 (6,376, 4,500, 12,012 and 4,646 species for Turbellaria, Monogenea, Digenea and Cestoda, respectively, Table 1.2), only 0.36% of platyhelminths are capable of regeneration. This low value may be compared with 1.98% (204 out of 16,911 species) and 2.91% (247 out of 7,000 species) for annelids and echinoderms, respectively (Pandian, 2018, 2019). In platyhelminths, the costlier epimorphic regeneration has limited it to fewer species.

## B 2.8  Cell Counts and Mitosis

With recognition of the importance of neoblasts in regeneration by the 1960s, histological techniques were used to count the number of neoblasts to understand the effects of body length and antero-posterior regions of the body. By then, it was well known that regeneration potency decreases in antero-posterior direction faster in a smaller worm (4 mm) than in a larger one (18 mm) (Fig. 2.9A). Of course, increase in temperature up to a point also accelerates regeneration rate (Brondsted and Brondsted, 1961). Improving David's technique of whole body maceration in solution containing methanol, glycerol acetic acid, glycerol and distilled water (3:1:2:14 ratio) at 8–10°C, Baguna and Romero (1981) quantified for the first time cell types and their numbers in **clonal strain** of *Schmidtea mediterranea* and *Dugesia tigrina*: (i) In them, 14 cell types have been recognized and quantified. (ii) In *S. mediterranea*, the cell number increases from 110,000 in a smaller (4 mm) worm to 1,800,000 in a larger (16 mm) one, (iii) With increasing body length, the proportions of neoblasts, nerve and epidermal cell types decrease but those of parenchyma,

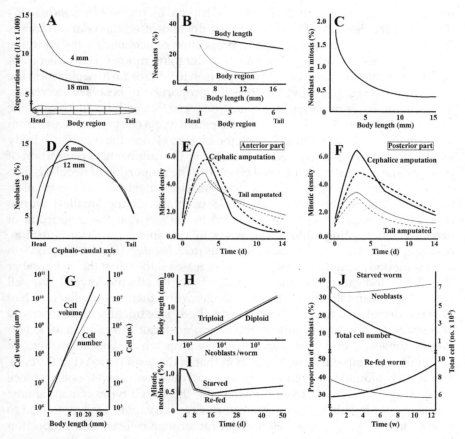

### FIGURE 2.9

A: Regeneration rate as a function of body regions in small and larger worms. B: Decrease in proportion of neoblasts as functions of body length and body regions. In it, 1, 3 and 6 represent the head, pharynx and tail, respectively. C: Neoblasts in mitosis as a function of body length. D: proportion of neoblasts as a function of body length in triploid worms of different body lengths. Mitotic density as a function of time in anterior (E) and posterior (F) parts (in dark continuous or dotted line for 4 mm size, in light continuous or dotted line for 10 mm size) following cephalic or tail amputation. G: Cell number and cell volume as a function of body length. H: Relation between number of neoblasts and body length. I: Mitotic neoblasts as function of starved and re-fed worms. J: Total number of cells and proportional neoblasts in 12-w starved and subsequently fed worms. These worms belong to *Schmidtea mediterranea* and *Dugesia lugubris* (all figures are simplified and redrawn from Dubois, 1949, Lange, 1967, Baguna, 1976a,b, Baguna and Romero, 1981).

gastodermal, acidophilic and basophilic cell types increase. For example, the proportion of neoblasts decreases from 32% (i.e. 35,200 neoblasts) in a small worm to 18% (i.e. 324,000 neoblasts) in a larger one (Fig. 2.9B). This is also true of nerve and epidermal cell types. As a result, the proportion of neoblasts is less by an order of magnitude, although the total number of neoblasts increases

74    *Reproduction and Development in Platyhelminthes*

with increasing body length in diploid and triploid biotypes of *Dugesia lugubris* (Fig. 2.9H). Incidentally, elevation in ploidy does not affect this relation (Lange, 1967). The reason for the decrease in proportion of neoblasts is traced to the decrease in the proportion of neoblasts undergoing mitosis, i.e. ~ 2% or 704 neoblasts in the smallest worm, in comparison to 0.5% or 1,620 neoblasts in the largest worm (Fig. 2.9C). (iv) Whereas the proportion of neoblasts decreases, there are increases in parenchyma, gastrodermal, acidophilic and basophilic cell types. For example, the proportion of parenchyma cell type increases from 11.5% in the smallest worm to 25% in the largest. Hence, the increase in body length of planarians is realized by increase in the number and proportion of parenchyma and gastrodermal cell types. (v) The proportion of neoblasts also decreases from 32% in the head region (i.e. 11,264 neoblasts in the smallest worm) to 24% in the tail region (8,448 neoblasts in the smallest worm) (Fig. 2.9B). A reason for it can be traced to the fact that the proportion of neoblasts undergoing mitosis decreases in the antero-posterior direction, covering the entire range of body length, as in *D. lugubris* (Fig. 2.9D). (vi) With increasing body length, the cell volume increases rather than the cell number (Fig. 2.9G), indicating that the longer the body length, the lower is the cell density, i.e. mean cell size is larger in a longer worm than in a smaller one. Not only cell density decreases with increasing body size but also mitotic density decreases as a function of time in the antero-posterior direction in all length classes; the levels of decrease are lower in a large worm than in smaller one and in the tail amputated region than in that of the head region (Fig. 2.9E, F). This has a profound implication to growth and degrowth as well as senescence and ageing. (vii) The relative proportions of the 14 cell types remain almost constant but are substantially changed during starvation. Degrowth due to starvation changes the total cell number rather than cell size and proportion of neoblasts. In a 7 mm long *S. mediterranea*, the total number of cells decreases from $6.8 \times 10^5$ to $0.5 \times 10^5$ after 12 weeks of starvation. However, following subsequent feeding for 12 weeks, it is increased to $9.6 \times 10^5$ cells (Fig. 2.9J). Owing to the changes in cell number, the proportion of neoblasts increases to 37% after starvation but decreases to 27% after feeding (Baguna and Romero, 1981). Baguna et al. (1990) undertook a year long experiment to study the effects of different feeding frequencies on the rates of cell birth and death in a small (3 mm) and a larger (11 mm) *Girardia tigrina*. With proportionately more neoblasts, rates of birth and death of cells are much faster in a young worm than in an older worm. To maintain a study state, the young worms have to be fed at least once every two weeks and the older ones once every three weeks, respectively (see also Romero, 1987). The terricolid triclad *Artioposthia triangulata* undergoes natural periods of growth and degrowth in relation to availability of its prey, the earthworm. Upon starvation, the adult animal resorbs tissues and depletes body resources (Blackshaw, 1992). These starved worms cannot be distinguished from their juveniles (Boag et al., 2006). Incidentally, planarians grow and degrow by changing total cell numbers depending on food availability leading to a 40-fold change in body size between 0.5 mm and 20 mm in *S. mediterranea* (see Rink, 2013). Using a new staining technique, Baguna (1974) reported the dramatic mitotic response in total cell number within 1 hour after feeding and the maximum between 3 and 8 hours. In planarians, ageing has been considered in the context of changes in (a) morphological and (b) physiological as well as (c) regenerative potency. In *D. lugubris*, (i) the supernumerary ocelli continue to form throughout the longest life span of 2.5 years and (ii) different parameters of sexual reproduction have been considered to decline. For example, cocoon production peaks between 3 and 12 months of its life span. Subsequently, 30% reduced level is maintained until 2.5 years. The number of embryos/cocoons also remains constant between 4 and 20 months but fertility decreases at the rate of 4.4%/month. Regenerative potency progressively decreases from 4-week old worm throughout the life time. As planarians grow in body length and age, neoblast density decreases. Starvation and regeneration are rejuvenators, due to reduction in cell number and tissue volume accompanied by increase in neoblast density (see Lange, 1968).

Using molecular markers, Takeda et al. (2009) investigated the maintenance of cell number during body growth in *D. japonica*. Their findings are summarized below: (1) In planarians, the number of cells increases with increasing body length. (2) Planarians have a constant ratio (1:3) of the head to the whole body length at a cellular level, irrespective of body length ranging from 2 to 9 mm. (3) The ratio between the specific neurons in the eye and brain also remains constant, regardless of the brain and body size, suggesting the planarians maintain a constant ratio of different cell types in an organ. From these results, Takeda et al. have concluded that a 'counting mechanism' regulates both the absolute and relative number of different cell types in complex organs such as the brain, turnover of cells, starvation and regeneration.

At this junction, it must be indicated that further investigations are required to note the number of cell types present in flukes and tapeworms, including their larval forms. In some teaniids, uterus alone occupies the ripe proglottids implying that fertilization of eggs switches off the neoblasts, which renew all other cell types.

## B 2.9 Blastema and Regeneration

In planarians, regeneration commences with the spreading of an epidermis over a wound surface followed by signaling that triggers the initiation of regeneration. Neoblasts are maintained in the parenchyma in adequate numbers, where they respond by proliferation. Neoblast's progenies 'migrate' and generate a blastema. The cells within the regenerative blastema differentiate and generate organ primordia (Reddien et al., 2005). These events are organized in a sequence of (i) wound-healing, (ii) mitosis and

76  *Reproduction and Development in Platyhelminthes*

'migration', (iii) blastema and differentiation, and (iv) pattern formation. Wound closure commences with strong contraction of the muscular layer and the consequent passive stretching of the pre-existing epidermal cells. As a shelter, the wound epidermis protects developing tissues but degenerates gradually, when the new surface closes around it and eventually falls off. The new epidermis is formed below the wound epidermis by a process of morphallactic reorganization of old tissues and by 'migration' of new cells from the parenchyma (Palmberg, 1986). In *Dugesia subtentaculata*, Chandebois (1980) reported an asymmetrical covering by the wound epidermis, namely dorsal epithelium covering the stump in the anterior regeneration and ventral epithelium in the posterior regeneration. From this observation, he suggested that determination of antero-posterior axis is caused by the differential interaction between one or other epithelium and underlying parenchyma. Even a simple needle poke up to the dorsal epithelium is adequate to induce robust mitotic response (Wenomoser and Reddien, 2010). Nevertheless, the hypothesis of Chandebois that dorso-ventral epidermal interaction may initiate antero-posterior axis may still remain valid.

*Proliferation and expansion*: While wound epidermis does not proliferate, groups of undifferentiated cells soon appear to form a few layers of cells that grow by addition of new undifferentiated cells formed by cell division in the underlying parenchyma. In *Microstomum lineare* with limited regenerative potential (see Table 2.8), $^3$H thyrimidine labeled neoblasts proliferate adequately and migrate from the post-pharyngeal part of the body to participate in regeneration of the head (Palmberg, 1986, 1990). Measurements by Salo and Baguna (1985a) showed that neoblasts 'migrate' rather expand (see Newmark and Alvarado, 2001) from 40 μm/day in intact planarians to 120–140 μm/day in those that are regenerating. By means of pulse-chase experiments, Ladurner et al. (2000) estimated the average 'migration' of neoblasts in *Macrostomum* as 6.5 μm/hour, which amounts to 156 μm/day. On the other hand, pre-existing local neoblasts proliferate in adequate number to commence blastema formation (Baguna, 1976b). However, 'migration' of neoblasts from other parts of the body is also reported (e.g. Gremigni and Miceli, 1980). For example, neoblasts from other regions expand to the amputated photoreceptors, a region devoid of neoblasts (Wenomoser and Reddien, 2010).

*Blastema and differentiation*: The blastema cells arise from a permanent population of undifferentiated stem cells, the neoblasts. In planarians, a number of factors, substances, and chemicals that accelerate or inhibit cell proliferation have been reported. At nanomolar concentrations, the neuropetides, Substance P, Substance K, bradykinin and *Hydra* peptide as well as the Epidermal Growth Factor are potent mitogens. The simultaneous use of Substance P and Substance K and an antagonist, the Spanide have a specific mitogenic role, especially the last one abolishes the mitotic response. The other mitogens bind to specific membrane receptors through inositol triphosphate/diacil glycerol (IPs/DG) mitogenic signal pathway. This leads to DNA synthesis and mitosis by stimulation of cytosolic calcium release and protein kinase activation (Baguna et al., 1989). To study the neoblast density affecting the cell proliferation rate, Salo and Baguna (1985a) injected different doses of neoblasts into irradiated host planarians. The results suggested that some kind of autocrine inhibitory substance is produced by the neoblasts to inhibit their proliferation, when their density is very high in the blastema.

Two patterns of neoblast proliferation are recognized: (i) mono S-phase proliferation occurring in non-triclad turbellarians and (ii) biphasic proliferation in triclads. In *Macrostomum lignano*, proliferation rate within blastema increases rapidly and goes through a (monophasic) S-phase until 72 hours after amputation, when differentiation is commenced (Egger et al., 2009). This type of S-phase proliferation is also reported in other non-triclad turbellarians (e.g. *Isodiametra pulchra*, De Mulder et al., 2009). In a detailed description of tail regeneration in *M. lignano*, Egger et al. (2009) reported that the tail accommodates the male genital apparatus and consists of ~ 3,100 cells, about half (1,560 cells) of which are epidermal cells. A distinct blastema consisting of 420 cells is formed 24 hours after amputation by the accumulation of proliferating neoblasts. Two days after amputation, differentiation is commenced and the male apparatus is developed in 4–5 days. In pulse-chase experiments, dispersed distribution of the label suggests that S-phase labeled progenitor cells of the male apparatus undergo further proliferation before differentiation. After 3 days, the blastemal 1,420 cells are gradually transformed into organ primordia, while proliferation is progressively decreased.

In the triclad *Schmidtea meditteranea*, a detailed description of biphasic neoblast proliferation is reported. Engaging a number of markers, Wenemoser and Reddien (2010) carried out a series of experiments in clonal strain of *S. meditteranea*. Their findings are briefly summarized: (1) The antibody that recognizes Histone H3phosphorylated at serine (anti-H3P) allows quantification and spatial resolution of neoblast mitosis in the entire animal fragment. Following amputation, a temporally biphasic mitotic pattern occurs and confirms the similar observation by Salo and Baguna (1984a), who counted the mitotic figures in successive tissue stripes. A rapid 5-fold increase in mitosis number in the first peak within 6 hours is followed by decrease to 2-fold level between 8 and 18 hours, and the second higher peak between 48 and 72 hours after the amputation. The magnitude of the first mitotic peak scales with the wound size. (2) Whereas the first peak occurs throughout the entire planarian fragment, the second one is localized to the wound area. In triclads, initiation of regeneration involves two phases of wound response. In the first, the wound triggers entry into mitosis at the body-wide long distances and detection of tissue loss rather than injury. And the second peak induces localized mitotic peak. (3) Cycling cells are *smedwi-1*[+]/SMEDWI-1[+]. But the cells that cease expression of *smedwi-1*[+]/SMEDWI-1[+] exit the cycle and transiently express *smedwi-1*[-]/SMEDWI-1[-], due to protein endurance. Engaging *smedwi-1mRNA* and antibody to recognize SMEDWI-1, it is possible to distinguish cycling neoblasts from those that have exited the cycle to differentiation. Before and during the second mitotic peak, a proportion of neoblast descendants exits the cycle and gives rise to a layer of non-cycling cells at the wound site. The signals from the wound site initiates progressively increased differentiation.

Pattern formation is characterized by the sequential or simultaneous determination of different groups of cells that are to become a particular tissue or organ in the final form of the regenerating flatworm. In planarians, it is a very early event (cf Chandebois, 1980) occurring in a narrow strip of cells below the wound, when the blastema is not yet formed (Salo and Baguna, 1984b). The determination of different regions along the antero-posterior axis seems to be simultaneous. The overall pattern is first manifested by morphallaxis in a very narrow strip of tissues below the wound during the first day of regeneration and is subsequently amplified and refined by epimorphosis. Classical experiments have shown that in general, planarians amputated at any level along the Antero-Posterior (AP) axis can regenerate the head. However, the potency for the head regeneration decreases posteriorly resulting in a 'time-graded regeneration rate' along the AP axis. In another classical experimental series, narrow, thin bipolar regeneratory pieces generate (two) Janus headed planarians (Fig. 2.11B1), indicating the need for a minimal dorsal-ventral distance in the AP axis for proper patterning of regeneration. Considering the dynamic gradient along the AP axis, Morgan (1905) proposed that the polarity is determined by structural (or substance) differences along the axis but Child (1941) proposed a physiological gradient hypothesis. However, sequencing the planarian genome and the possibility of performing gene functional analysis of RNA interference (RNAi) have led to the isolation of the bone morphogenetic protein (Kobayashi et al., 2007), *Wnt* (Molina et al., 2007) and Fibroblast Growth Factor (FGF). Their pathways control patterning and axial polarity during regeneration and homeostasis of planarians. A series of experiments have shown that *Wnt* and *BMP* determine the fate of cells antero-posterior and Dorso-Ventral (DV) axes, respectively (Adell et al., 2010, 2015). The *Wnt* signaling is required for posterior regeneration (De Robertis, 2010) but not for the anterior regeneration. This conclusion is also confirmed by a series of reports published in Nature. *β-catenin* signaling is responsible for the deficiency in head regenerative potency of posterior fragments in *Phagocata kawakatensis* (Umesono et al., 2013), *Procotyla fluriatilis* (Sikes and Newmark, 2013) and *Dendrocoelum lacteum* (Liu et al., 2013). Reduced activity of *β-catenin-1* forces the head regeneration, irrespective of the wound context (Gurley et al., 2008, Iglesias et al., 2008, Petersen and Reddien, 2008). The patterning of the medio-lateral axis and bilateral symmetry of central nerve system requires *natrin, slit, robo* (Cebria, 2007, Cebria and Newmark, 2005, Cebria et al., 2007) and *Wnt-5* (Adell et al., 2010, Gurley et al., 2010). Several elements of the

BMP pathways are required for the reestablishment of proper DV axis during regeneration (Molina et al., 2007, Reddien et al., 2007). The RNAi silencing of the planarian homologs of *BMP, Smad 1 and Smad 4* transforms the dorsal side into ventral one in both intact and regenerating planarians.

## B 2.10  Genes in Regeneration

A comparison of microarray based expression analysis of intact and irradiated worms in two planarian species has produced a catalog of genes potentially involved in neoblast biology (Eisenhoffer et al., 2008, Blythe et al., 2010). In *Schmidtea meditteranea*, transcriptome has identified 2,300 transcripts (out of 25,000) that are significantly down-regulated by irradiation (Newmark and Alavarado, 2001). Many of these genes with roles in stem cells and germ cells are associated with RNA binding and metabolism, and are localized in neoblast's Chromatoid Bodies (CBs). The three planarian orthologs, *smedwi-1* (*piwi-1*, see Rink, 2013), *-2* and *-3* are all expressed in neoblasts (see Aboobaker, 2011). Both *smedwi -2* and *-3* are required to accomplish regeneration and maintenance of homeostasis.

Selecting a representative sample of 1,065 genes from *S. mediterranea* genome, Reddien et al. (2005) found that the defects associated with the RNAi of 240 genes define regeneration and homeostasis. Of them, 205 are predicted to encode proteins with significant homology. Among 143 genes that are associated with dsRNA-induced regenerative defects, *SMAD4* is a gene for proper blastema formation. Of the selected 139 genes in the 24 hours H3P data set, eight genes allow a normal number of mitotic neoblasts after wounding. Blastemas in HE.2.07D BMP1 (RNAi) worms indicate that *BMP* signaling, which regulates DV axis, may control regeneration of midline tissues. Since 85% of the genes have homology to genes found in other organisms including 35 that are similar to human disease genes, Friedlander et al. (2009) reported that piRNA of *S. mediterranea* are in part organized in genomic clusters and share characteristic features with piRNAs of mammal and drosophila. These miRNAs are downregulated in worms, in which stem cells have been abrogated by irradiation. Thus, they are associated with specific stem cell function. Cowles et al. (2013) identified 44 genes predicted to code for a basic Helix-Loop-Helix (bHLH) domain, of which 12 are expressed in the Central Nervous System (CNS) and neoblasts. Some of these genes are co-expressed with cholinergic-, GABAergic-, octapaminergic, dopaminergic or serotogeneric neuronal population. Strikingly, two of these genes *coe* and *sim* are co-expressed with proliferating neoblasts and contribute to the regenerative blastema and thereby support the existence of neural progenitor cells in planarians. Further, silencing of the homologs of either *netrin* or *netrin* receptor disrupts neural architecture in intact and regenerative flatworms (Cebria et al., 2015).

# Part C: Clonal Reproduction

## Introduction

Sex is costly and demands time and resource. But clonal reproduction saves time and avoids the risks involved in sexual reproduction. It also provides a mechanism for potential rapid amplification of a genotype known for its fitness, for example, oncomiracidium (in *Entobdella* sp), miracidium and oncosphere/coracidium that have successfully infected the appropriate definitive/intermediate hosts. However, it involves no gametogenesis and recombination. In the absence of recombination and fusion of gametes in clonally reproducing flatworms, adaptation can be impeded and deleterious mutations may be accumulated, due to Muller's ratchet (Pandian, 2019). Understandably, the incidence of clonal reproduction is limited to adults of a few turbellarians and the monogenean *Gyrodactylus* as well as intra-molluscan larval stages of all digeneans and larval stage of ~ 0.5% cestodes.

Embryogenesis, regeneration and clonal reproduction are similar but not identical developmental processes. For example, wound healing and blastema formation are not part of embryogenesis. No blastema is formed during clonal reproduction. In the light of the definition for clonal reproduction, the scope of the term regeneration may have to be extended. In some triclads (Baguna, 1998 and *Macrostomum ligano*), regeneration of the small anterior segment goes through stages similar to post-embryonic development. From their observations, Cardona et al. (2005) noted the recapitulation of embryonic pathways during regeneration. *Vasa* expression is also similar in hatchlings and gonad-restoring regeneration (Pfister and Ladurner, 2005). In the presence of the brain, the decisive organ for regeneration, the paratomic and budding acoels 'pre-generate' many organs of the newly developing zooids, while still in the presence of parental organs. The head may dictate the newly developing organs to recapitulate the same pathways, as in embryogenesis (see Egger et al., 2006).

Clonal reproduction is bi- (or multi-) directional, when an animal (genet) divides to produce two or more fully functional progenies (ramets). But it is unidirectional, when only a single (half of the ramets) progeny is developed (Rychel and Swalla, 2009), as in enteropneusts and solitary ascidians (Pandian, 2018). In principle, reproduction, whether sexual (including self-fertilizing hermaphrodite involving a single parent) or asexual (clonal), is expected to generate more than one offspring. Hence, a fission or amputation that generates a single ramet, as a product of unidirectional cloning, can be considered more as regeneration than reproduction. Unlike annelids, the high frequency of bidirectional clonal multiplication in turbellarians leads to increase in progeny output. The turbellarian species, for which reports are

*Turbellaria* 81

**TABLE 2.10**

Clonal potency in turbellarian taxa. A = architomy, P = Paratomy, B = Budding, DA = double architomy

---

**Acoela**: *Adenopea canata* (A), *Amphiscolops langerhansi* (A), *Convolutriloba longifissura* (DA), *Paratomella rubra* (A), *Pseudohaplogonaria macnaei* (A), *Pa. unichaeta* (P), *C. retrogemma* (B), *C. okinawa* (B) (Akesson et al., 2001), *Isodiametra pulchra* (De Mulder et al., 2009)       **(9)**

**Catenulida**: *Stenostomum grande* (P) (Egger et al., 2006), *Catenula* (Moraczewski, 1977), *Stenostomum leucops* (P) (Palmberg, 1990), *S. unicolor* (Hartmann, 1922), *S. tenuicauda* (see Table 2.11)       **(4)**

**Macromorpha**: *Alaurina aethiopica* (P) (Stocchino and Manconi, 2013)       **(1)**

**Tricladida**: *Cura foremanii* (Rivera and Perich, 1994), *Dugesia aethiopica*, *D. afromontana* (Stocchino and Manconi, 2013), *D. anderiani* (Hauser, 1987), *D. benazzii*, *D. entrusca* (Stocchino and Manconi, 2013), *D. dorotocephala* (Rivera and Perich, 1994), **D. fissipara (P)** (Stocchino and Manconi, 2013), *D. gonocephala* (Brondsted, 1969), *D. japonica* (Shibata et al., 2012), *D. japonica japonica* (Sakurai, 1981), *D. lugubris* (Grasso and Benazzi, 1973), *D. maghrebiana*, **D. paramensis (P)** (Stocchino and Manconi, 2013), *D. ryukyuensis* (Hase et al., 2003), *D. sanchezi* (Benazzi, 1981), *D. sicula* (3n only) (Lazaro et al., 2009), *D. tigrina* (Sheiman et al., 2006), *D. subtentaculata* (Stocchino and Manconi, 2013), *D. tahitiensis* (P) (Peter, 2001). *Dendrocoelopsis vaginatus* (Rivera and Perich, 1994), *Euplanaria tigrina* (Kenk, 1937), *Fonticola morgani* (Benazzi and Ball, 1972), *Girardia tigrina* (Benazzi, 1993), *Planaria maculata* (Curtis, 1902), *Phagacata velata*, *P. vivida*, *P. vitta*, *Polycelis cornuta* (Brondsted, 1969), *Schmidtea mediterranea* (Tan et al., 2012)       **(30)**

---

available for clonal potency, are listed in Table 2.10, which is not exhaustive. It seems that with increase in size, the fission frequency is decreased. Further, a number of substances (e.g. hormonoids, neuropharmalogical agents) may prevent regeneration and stimulate bi-directional/multi-directional reproduction (see Krichinskaya, 1986). Hence, this account also includes the uni-directional clonal multiplication as clonal reproduction.

# C 2.11  Surgery and Potency

In turbellarians, the relatively small size (0.3 mm–27 mm, rarely 60 cm, see Fig. 1.9) rarely allows field observations on natural clonal reproduction (e.g. *Dugesia benazzii*, Stocchino and Manconi, 2013). However, the turbellarians are readily amenable to rearing and to surgical amputations and grafting. Hence, our understanding of their regeneration and clonal potency is mostly based on laboratory observations and experimentations. The typical morphology and position of the pharynx, an organ with no neoblasts, are shown in Fig. 2.10A. Transverse (Fig. 2.10A2, B, C, e.g. *D. japonica*), longitudinal (Fig. 2.10D) or oblique (Fig. 2.10E) amputations at any level results in complete regeneration of all missing body parts, *albeit* in different sizes and shapes but with retention of original polarity in many planarians. In fact, a

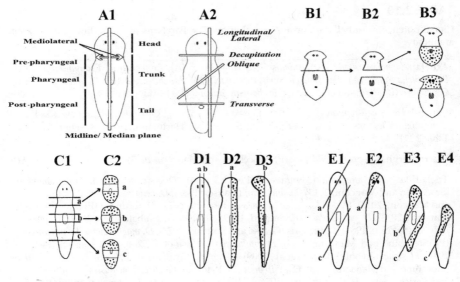

**FIGURE 2.10**

Surgical amputation in planarians: A1 shows the morphology and location of pharynx and A2, the directions of amputations. B and C transverse, D longitudinal and E oblique amputations at different levels; the dotted area indicates the newly regenerated body and the white area remaining parental body (free hand drawings from Reddien and Alvarado, 2004).

small fragment as small as 1/279 of its body can successfully regenerate a complete worm (see Collins, 2017). However, the size and morphology of regenerated head, regeneration rate and tail length are progressively reduced with more and more posterior fragments (e.g. *D. derotocephala*). There are others, in which the middle and posterior fragments cannot regenerate the head and/or tail (e.g. *Macrostomum* spp, see Egger et al., 2006). Therefore, the head controls the pattern formation namely the spatial arrangement and proportions of organs and other body parts. In planarians, the smaller fragments and appropriate splits can generate the formation of Janus (with two) heads, crotech head and so on (see Fig. 2.11B series) (for more details, consult Hyman, 1951).

Grafting is another surgical technique to understand the regenerative and clonal potencies. For example, when a rectangular or triangular piece from a donor planarian is grafted either in a normal or reversed position into a recipient, from which the equivalent has already been removed, the polarity is restored (Fig. 2.11A). Interestingly, right and left longitudinal fragments from two planarians can also be fused (Fig. 2.11B). A graft from the cephalic region induces the formation of the head at the pre-pharyngeal, post-pharyngeal and the tail region as well as Janus, crotch and multiple heads (Fig. 2.11B). Kenk (1941) had successfully induced the development

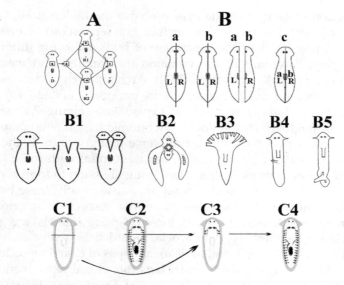

**FIGURE 2.11**
Grafting experiments in planarians. (A) Retention of polarity, when a graft removed from a donor (D) is grafted to recipients (R1, R2), in which the direction of the grafts has been changed and is corrected in P. (B) Splitting and grafting lead to the development of (B1) Janus heads, (B2) crotch head, (B3) multiple heads and (B4) formation of post-pharyngeal head and (B5) caudal head. (C) *Dugesia derotocephala*: C1 and C2 are asexual and sexual planarians, respectively. Grafting anterior fragment from sexual donor to asexual recipient results in sexualization of the recipient (Figures A and B are redrawn from an unknown source; B1 to B5 are free hand drawings from Hyman, 1951; C1 to C4 are drawn from information reported by Kenk, 1941).

of gonads and copulatory organ in the body of clonal *D. derotocephala* by grafting an anterior fragment of a sexual worm, after its anterior body has earlier been removed (Fig. 2.11C).

## C 2.12 Clonal Types

Clonal reproduction includes different types: (1) Autotomy or spontaneous fragmentation represents a stress-induced reaction to an unfavorable condition; (2) Architomy, which means fission preceding the differentiation of the new organs of the daughter flatworms (cf annelids, Pandian, 2019); (3) Double fission is a form of architomy that occurs in two steps, as in the acoel *Convolutriloba longifissura* (Akesson et al., 2001); (4) In paratomy, differentiation of ganglia or the eyes precedes the fission itself, i.e. fission into new worms after the formation (but not separation) of new organs; and (5) Budding involves the separation of buds following differentiation of the organs. Paratomy is also common in catenulid and macrostomid species (Reuter and Kreshchenko, 2004). Architomy characterizes triclads, although *D. fissipara* and *D. paramensis* are paratomic (Table 2.10). Of these types, double architomy requires an explanation. Figure 2.12 shows the sequence, through which the double architomic fission is accomplished in *C. longifissura*. A post-pharyngeal transverse fission occurs in the parental worm; the anterior fragment regenerates the caudal with three lobes. But the caudal one undergoes a second longitudinal fission to form a W-shaped butterfly stage attached to the substratum. Subsequently, these two caudal fragments complete the regeneration of the respective anterior region. Briefly, from a single parent (genet), three offspring (ramets) are generated in the double architomic type. In acoela, regarded as a very early offshoot of the Bilateria (Littlewood et al., 1999a), all types of clonal reproduction are reported (Akesson et al., 2001). In triclads too, clonal reproduction occurs in some species of the primitive families of Dugesiidae, Planariidae and Dendrocoelidae. Even among them, dendrocoelid can regenerate the anterior only following pre-pharyngeal amputation (Newmark and Alvarado, 2001). Incidentally, regarding the origin of clonal potency, there are alternative views. Ax and Schulz (1959) considered that clonals arose from flatworms with high regenerative potency. But Benazzi (1974) suggested the 'fissiparous genes' induced clonal reproduction. He may be correct, as the presence of sexual and asexual neoblasts have been reported (Kobayashi et al., 2008, 2009).

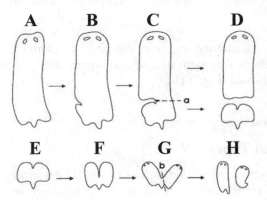

**FIGURE 2.12**

Double architemy in *Convolutriloba longifissura*. The first row from A to D shows the first trasverse fission. The second row E to H shows the second vertical fission (free hand drawing from Akesson et al., 2001).

## C 2.13 Switching Sexuality

In many planarians (*Girardia tigrina*, *D. japonica*), the clonal strain reproduces exclusively by fission and does not express sexuality at all. The sexual strain can switch to the clonal state spontaneously and subsequently revert to sexual state. These are named as exfissiparous strain (Benazzi, 1974). In others, the clonal strain can switch to a sexual state, when they are fed with sexual planarians as food, as in *D. ryukyuensis*. Clonal and sexual reproductive modes have their own advantages of low cost production and increase in genetic diversity, respectively. Switching from sexual to clonal and *vice versa* is an effective strategy to gain advantage of both in response to change in environmental conditions.

Despite numerous publications, clonal reproduction seems to be limited to 50 + species (Table 2.10). The absence of external (e.g. genital apparatus, a key structure for species level identification) and internal (e.g. ovary, testis, yolk glands) structures make it more difficult to identify clonal turbellarians. The presence of triploids, chromosomal heteromorphism (e.g. *Schmidtea mediterranea*, Newmark and Alvarado, 2002), anuploid and variable number of B chromosomes (e.g. *S. mediterranea*, Pongratz et al., 2003) make the task more onerous. In recent years, a small fragment of the mitochondrial COI sequence is being used as a universal marker for species identification in the DNA bar-coding system. Despite some drawbacks, the COI and ITS (internal transcribed spacer) sequences are useful and it has been possible to identify and assign 29 asexual dugesiid populations to their sexual species counterparts (Lazaro et al., 2009). Hence, the list summarized in Table 2.10 is not exhaustive. More species may be added but the number of clonal species may not increase beyond 100 species.

Barring 9 acoelid species, all others are inhabitants of freshwater, especially lentic water. To overcome unfavorable conditions like freezing or drying, the cloners switch to sexual (except *Phagocata velata* producing encysting clones) reproduction. Reared under optimal conditions (20°C for *P. vitta*), half a dozen turbellarian species are reported to sustain clonal reproduction for 3–15 years and pass through over 1,000 generations without any sign of sexual organ/apparatus (Table 2.11). However, sexualization does occur in 30–40% *Dugesia* spp after a period of laboratory rearing for 7 months in *D. aethiopica* and 2 years in *D. afromontana* (Table 2.11). Yet, these dugesiids retain clonal potency. Notably, the exfissiparous sexualized dugesiids remain mostly sterile (e.g. *D. maghrebiana*); those produce a few cocoons (i.e. *D. aethiopica*, *D. afromontana*), the cocoons are sterile and fail to hatch. The number of cocoons (3–8 cocoons/week) and their hatchability (1–6 hatchlings/cocoon) may be compared with the production of 10–20 cocoons/week, 98% hatching success and 10 hatchlings/week by the semelparous exclusively sexual species *D. hepta*. Incidentally, the fission frequency in the clonal turbellarians

# TABLE 2.11

Reported observations on sustenance of clonal reproduction in turbellarians and life cycle of selected *Dugesia* species (condensed from Stocchino and Manconi, 2013). coc = cocoon, w = worm

| Sustenance of clonal reproduction in laboratory | |
|---|---|
| Species, reference | Reported observations |
| *Stenostomum grande,* | Clonal reproduction by fission lasted for 4 years |
| *S. tenuicauda* (Nuttycombe and Waters, 1938) | Over 1,000 clonal generations lasted for 11 years without appearance of sexual |
| *S. unicolor, S. leucops* (Hartmann, 1922) | Clonal reproduction for > 2 years. Following 50 amputations clonal reproduction continued |
| *Microstomum* (Sekera, 1906) | Paratonic colonies production lasted for 3 years |
| *Phagocata velata* (Child, 1914) | 13 generations of fragmentation, encystment and hatching lasted for 3 years |
| *P. vitta* (see Brondsted, 1969) | Clonal reproduction lasted for 15 years at 20°C |
| *D. derotocephala* (Kenk, 1937) | Clonal reproduction lasted for 6 years in the asexual strain |

| Life cycle of asexual *Dugesia* spp in laboratory | | | |
|---|---|---|---|
| Species, source, rearing period | Sexualization in lab | Reproduction | |
| | | Sexual | Cloanl by ex-fissipare |
| **Natural sexual population** | | | |
| *D. hepta*, Sardina, 10 years | Univoltine | 10–20 coc/w, 98% hatch, 10 hatchlings/w | – |
| **Natural clonal populations** | | | |
| *D. aethiopica*, Lake Tanaka, Ethiopea, 6 years | After 7 months, 30% sexualized | 50–90% sterile coc; of remaining, 5–7 coc/season, 1–6 hatchlings/coc | Following gonadal loss, caudal remains fissiparous, cephalic with or without copulatory organ but fissiparous |

*Table 2.11 contd. ...*

| Natural clonal populations | | | |
|---|---|---|---|
| *D. afromontana*, South Africa, 6 years | After 2 years, 30% sexualize in spring, but retains fission potency | 3–8 coc/w, 60–80% sterile coc, of remaining, 20–40% hatch, 1–2 hatchling/coc | Resorbs gonads and resumes fissioning |
| *D. maghrebiana* | After 1 year, 40% sexualized, but retains fission potency | No cocoon produced | Resumes fissioning |
| Mixed sexual and fissiparous populations | | | |
| *D. benazzii, D. entrusca*, Saradinia, > 13 years | Sexual reproduction during autumn | 10–20% fertile coc/w, 8–10 hatchlings/coc | Caudal tips become fissiparous, but cephalics become ex-fissiparous |

88    *Reproduction and Development in Platyhelminthes*

ranges from 0.1–0.4 times/worm/day in triclads to 0.32–0.52 times/worm/ day in the acoels (Table 2.12). Hence, the acoels undergo fission more frequently than triclads. Experiments have revealed that the frequency also depends on population density (Brondsted, 1969).

Many tricladid species are characterized by the simultaneous existence of sexual and clonal strains (e.g. *D. ryukyuensis*). In sexual strain, cloning can be induced by the season (e.g. Kobayashi et al., 2002b), temperature or food. Clonal fission can be induced in sexual strains of *Phagocata vivida* at 20°C. In *D. gonocephala* and *D. tigrina*, it is induced at 17–20°C and 13–16°C in mature and immature worms of sexual strain, respectively; however, clonal strain commences fission even at 11–13°C (see Brondsted, 1969). Perhaps, Grasso and Benazzi (1973) were the first to sexualize the clonals. They reported that sexualization of *D. gonocephala* by feeding sexual *Polycelis nigra* resulted in the development of hyperplasic ovaries, testicular follicles and copulatory apparatus but production of sterile cocoons. Feeding *P. nigra* also sexualized *D. tigrina*, *D. derotocephala* and *D. mediterranea*, but perhaps with similar development, as in *D. gonocephala* (Benazzi and Grasso, 1977). Clearly, the exfissiparous sexual dugesiids are not functionally sexual.

Through a series of publications, Kobayashi and his colleagues brought to light a holistic picture of sexualization in clonals of *D. ryukyuensis*. In exclusively clonal OH strain, feeding with sexually mature worms of *Bdellocephala brunnea*, an exclusively oviparous species, results in the (i) development of gonopore, (ii) a pair of ovaries and (iii) complete inhibition of fission on the 2nd, 2–3rd and 3rd week after feeding, respectively. The level of sexualization can be recognized at five stages: the appearance of the testes primordia and copulatory apparatus at stage 3, genital pore, primordia of yolk glands and spermatocytes at stage 4, and mature sperm at stage 5. Clearly, the oviparous worms contain a putative sexualizing substance, which is not digestible and species-specific (Kobayashi et al., 1999). Firstly, from stages 1 and 2, the treated worms can return to a clonal state. But stage 3 represents a point of no return. Hase et al. (2003) isolated a novel gene *Dryg* encoding 655 amino acids and a predicted molecular mass of 79 kDa. It is expressed specifically in yolk glands. Following fission, the original yolk glands disappear but a new set of neoblasts induces the development of new series of yolk glands. However, the true sexualizing substance is not the product of yolk glands, as *B. brunnea* with mature yolk glands are unable to induce complete sexualization (Kobayashi et al., 2002a). Secondly, sexualization is a time-dependent process. For example, 0, 43 and 94% of the worms, grouped in A, B and C are sexualized, when feeding is continued up to 1, 3 and 6 weeks, respectively. Hence, a period of 6–7 weeks is required to ensure 100% sexualization of the treated worms. Remarkably, 43 and 94% of the worms belonging to B and C groups, respectively, remain sexualized, even when feeding is ceased. Apparently, all the worms in group C remain sexualized. However, a few (43%) worms in B group are able to respond,

## TABLE 2.12

Fission frequency in some turbellarians

| Species, Reference | Fission frequency (no./worm/d) |
|---|---|
| Acoela | |
| *Convolutriloba longifissura* (Akesson et al., 2001) | 0.32, Architomy |
| *C. retrogemma* (Ishikawa and Yamasu, 1992) | 0.52, Budding |
| *Ampiscolops langerhansi* (Hanson, 1960) | 0–0.5, Architomy |
| Macrostomorpha | |
| *Macrostomum lignano* (Mouton et al., 2018) | 3 weeks for posterior regeneration |
| Tricladida | |
| *Dugesia ryukyuensis* (Kobayashi and Hoshi, 2002) | 0.1 |
| *D. derotocephala* (Best et al., 1969) | 0.2–0.4 |
| *D. tigrina* (Davison, 1973) | ~ 0.35 |
| *Euplanaria tigrina* (Kenk, 1937) | 0.4 |

while others are unable to do it (Kobayshi et al., 1999). Thirdly, the clonal worms are not completely sexualized during summer, *albeit* developing a pair of ovaries, when feeding is stopped. During winter, the worms, however, are completely sexualized with the development of ovaries, even when feeding is stopped (Kobayashi et al., 2002b). Hence, sexualization is a season- and dose-, i.e.—treatment duration dependent process (see also Hoshi et al., 2003).

Fourthly, Kobayashi et al. (2002a) also established an OH strain, an exclusively fissiparous strain as distinctly different from the sexual strain. An experimental bioassay system for the OH strain has also been developed (Ishizuka et al., 2007). Inbreeding of acquired sexual in the OH strain of *D. ryukyuensis* generates both clonals and innate sexual at the ratio of 2:1 (Kobayashi et al., 2009). This is also true of *D. benazzi* (see Benazzi et al., 1988). Besides, transverse amputations at the Head (H), Middle (M) and Tail (T) levels are followed by complete regeneration and reproduction. Notably, the H segment regenerates into clonal, whereas M and T segments into sexual. Evidently, the sex-inducing substance is present in posterior fragments but not in anterior (Kobayashi and Hoshi, 2002). Interestingly, sexualization of clonal worms also induces supernumerary ovary pairs along the ventral nerve cord up to pharyngeal level. However, the main ovaries alone are noticeably larger and harbor increased number of germ cells. Benazzi (1981) also reported that exfissiparous *D. sanchezi* is more fecund than sexual. The innate sexual do not produce supernumerary germ cells, even when fed *B. brunnea* (Kobayashi et al., 2012). The *nanos* homolog *Dr-nanos* is also expressed in the germ cells. Additionally, it is also strongly expressed in the brain of innate sexual (Nakagawa et al., 2012a). Incidentally, the *Smed-nanos* is also

90  *Reproduction and Development in Platyhelminthes*

expressed in germ cells as well as eye primordium of *Schmidtea mediterranea* (Handberg-Thorsager and Salo, 2007). Based on clonal and sexual progenies during successive generations, five types can also be recognized. It requires 4, 3, 3, 4 and 2 clonal cycles to generate completely sexualized progenies from the sexual worm-fed $F_o$ *D. ryukyuensis* in type 1, 2, 3, 4 and 5, respectively, i.e. individual clonal worms considerably differ in their ability to completely switch from clonal to sexual (Kobayashi and Hoshi, 2002).

In nature, *D. ryukyuensis* exhibit diploid (2n = 14), triploid (3n = 21) and mixed ploid (2n + 3n) (Tamura et al., 1991). Inbreeding of acquired sexual in the OH strain produces both diploid and triploid offspring at the ratio of 1:2 or 1:3, resulting in (i) diploid clonal, (ii) triploid clonal, (iii) diploid sexual and (iv) triploid sexual (Kobayashi et al., 2008). Using chromosomal mutation in Chromosome 4 as a marker, Kobayashi et al. (2008) showed that the Chromosome 4 is inherited in subsequent generation of both diploids and triploids. However, triploidy is considered as an evolutionary dead end, due to problems in chromosomal pairing and segregation during meiosis. This problem seems to have been solved in triploid *D. ryukyuensis*. In the germline of triploid neoblasts, the elimination of one set of chromosomes results in diploid germ cells in both female and male. Cytological studies have indicated the occurrence of chiasma between homogeneous chromosomes during both oogenesis and spermatogenesis, indicating that meiosis does occur in the triploids. With the production of haploid sperm and eggs, diploids are generated. However, triploid production involves, of course, a haploid sperm and presumably diploid egg produced by the fusion of nuclei following meiotic division (Kobayashi et al., 2008).

To understand the genetic differences between the neoblasts of clonal (AS), acquired sexual (Aqs) and innately sexual (InS) in *D. ryukyuensis* (Kobayashi et al., 2012), Nodono et al. (2012) transplanted the neoblasts from InS into lethally irradiated AS worm. The resultant transplant consisted entirely of donor-derived neoblasts. The transplanted neoblasts were engrafted successfully into the recipient, proliferated and ultimately replaced all recipient cells. Nodono and Matsumoto (2012) also transplanted InS neoblasts into a non-lethally irradiated AS worm and thereby produced chimeras consisting of neoblasts from both AS and InS. The AS became sexual revealing that InS can initiate a sexual state autonomously even under a chimeric state with coexisting AS neoblasts. Unlike AqS, only a pair of ovaries was developed in the chimeric AS with transplanted neoblasts from InS.

## C 2.14 Telomere and Senescence

In sexual eukaryotes, cell lineages have a definite life span, as in other organisms (Roger and Hug, 2006). The defined life span is linked to telomere shortening during each cycle of eukaryotic DNA replication (see Tasaka et al.,

2013). Sexual eukaryotes achieve telomere elongation during embryogenesis and development of the germline. However, their somatic cells do not maintain telomere sequence (length). Consequently, they become senescent, as teleomere shortens to a critical length. This is to avoid the possible genome instability and emergence of cancer cells. In fact, this protective senescence mechanism is a central part of the aging process (see Tan et al., 2012).

On the other hand, clonal eukaryotes like the clonal planarians seem not to suffer from telomere shortening and consequent defined life span (see Tasaka et al., 2013). Not surprisingly, many planarians serve as excellent models to understand their potency to maintain telomere length. Typical (not involving any translocation) karyotypes of innately sexual (InS) and clonal (AS) triploid *Dugesia ryukyuensis* show shortening of chromosomal length in InS (Fig. 2.13A). The use of telomere probe TAAGGG in FISH analysis fluorescently labels the chromosomal ends, indicating no reduction in chromosome length (Fig. 2.13B). For more figures on similar labeling of chromosomal ends of in *Polycelis tenuis*, Joffe et al. (1996) may be consulted. *In situ* Southern hybridization using TAAGGG probe has indicated that the mean length is

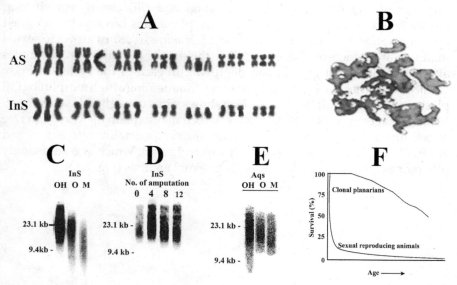

**FIGURE 2.13**

*Dugesia ryukyuensis*: (A) Karyotype of clonal and sexual triploids. (B) Fluorescence *in situ* hybridization showing telomere repeats in white against dark background. (C) Effect of ageing in telomere length in innate sexual (InS) worms. O = new born worms, M = worms with repeated amputations and regeneration for 2 years, OH = fissiparous OH strain. (D) Telomere length in InS following a series of amputations up to 12 times. (E) Effect of ageing in telomere length in acquired sexual worms (AqS) OH, O and M worms. (F) Survival as function of clonal and sexual planarians (from Tasaka et al., 2013, with permission of International Journal of Developmental Biology). (G) Survival as function of age in clonally and sexually reproducing animals (drawn partly from Fig. 3.9).

## 92 Reproduction and Development in Platyhelminthes

24 kb for AS. But it is only 18 and 14.5 kb for the fresh hatchling and 2-years old InS worms, respectively. Clearly, the length is not only shorter in InS worm but also is progressively shortened with increasing number of DNA replication cycle during 2-years life span (Fig. 2.13C). Conversely, the length is not shortened even after 12 cycles of DNA replication in AS clonal worms (Fig. 2.13D). This is true of acquired sexual (AqS) worms; the length remains the same both in the fresh hatchling and 2-years old AqS.

Karyotypically, *S. mediterranea* is a little different from *D. ryukyuensis* in that InS and AS can be distinguished by the presence of a chromosomal translocation only in AS (Newmark and Alavarado, 2002). In *S. mediterranea* too, the length is 28 and 21.2 kb for AS and InS, respectively. In InS, it decreases from 21.2 kb to 1.67 kb in a 3-years old worm. Remarkably, it is also decreased from 28 to 26 and 22.6 kb in AS worms that have not undergone clonal multiplication for the past 3 months and > 3 months (195 days). Clearly, AS can maintain telomere length somatically through fission and regeneration, but InS can achieve it only through sexual reproduction (Tan et al., 2012).

*Senescence*: In planarians, three processes namely (i) routine renewal and turnover of somatic cells, (ii) regeneration and (iii) clonal reproduction induce rejuvenation. The life span of these planarians can also be extended by regeneration and starvation (p 74–75). *Schmidtea polychroa* grows to sexual maturity at the age of 2 months. After the age of 5 months, it undergoes alternating phases of growth and degrowth. In each of the cycle, the body surface area varies between 35 and 100 mm². Maintaining a high morphological plasticity during adult life, *S. polychroa* shows signs of negligible senescence. Metabolic ageing is also not detectable, which may explain the very high rate of cell turnover and renewal. As a consequence, mortality rate is low and gradually increases resulting in their survival trend, which is contrastingly different as L-shaped one for the ageing sexual animals (Fig. 2.13E).

# 3

## Monogenea

## Introduction

Monogeneans are mostly ectoparasites with a digestive system but without an epidermis. They are covered externally by a specialized integument (Fig. 1.12B). In them, an oral sucker is absent or weak, when present; they are provided with anterior adhesive structure and posterior opisthaptor with an adhesive disk bearing hooks (Hyman, 1951). Their indirect life cycle involves non-feeding infective ciliated (e.g. *Entobdella soleae*, Kearn, 1974) or unciliated (e.g. *Pseudodactylogyrus bini*, Chan and Wu, 1984) oncomiracidium but with no intermediate host. Incidentally, many authors describe the monogenean life cycle as direct, meaning that they do not involve an intermediate host (e.g. Reed et al., 2012). The polystomatid monogeneans (~ 150 species, Badets and Verneau, 2009) are endoparasites, mostly on the urinary bladder of lungfish (e.g. *Concinnocotyla australensis*), frogs, toads (e.g. *Pseudodiplorchis americanus*, Tinsley and Earle, 1983), pharyngeal cavities of freshwater turtles like *Stenotherus odoratus* (Oglesby, 1961), but rarely on cephalopods and arthropods (Harris et al., 2004).

Monogeneans can become serious pests of fish and cause heavy mortality in aquaculture farms (Ogawa, 2015). In general, hyperplasia and hemorrhage of infected gills are frequently reported (e.g. *Dicentrarchus labrax* infected with *Diplectanum aequans*, see Gonzalez-Lanza, 1991). *Pseudodiplorchis americanus* feeds exclusively on the blood of its host *Scaphiopus couchii* ingesting 5 µl blood/d (see Tocque and Tinsely, 1994). Not surprisingly, *Seriola lalandi* infected with *Zeuxapta seriolae* remain lethargic, emaciated and anemic, and may die from asphyxiation (Mooney et al., 2006). On being heavily infected with *Neobenedenia girellae*, *Seriola dumerili* stops feeding, darkens its body, swims erratically and suffers from dermal ulceration and branchial hyperplasia (Hirazawa et al., 2010). One hundred percent mortality of amberjack *Seriola quinqueradiata* infected with *N. girellae* is reported from Okinawa, Japan (Ogawa, 2015). Regarding the damages inflicted and consequent economic impact by the gyrodactylids on cod, flatfish, eel, wolfish and cyprinids, Poppe (1999), Ernst et al. (2000), Solomatova and Luzin (1988), Buchmann and Lindenstrom (2001) may be consulted.

## 3.1 Taxonomy and Distribution

### 3.1.1 Taxonomy

Monogeneans are classified on the basis of adult features, especially the haptor; they are divided into two suborders: (1) Monopisthocotylea consisting of ectoparasites with a divided or undivided adhesive organ, the haptor (Fig. 1.4E, F) and (2) Polyopisthocotylea comprising ecto- and endo-parasites with a larger posterior disk-like opisthaptor bearing many suckers adorned with hooks and clamps (Hyman, 1951). For a description of the haptor in larvae and adults, Kearn (1999) and Reed et al. (2012) may be consulted. Reed et al. (2012) indicated the inclusion of 13 monogenean families, of which parasite species belonging to four families Gyrodactylidae, Dactylogyridae, Ancyrocephalidae and Capsalidae (Fig. 3.1A–D) are more frequently reported from aquaculture farms. However, in their respective analysis, Poulin (1996b) considered 39 families and Jianying et al. (2003) 33 families. Within these two analyses, 13 families considered by Poulin are not found in Jianying et al. and 8 families listed by Jianying et al. are not included by Poulin. Hence, monogeneans may include ~ 53 families (Littlewood and Bray, 2001). Further, these authors presented 228 and 337 genera in their lists. Considering 831 monogenean species, Poulin (1996b) found that species number decreases from 38/genus in smaller parasites of < 1 mm size to

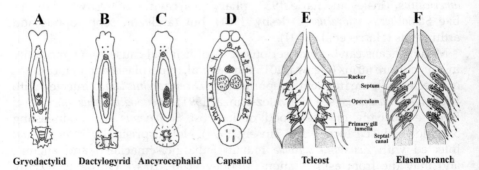

### FIGURE 3.1

A–D: Common families of monogeneans (free hand drawing from Reed et al., 2012). Diagrammatic sections in a horizontal plane through the buccal and gill chambers of a (E) teleost and (F) an elasmobranch show the arrangement of gills. I, II, III and IV represent gill arches. Arrows indicate the direction of water current entering and leaving the gill chamber (from Kearn, 2014).

Monogenea 95

2/genus in large parasites of 50 mm size. Until 1983, 4,000–5,000 monogenean species have been described (Schmidt and Robertes, 1985). Since then, reports on description and erection of more and more species appear (Fig. 1.3A), *albeit* not many more from dactylogyrid and gyrodactylid families (Poulin, 2002). Bakke et al. (2007) indicated that of 20,000 species of the mega-biodiverse genus *Gyrodactylus*, only 402 (2%) species have been described. The DNA barcoding technique is inadequate to distinctly identify new species. For example, Hansen et al. (2007) found that the use of mitochondrial cytochrome oxidase I (COI) is inadequate for biological characterization and to distinguish clonally reproducing *G. salaris* from *G. thymalli*. Lim (1995) assumed that there were 9,000 species waiting to be described from the Chinese marine fish species alone. Not surprisingly, monogeneans are expected to include 25,000 species (e.g. Whittington, 1998).

### 3.1.2 Distribution

Regarding the distribution of monogeneans, the following key findings may briefly be listed: (1) Considering Poulin's (1996b) analyses as an example, 6, 35 and 59% of 613 monogenean species are from freshwater, marine and freshwater cum marine fishes, respectively. Of 56 monogenean genera reported from the Australian waters, 46 are exclusively marine, one estuarine and another one both freshwater as well as marine habitats (Young, 1970, see also Rohde, 1993). Hence, monogeneans are mostly from marine fish. (2) An analysis of the publication by Jianying et al. (2003) reveals that of 337 marine fluke species from China, only 17, i.e. 5.0% elasmobranchs serve as host for fluke species (Table 3.1). Clearly, the monogeneans prefer teleosts over elasmobranchs. An important reason for the inability of the fluke to parasitize elasmobranch is elaborated in pp 245–246. A second reason is the presence of an operculum in teleosts protecting the parasite from predators and from being carried away by relatively stronger outgoing branchial current (Fig. 3.1E, F). Unlike the slippery wet scales of teleosts, those in elasmobranchs, for example, the dogfish *Styliorhinus canalicula* provide a hard substratum unsuitable for the deployment of hooks by *Leptocotyle minor* (Kearn, 2014). (3) Prevalence and intensity of monogenean infection are higher in tropical marine fish than in their temperate counterpart (Rohde, 1982).

In an admirable contribution, Jianying et al. (2003) listed the incidence and intensity of monogenean species from the Chinese marine fishes. Their list provides an opportunity to analyze and draw the following new inferences: (1) From the 18,000 km of mainland coast line and 14,000 km of island coast line, a total 337 monogenean species belonging to 147 genera and 33 families are reported (Table 3.1). (2) The incidence covers a very wide range of 343 teleost species from 203 genera, 60 families and 15 orders infected by 320 fluke species belonging to 139 genera and 57 families. In elasmobranchs, it is limited to 17 fluke species belonging to 8 genera and 7 families parasitic on 17 host species (i.e. 5%) in 10 genera and 7 families. Briefly, there are 337 fluke

# TABLE 3.1

Monogenean parasites infecting Chinese marine fishes (reorganized from Jianying et al., 2003)

| | Host: Order | Parasite: Host Family | Parasite (no.) | | Host fish (no.) | |
|---|---|---|---|---|---|---|
| | | | Genus | Species | Genus | Species |
| | **1. Monocotylidae** | | 6 | 14 | 7 | 14 |
| 1 | Myliobatiformes | Dasytidae 1, Gymnuridae 2 | | | | |
| 2 | Rhinopristiformes | Rhinidae 3, Rhinobatidae 4 Platyrhinidae 5 | | | | |
| 3 | Torpediniformes | Rajidae 6 | | | | |
| 4 | Rajiformes | | | | | |
| | **1. Capsalidae** | | 10 | 13 | 12 | 13 |
| 5 | Perciformes | Holocentridae 7, Terapontidae 8, Lutjanidae 9, Ephippidae 10, Scombridae 11, Serranidae 12, Caproidae 13, Lethrinidae 14, Haemulidae 15 | | | | |
| | **3. Dionchidae** | | 1 | 6 | 4 | 4 |
| | Perciformes | Rachycentridae | | | | |
| | **4. Gyrodactylidae** | | 1 | 2 | 2 | 2 |
| | Perciformes | Lateolabricidae 17, Mugilidae 18 | | | | |
| | **5. Tetraonchoididae** | | 5 | 7 | 7 | 9 |
| | Perciformes | Uranoscopidae 19, Labroidae 20, Ammodytidae 21 | | | | |
| 6 | Auliformes | Synodontidae 22 | | | | |
| | **6. Neocalceostomatidae** | | 1 | 1 | 1 | 1 |
| 7 | Siluriformes | Ariidae 23 | | | | |

| | | | | | | |
|---|---|---|---|---|---|---|
| | **7. Calceostomatidae** | | 3 | 3 | 3 | 3 |
| | Perciformes | Serranidae, Haemulidae, Drepaneidae 24 | | | | |
| | **8. Amphibdelliatidae** | | 1 | 2 | 1 | 3 |
| | Torpediniformes | Nakidae 25 | | | | |
| | **9. Dactylogridae** | | 1 | 3 | 1 | 1 |
| | Perciformes | Lateolabricidae | | | | |
| | **10. Ancyrocephalidae** | | 26 | 104 | 44 | 109 |
| | Perciformes | Drepaneidae, Gerreidae 26, Leiognathidae 27, Lutjanidae, Chaetodontidae 28, Lethrinidae, Serranidae, Pempheridae 30, Haplogenyidae 31, Sciaenidae 32, Haemulidae, Apogonidae 33, Scatophagidae 34, Pomacanthidae 35, Priacanthidae 36, Sparidae 37, Gobiidae 38, Mugilidae | | | | |
| 8 | Anguilliformes | Congridae 39 | | | | |
| 9 | Beloniformes | Belonoideae 40, Hemiramphidae 41 | | | | |
| 10 | Pleuronectiformes | Paralichthyidae 42 | | | | |
| 11 | Scorpaeniformes | Platycephalidae 43 | | | | |
| 12 | Siluriniformes | Ariidae | | | | |
| 13 | Syngnatheformes | Mullidae 44, Dactylopteridae 45 | | | | |
| 14 | Tetrodontiformes | Triacanthidae 46 | | | | |
| | **11. Protogyrodactylidae** | | 1 | 13 | 2 | 7 |
| | Perciformes | Gerreidae, Terapontidae | | | | |

*Table 3.1 contd. ...*

| | Host: Order | Parasite: Host Family | Parasite (no.) | | Host fish (no.) | |
|---|---|---|---|---|---|---|
| | | | Genus | Species | Genus | Species |
| | **12. Diplectanidae** | | 8 | 31 | 23 | 35 |
| | Perciformes | Tetrodontidae 47, Sillaginidae 48, Sparidae, Sciaenidae, Terapontidae, Latidae 49, Polynemidae50, Lutjanidae, Serranidae | | | | |
| | **13. Hexabothridae** | | 2 | 3 | 3 | 3 |
| | Myliobatiformes | Dasytidae | | | | |
| | Rajiformes | Rajidae | | | | |
| 15 | Carcharhiniformes | Triakidae 51 | | | | |
| | **14. Plectanocotylidae** | | 3 | 3 | 3 | 4 |
| | Perciformes | Trichiuridae 52 | | | | |
| 16 | Scorpaeniformes | Peristeiidae 53 | | | | |
| | **15. Mazocraeidae** | | 10 | 22 | 13 | 18 |
| 17 | Clupeiformes | Clupeidae 54, Engraulidae 55, Dussumieridae 56, Scombridae, Carangidae, Rachycentridae, Sphyraenidae 57, Lactariidae 58 | | | | |
| | **16. Pseudodiclidophoridae** | | 1 | 1 | 2 | 2 |
| | Perciformes | Rachycentridae, Carangidae | | | | |
| | **17. Protomicrocotylidae** | | 5 | 8 | 4 | 5 |
| | Perciformes | Carangidae, Sphyraenidae, Lactariidae | | | | |
| | **18. Allodiscocotylidae** | | 3 | 4 | 5 | 7 |
| | Perciformes | Carangidae | | | | |

| | | | 4 | 4 | 1 | 3 |
|---|---|---|---|---|---|---|
| **19. Chauhaneidae** | | | | | | |
| Perciformes | Sphyraenidae | | | | | |
| **20. Bychowskicotylidae** | | | 3 | 3 | 2 | 2 |
| Perciformes | Haemulidae | | | | | |
| **21. Gastrocotylidae** | | | 6 | 14 | 12 | 20 |
| Perciformes | Carangidae, Scombridae | | | | | |
| **22. Neothoracocotylidae** | | | 2 | 2 | 1 | 2 |
| Perciformes | Scombridae | | | | | |
| **23. Gotocotylidae** | | | 2 | 4 | 1 | 4 |
| Perciformes | Scombridae | | | | | |
| **24. Hexostomatidae** | | | 2 | 2 | 2 | 2 |
| Perciformes | Scombridae | | | | | |
| **25. Axinidae** | | | 5 | 6 | 5 | 8 |
| Beloniformes | Hemiramphidae, Exocoetidae 59 | | | | | |
| **26. Heteraxinidae** | | | 8 | 9 | 9 | 11 |
| Perciformes | Stromateidae 60, Carangidae, Haplogenyidae, Lethrinidae | | | | | |
| **27. Microcotylidae** | | | 13 | 26 | 23 | 38 |
| Perciformes | Malacanthidae 61, Terapontidae,  Labridae, Neripteridae 62, Carangidae,  Haemulidae, Sciaenidae, Sparidae, Mugilidae, Gerreidae, Lutjanidae, Polynemidae, Leiognathida | | | | | |
| Syngnatheformes | Fistulariidae 63 | | | | | |

*Table 3.1 contd. ...*

*... Table 3.1 contd.*

| | Host: Order | Parasite: Host Family | Parasite (no.) | | Host fish (no.) | |
|---|---|---|---|---|---|---|
| | | | Genus | Species | Genus | Species |
| | **28. Diclidophoridae** | | 8 | 16 | 12 | 20 |
| | Clupeiformes | Engraulidae, Pristigasteridae 64 | | | | |
| | Perciformes | Sciaenidae | | | | |
| 18 | Tetradontiformes | Tetradontidae | | | | |
| | **29. Heteromicrocotylidae** | | 2 | 7 | 2 | 6 |
| | Perciformes | Carangidae, Gerreidae | | | | |
| 19 | Alubuliformes | Alubulidae 65 | | | | |
| | Clupeiformes | Chirocentridae 66 | | | | |
| 20 | Pleuronectiformes | Cynoglossidae 67 | | | | |
| | **30. Paramonaxinidae** | | 1 | 1 | 1 | 1 |
| | Perciformes | Gerreidae | | | | |
| | **31. Pterinotrematidae** | | 1 | 1 | 1 | 1 |
| | Alubuliformes | Alubulidae | | | | |
| | **32. Anchorophoridae** | | 1 | 1 | 1 | 1 |
| | Pleuronectiformes | Cynoglossidae | | | | |
| | **33. Megamicrocotylidae** | | 1 | 1 | 1 | 1 |
| | Clupeiformes | Chirocentridae | | | | |
| | | Total | 147 | 337 | 211 | 360 |

*Monogenea* 101

species parasitic on 360 fish species in 211 genera and 67 families. This may indicate that monogeneans are host specific, i.e. for every parasite species, there are only 1.07 host species (see Table 3.13). (3) Among parasite families, the incidence of parasite species decreases in the following descending order: Ancyrocephalidae (104 species) > Diplectanidae (31 species) > Microcotylidae (26 species) > Mazocraeidae (22 species). (4) Among host species, carangids are most susceptible to parasite species belonging to 8 families; these values are 6 and 4 for scombrids and gerrids, respectively.

The distribution of monogeneans is influenced by a number of internal and external factors. The internals include (i) host susceptibility, (ii) host size (iii) host's condition factor/immunity (iv) sex and (v) microhabitat. The external factors comprise (i) lotic or lentic habitat, (ii) latitude and season and (iii) co-existence or competition between intra- and inter-specific individuals/species.

*Susceptibility*: Llewellyn (1956) estimated the distribution of a dozen fish species brought to Plymouth Laboratory, UK. Out of 507 and 509 specimens of *Gadus merlangus* and *G. luscus*, (prevalence or) susceptibility to monogenean infection was 8.7 and 21.0%, respectively. In Lake Ayame 2, Cote d'Ivoire, Africa, Adou et al. (2017) reported that in *Tilapia guineensis*, it susceptibility decreased from 15.4% for *Cichlidogyrus kouassii* to 96.2% for *C. vexus*. The intensity for these parasites also decreased from 33.3/host to 15.4/host. In aquacultural farms of Uganda, it was 52% for *Cichlidogyrus* sp on *Oreochromis mossambicus*, in comparison to 0.8% on *Clarias gariepinus* (Akoll et al., 2011). Evidently, *G. merlangus* and *C. gariepinus* are less susceptible, and *T. guineensis* is less susceptible to *C. kouassii* than to *C. vexus*.

*Body size*: Increase in body size/age may also increase the gill surface area and provide a larger area for infection (see Silan et al., 1987). On the other hand, with increasing size/age, the infected host may also develop immunity and resistance. Incidentally, both the serum and dermal mucus of the infected sole *Pleuronectes vetulus* contain resistance factors against *Gyrodactylus stellatus* (Moore et al., 1994, however, see also Faliex et al., 2008). Hence, the reports on the relation between prevalence and intensity of infection vs host body size/age may not yield a common generalized trend (Fig. 3.2). Among the susceptible host fish, body size has been measured in length (e.g. *Cichlidogyrus vexus* on *Tilapia guineensis*, Adou et al., 2017), weight (e.g. *Microcotyloides* sp on *Terapon puta*, Fig. 3.2D1) or age (e.g. *Gastrocotyle trachuri* on *Trachurus trachurus*, Fig. 3.2B1). With increasing body size, not only prevalence but also intensity of branchial flukes increase positively almost linearly on three cichlid species, for which information is available. Likewise, both prevalence and intensity of infection by *Diplectanum aequans* and *D. laubieri* on sea bass *Dicentrarchus labrax* also increase with advancing age up to 4+ years (Gonzalez-Lanza et al., 1991). Positive relationships between host size and prevalence as well as intensity of fluke are also reported for *Halitrema* sp on *Dascyllus aruanus* and *Benedenia* sp on *Cephalopholis argus* (Lo et al., 1998). Interestingly,

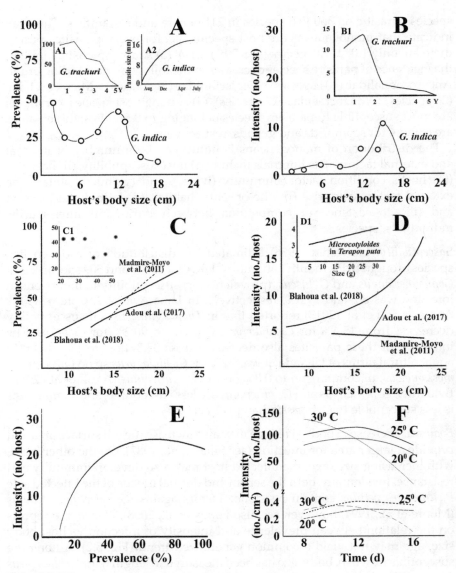

**FIGURE 3.2**

Effect of the host's body size on prevalence of (A) *Gastrocotyle indica* in *Caranx kalla* and (A1) *Gastrocotyle trachuri* on *Trachurus trachurus* (A1) and growth of *G. indica* as function of time (A2). The same on the intensity of infection (B) (modified and redrawn from Llewellyn, 1956, Radha, 1971). (C) Prevalence of *Cichlidogyrus* spp in *Tilapia zillii*, *T. guineensis*, *Oreochromis mossambicus* and (C1) *Cyprinus carpio* (drawn from data reported by Blahoua et al., 2018, Adou et al., 2017, Madanire-Moyo et al., 2011, Tekin-Ozan et al., 2008, respectively). (D) The same as function of intensity. (D1) Intensity of infection as a function of body weight (from Khidr et al., 2012). (E) Relationship between prevalence and intensity of infection by monogenean parasite species in finfish (drawn after ploting values reported by many authors). (F) Increase in intensity of *Neobenedenia girellae* with time and temperature. Note the increases per fish and per unit surface area of *Seriola dumerili* (compiled, modified and redrawn from Hirazawa et al., 2010).

*Monogenea* 103

intensity of the fluke *Neobenedenia girellae* increases with increasing time up to 12 days and temperature not only for an individual fish but also for a unit area of body surface in *Seriola dumerili* (Fig. 3.2F). Hence, surface area available for monogenean infection on the body increases with increasing size and age. It may also hold true for branchial flukes. Conversely, both prevalence (Fig. 3.2A1) and intensity (Fig. 3.2B1) decrease with advancing age in *Trachurus trachurus* infected by *Gastrocotyle trachuri*. Many other trends like those of *Gyrodactylus indica* (Fig. 3.2A,B) and *Discocotyle sagittata* (Paling, 1965, not shown in figure) reported for the relationship between body size and prevalence/intensity fall between the trends known for the cichlid at one end of spectrum and *T. trachurus* at the other end of it. The trends reported in Fig. 3.2C, D are based on (i) *Oreochromis mossambicus* infected with four *Cichlidogyrus* species + *Scutogyrus longicornis* in the Niwanedi-Luphephe dam (Madanire-Moyo et al., 2011), (ii) *Tilapia guineensis* infected with 11 *Cichlidogyrus* species in Lake Ayame 2 (Adou et al., 2017) and (iii) *T. zillii* infected with three *Cichlidogyrus* species in Lobo River—all in Africa. Interestingly, Blahoua et al. (2018) recorded a significant relation between the condition factor of *T. zillii* and prevalence of *C. vexus* (host condition factor: 0.41), *C. aegypticus* (0.62) and *C. digitatus* (0.83). Relatively healthier *T. zillii* were more susceptible to *C. digitatus* (cf Fig. 5.8C) than *C. vexus* with low condition factor. Incidentally, *Diclybothrium armatum* is not found in young sturgeon but infects 40–80% of adults. In < 1-year old *Hypophthalmichthys*, prevalence and intensity of *Dactylogyrus skrjabini* are 25% and 3/host; these values are 70% and 20/host in 2-year old ones. *Microcotyle spinicirrus* does not infect the young (< 1-year old) *Aplodinotus grunniens* but infects most of 2-year old ones (see Rohde, 1979).

In Ancyrocephalidae, Gastrocotylidae, Capsalidae and Microcotylidae, the incidence frequency of fluke species progressively decreases with increasing host body size. But the decrease is observed after an initial peak in Diplectanidae and Diclidophoridae. It is likely that anchyrocephalid group may also pass through an initial peak, however smaller sized specimens have been examined. From his analysis of 65 host species, Poulin (2002) also noted an initial peak in the host species frequency-body size relation. Hence, smaller host species are more frequently infected than the larger ones.

From data reported by Radha (1971), it is possible to trace the growth trend for *Gastrocotyle indica*. With commencement of North-east monsoon in the Madras coast (South India) during late August, *G. indica* larva infects *Caranx kalla*. It grows to juvenile and adult stages during October and December, respectively (Fig. 3.2A2). Both *G. indica* and its host *C. kalla* are apparently annuals. This asymptotic trend also holds true for *Pseudothoracocotyla gigantica* that parasitizes *Scomberomorus commersoni* (but as a function of host size, Rohde, 1976) that lives longer than 13 years (Lee and Mann, 2017). Hence, it appears that the life span and body size of the host species may also influence those of monogenean species. However, an analysis of 39 monogenean families has revealed no relationship between

104    *Reproduction and Development in Platyhelminthes*

host size and parasite size (Poulin, 1996b). Nevertheless, of 35 relations "22 positive suggested a tendency for host size to vary in the same direction as monogenean body size" (Poulin, 1996b). The same holds true for latitudes suggesting that higher the latitude the larger the body size of monogenean parasites. Incidentally, Rohde (1985) reported that viviparous monogeneans are more common in higher latitudes.

*Host sex*: A higher prevalence of *Discocotyle sagittata* on male trout *Salmo trutta* was first observed by Paling (1965). Radha (1971) confirmed this observation and went on to show that both prevalence and intensity of infection by *Gastrocotyle indica* are higher in male carangid *Caranx kalla*. A significant host's sex specific prevalence has also been reported for *Polystoma gallieni* infecting *Hyla meridionalis* (Badets et al., 2010). The subsequent reports on sex-dependent differences are summarized in Table 3.2. Buchmann (1997b) injected Testosterone (T) into naïve rainbow trout and allowed them to co-inhabit with trouts that were already infected with *Gyrodactylus derjavini*. The T-treated trouts were also infected and the density of *G. derjavini* was higher within a few days in T-treated trouts than in the sham-treated ones. He suggested a possible immune-suppressive effect of T. However, two facts go against Buchmann's suggestion. Firstly, the prevalence and intensity of infection of many monogeneans are higher, for example, in females of *Astyanax altiparanae* and *Rhamdia quelen* than in males. Secondly, the female fish also secrete T, which is then converted into estradiol by an aromatase enzyme (see Pandian, 2013). Briefly, none of these authors have ever subjected their respective data to statistical analysis to conclusively show that the observed sex-dependent differences are indeed significant.

*Ploidy level*: Posing a question whether polyploid hosts are attractors of monogenean infection, Guegan and Morand (1996) considered 29 cyprinid species; of them 13 were with 2n = 50 chromosomes (e.g. *Barbus issenensis*), 10 with 2n = 100 chromosomes (e.g. *B. nasus*) and 6 with 2n = 150 chromosomes (e.g. *B. sacratus*). His analysis suggested that with elevation of ploidy, susceptibility is increased.

*Prevalence vs intensity*: To know the interaction between prevalence and intensity of monogenean infection, available values on these components were plotted. Intensity increased only up to 30% prevalence, beyond which it leveled off at the intensity of 25/host (Fig. 3.2E).

*Microhabitat*: Like in other vertebrates, the fish skin consists of an outer epidermis and an inner dermis. Unlike mammals, fish epidermis, however, consists of 10 or more layers of live epidermal cells, with continuing cell divisions in the deeper layers. Whereas a cornified protective layer is formed on death of epidermal cells in mammals and it is difficult to digest by parasites, the wound of fishes is healed rapidly with the production of more and more epidermal layers. Thus, fish epidermis is readily accessible and constitutes a nutritious food for skin parasites. Gut contents of recently-fed *Entobdella*

# TABLE 3.2

Effect of host sex on prevalence and intensity of monogenean parasitic infection in selected fishes

| Host species | Parasite species | Prevalence (%) | | Intensity (no./host) | | Reference |
|---|---|---|---|---|---|---|
| | | ♀ | ♂ | ♀ | ♂ | |
| *Salmo trutta* | *Discocotyle sagittata*† | 48 | 68 | 1.3 | 2.6 | Paling (1965) |
| *Caranx kalla* | *G. indica* | 25.6 | 29.7 | 4.1 | 5.5 | Radha (1971) |
| *Polystoma gallieni* | *Hyla meridionalis* | 18.7 | 41.5 | | | Badets et al. (2010) |
| *Astyanax altiparanae* | *Ampbitbecium* sp | 7.9 | 3.9 | 14.3 | 2.2 | |
| | *Notozotbecium* sp | 5.3 | 1.3 | 37.5 | 1.2 | Ferrari-Hoeinghaus et al. (2006) |
| *Rhamdia quelen* | *Scleroductus* sp | 15.8 | 1.0 | 4.2 | 1.0 | |
| | *Urocleidoides mastigatus* | 42.1 | 43.4 | 29.2 | 12.0 | |
| *Scarus* sp | | | | | | |
|   Jeddah, Red sea | | 0 | 65 | 0 | | Bakhraibah (2018) |
|   Rabigh, Saudi Arabia | | 84 | 27 | 6.4 | 19 | |
| *Sparus aurata* | *Spariocotyle chrysophrii* | | | | | |
|   I  Y Immature | | 48 | | 2.6 | | Antonelli et al. (2010) |
|   II  Y Male | | | 54 | | 2.1 | |
|   III Y Female | | 54 | | 1.9 | | |
| *Tilapia zilli* | *Cichlidogyrus digitatus* | 72.6 | 70.4 | 24.4 | 23.4 | Blahoua et al. (2018) |
| *Dicentrarchus labrax* | *Diplectanum aequans* | 70 | 89 | 75 | 135 | Gonzalez Lanza et al. (1991) |
| | *D. laubieri* | 65 | 204 | 35 | 71 | |
| *Acantholochus unisagittatus* | *Centropomus undeclimatis* | 82 | 96 | – | – | see Iyaji et al. (2009) |
| *Salmon* | *Ichthyophonus* | 33 | 26 | – | – | |
| *Oreochromis mossambicus* | *C. vexus* | 97.9 | 90.2 | 24.5 | 13.3 | Adou et al. (2017) |
| | *C. ergensis* | 82.5 | 81.5 | 20.4 | 9.1 | |
| | *C. anthenocolops* | 73.0 | 70.5 | 11.2 | 11 | |
| *Vimba vimba tenella* | *Dactylogyrus cornu* | 100 | 100 | 159 | 113 | Ozer and Ozturk (2005) |
| *Gasterosteus aculeatus* | *Gyrodactylus arcuatus* | 82 | 74 | 38 | 29 | Ozer et al. (2004) |

106  *Reproduction and Development in Platyhelminthes*

*soleae*, for example, are all colorless. As colored red cells are not found in the gut contents, grazing may be limited to superficial pigmented epidermal cells and not deep into unpigmented skin and blood vessels. Hence, some monogenean skin parasites may not attract an immune reaction from the host (Kearn, 1963). Topographical distribution of *Gyrodactylus salaris* on *Salmo salar* suggests that (i) the dorsal fin is the most preferred microhabitat and (ii) preference for right and left pectoral is equal and the same holds true for pectoral fins (Table 3.3).

Most monogeneans are found on the gills of fishes. The gills of fishes are soft and richly vascularized. They constitute a favorable substratum for the attachment of monogeneans. The branchial chamber, which is well aerated by a respiratory current and also sheltered/protected from environmental factors outside the fish, proves to be a very suitable ecological niche for not only monogeneans but also other parasites. Still, the relatively simple gill apparatus can be subdivided into a number of microhabitats like dorsal and ventral filament, proximal and distal part of the filament, first to fourth gill arches and so on. Regarding selection of these microhabitats, even congeneric monogeneans may select different microhabitats; some species select primarily the dorsal part of the gill filament, but others attach to the proximal portion. The attachment can be on anterior or posterior gill arches and outer or inner hemibranchs. The shape of the head and consequent space available for the different arches within the branchial chamber (Fig. 3.3) and respiratory water current may influence the selection and ultimate settlement of oncomiracidia in different microhabitats on the gills. Recent studies have indicated that a microhabitat selection and settlement of gyrodactylids on salmonids are quite dynamic processes and are partly regulated by immune response (see Buchmann and Lindenstrom, 2002).

Restriction to a microhabitat in host species is universal among parasite species. It increases the chances for selection and mating. However, 'the crowding effect' may lead to an intense competition for space and food. The

## TABLE 3.3

Topographical distribution of *Gyrodactylus salaris* on *Salmo salar* (modified from Mo, 1992)

| Site | (%) | Site | (%) |
|---|---|---|---|
| Dorsal fin | 34.4 | Head | 3.5 |
| Left pectoral fin | 13.5 | Gills | 2.6 |
| Right pectoral fin | 13.5 | Body | 7.8 |
| Left pelvic fin | 3.5 | | |
| Right pelvic fin | 3.6 | | |
| Tail fin | 7.5 | | |
| Anal fin | 6.8 | | |
| Adipose fin | 3.3 | | |

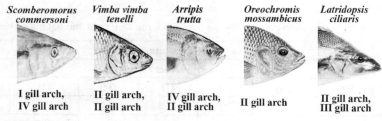

**FIGURE 3.3**

The heads of fishes showing the differences in snout shape and its possible effect on branchial chamber covered by the operculum. The upper lines show the maximum prevalence on the indicated gill arch and the lower lines the maximum intensity on the indicated gill arch.

interaction among competing parasite species may lead to site segregation (Rohde, 1979). In monogeneans, the segregation has restricted their microhabitats to the skin and/or gills. With their adhesive/clinging haptor, they are attached to the host's gill arches with anterior and nearer to the distal end of the primary lamellae. Thus, they lie upstream relative to the gill ventilating current of the host with the mouth of the parasite downstream (Llewellyn, 1956). The attaching organ may be suckers (e.g. *Cyclocotyla chrysophryi* on the gills of *Pagellus centrodontus*) or clamps (e.g. *Anthocotyle merluccii* on the gills of *Merluccius merluccius*). The development of additional suckers may tilt symmetry of the monogenean body from bilateral to asymmetry. For example, the number of adhesive suckers is increased to > 6–8 but in some cases to 50–70 in *Axine belone* on the gills of *Belone beloni*; the newly developing suckers are borne in a single oblique row on the right (87%) or left (13%) side of the parasite and thereby manifest asymmetry in it (Llewellyn, 1956). Hence, the symmetry may play a role on the microhabitat distribution of the monogenean parasite (Rohde 1993).

Earlier studies have reported that monogeneans like *Diplozoon paradoxum* and *Pseudodactylogyrus anguillae* prefer the attachment to either left or right set of gills (see Ozer and Ozturk, 2005). However, subsequent investigations on *Cichlidogyrus* spp and *Dactylogyrus cornu* have shown no such preference, albeit *Polystoma australis* prefers left gills of *Kassina senengalensis* tadpole (Kok and DuPreez, 1987). Llewellyn (1956) noted the preference of *Diclidophora merlangi* to the first fill arch of *Gadus merlangus* and *Anthocotyle merluccii* for the second gill arch of *Merluccius merluccius*. Subsequent studies have shown the preference for the first arch by five monogenean species on *Scomberomorus commersoni*, the second arch by *Mediavagina latridis*, *Cichlidogyrus* spp, *Dactylogyrus cornu*, and the fourth arch by *Kahawaia truttae* (Table 3.4). For intensity of infection too, there are differences in preference; however, the arch preference for intensity is not the same as for prevalence. For example, the highest preference is the first arch for prevalence but the fourth arch for intensity for 5 monogenean species on the gills of *S. commersoni*. Notably,

108  *Reproduction and Development in Platyhelminthes*

## TABLE 3.4

Distribution of monogenean parasite species in gill arch number, right and left gill arches

| Gill arch | Reference number † | | | | | | |
|---|---|---|---|---|---|---|---|
| | Prevalence (%) | | | | | | |
| | **(1b)** | **(1a)** | **(5)** | **(7)** | **(1c)** | **(2)** | **(4)** |
| I | 0 | 13 | 15 | **31.8** | **51.2** | 67.3 | 97.6 |
| II | 0 | **33** | 39 | 24.6 | 26.2 | **76.7** | **98.4** |
| III | 5 | 27 | 35 | 26.2 | 29.6 | 71.7 | 94.3 |
| IV | **21** | 15 | 11 | 17.3 | 19.8 | 50.0 | 95.1 |
| | Intensity (no./host) | | | | | | |
| | **(1b)** | **(2)** | **(1a)** | **(4)** | **(3)** | **(1c)** | |
| I | 5 | 4.4 | 8 | 20.7 | 19.4 | 22.8 | |
| II | **14** | **6.5** | 9 | **24.3** | **37.6** | 21.4 | |
| III | 4 | 5.5 | **35** | 18.3 | 31.2 | 29.6 | |
| IV | 0 | 1.8 | 4 | 13.1 | 11.8 | **31.4** | |
| | Left gill arches | | | Right gill arches | | | |
| (4) | 96.3% | 19.4/host | | 96.3% | | 18.8/host | |
| (2) | 77.7% | 17.3/host | | 77.7% | | 17.0/host | |
| (7) | 51% | | | 49% | | | |
| (6) | 13/17 samples | | | 9/17 samples | | | |

† (1a) *Mediavagina latridis* on *Latridopsis ciliaris*, (1b) *Kahawaia truttae* on *Arripis trutta*, (1c) 5 parasite species on *Scomberomorus commersoni* (Rohde, 1979). (2) *Cichlidogyrus* 3 species (Blahoua et al., 2018). (3) 10 species on 11 host species (Llewellyn, 1956). (4) *Dactylogyrus cornu* on *Vimba vimba tenelli* (Ozer and Ozturk, 2005). (5) *Cichlidogyrus* spp on *Oreochromis mossambicus* (Madanire-Moyo et al., 2011). (6) *Polystoma australis* on *Kassina senengalensis* (Kok and Du Preez, 1987). (7) 7 parasites on *S. commersoni* (Rohde, 1976).

none of the authors have traced any reason for the reported preference and intensity. Investigation on this aspect in pleuronectids with asymmetric twist of the head may provide new keys. Incidentally, the differences in the head shape, especially the narrowing snout depressing the branchial chamber may be a reason for the highest prevalence and intensity on different gill arches reported (Fig. 3.3). For example, the wider snout and the consequent broad branchial chamber of *S. commersoni* facilitate the maximum prevalence on the proximal first gill arch itself. But the narrow snout and consequent depression of the branchial chamber of *Latridopsis ciliaris* lets the maximum prevalence on the second gill arch.

*Habitat*: Infective period of oncomiracidium may depend on its ciliation and consequent motility as well as static or running waters to facilitate contact frequency between oncomiracidium and its appropriate host. The publications by Adou et al. (2017) and Blahoua et al. (2018) on the prevalence and intensity of infection by *Cichlidogyrus digitatus* and *C. vexus* in Lake Ayame and Lobo River in Cote d'Ivoire provide a rare opportunity to compare the effect of static and running waters on the infection ability of monogeneans. The higher prevalence and intensity of infection in running water of Lobo River than in static waters of Lake Ayame suggests that *C. digitatus* is more adapted to running water (Table 3.5). Conversely, *C. vexus* is more adapted to static waters.

*Latitudes and seasons*: In temperate zones, seasonal changes are usually related to temperature and photoperiod but in the tropics to rainfall . Intensity and prevalence of *Microcotyloides* sp on *Terapon puta* peak during spring and summer, respectively in the Mediterranean coast of Egypt (Table 3.6). Obviously, *Microcotyloides* eggs are hatched and become abundant during the warmer season. This is also true of *Gastrocotyle salaris* on *Salmo salar* (Mo, 1992). In *Dactylogyrus* spp, the response to temperature varies widely; for example, warmer temperature of 20–24°C is required for egg deposition and larval development in *D. vastator* and *D. skrjabini*. But *D. solidus* lays eggs even at 1°C (see Rohde, 1979). In Cote d'Ivoire, Africa, four seasons are recognized namely (i) Long Rainy Season (LRS) from March to June, (ii) Short Dry Season (SDS) from July to August, (iii) Short Rainy Season (SRS) from September to October and (iv) Long Dry Season (LDS) from November to February. For prevalence and intensity, SRS is the favorable season for *Cichlidogyrus* spp on *Tilapia zillii* and *T. guineensis* in Lobo River and Ayame Lake, respectively (Table 3.6). Eggs are hatched during September–October immediately following SRS. In the Madras coast, South India, *Gastrocotyle indica* eggs are hatched during August, immediately following the first spell of Northeast monsoon. Hence, rain following long dry season seems to trigger egg hatching in *Cichlidogyrus* spp and *G. indica*.

*Coexistence*: Monogeneans readily coexist not only with conspecific individuals (e.g. *Hatschekia* sp on *Cephalopholis argus*, Lo et al., 1998) but also with non-conspecifics. The maximum number of a fluke species, that can be accommodated, ranges from 33/host for *Cichlidogyrus* sp (22.5 cm) on the

**TABLE 3.5**

Effect of lotic and lentic habitats on infection ability of *Cichlidogyrus* in Cote d'Ivoire

| Parasite, Host, Reference | Lobo River | | Lake Ayame | |
|---|---|---|---|---|
| | Prevalence (%) | Intensity (no./host) | Prevalence (%) | Intensity (no./host) |
| *C. digitatus* on *Tilapia zillii*, Blahoua et al. (2018) | 71.0 | 24.4 | 54.3 | 2.7 |
| *C. vexus* on *T. guineensis*, Adou et al. (2017) | 31.0 | 7.1 | 90.2 | 13.3 |

## TABLE 3.6

Effect of seasons on prevalence and intensity of monogenean parasitic infection in selected fishes. SDS = Short Dry Season, SRS = Short Rainy Season, LDS = Long Dry Season, LRS = Long Rainy Season

| Prevalence (%) | | | | Intensity (no. of parasites/ host) | | | |
|---|---|---|---|---|---|---|---|
| SDS | SRS | LDS | LRS | SDS | SRS | LDS | LRS |
| *Tilapia zillii* infected by 3 *Cichlidogyrus* spp in Lobo River, Cote d'Ivoire (Blahoua et al., 2018) | | | | | | | |
| 7.2 | 61.9 | 6.0 | 14.5 | 4.2 | 7.4 | 3.2 | 8.6 |
| *T. guineensis* infected by 13 *Cichlidogyrus* spp in Ayame Lake, Cote d'Ivoire (Adou et al., 2017) | | | | | | | |
| 34.5 | 51.0 | 42.0 | 45.0 | 5.5 | 10.3 | 3.78 | 8.26 |
| *Terapon puta* infected by *Microcotyloides* sp in Mediterranean coast of Egypt (Khidr et al., 2012) | | | | | | | |
| Spring | Summer | Autumn | Winter | Spring | Summer | Autumn | Winter |
| 87 | 92 | 69 | 89 | 5.0 | 3.44 | 3.67 | 2.97 |
| *Caranx kalla* infected by *Gastrodactyle indica* from Madras coast, India (Radha, 1971) | | | | | | | |
| North east monsoon, Aug–Nov | | Post monsoon, Dec–Mar | | | Summer, Apr–Jul | | |
| 31% | 2.0 no./host | 37.9% | | 5.4 no./host | 5.4% | | 6.4 no./host |

gills of *Tilapia zillii* (Blahoua et al., 2018) to 159/host on *Vimba vimba tenelli* by *Dactylogyrus cornu* (Ozer and Ozturk, 2005). The estimate of intensity on a single host species suggests the coexistence of as many as 45 *Cemocotylella* sp on a single gill of *Caranx melampygus* and 158 *Allodiscocotyle* sp on a single gill of *Chorinemus tol* (Fig. 3.4). Rohde (1976, 1979) recorded the coexistence of 5 monogenean species (3 of them on smaller fish and 2 of them on large fish) on the gills of *S. commersoni* (see Fig. 3.9A). Firstly, segregation and consequent different spatial occupancy allow the coexistence of more than one individual of a monogenean species and one or more monogenean species on the gills of a single host (Fig. 3.5C, D). For example, *Pseudaxine trachuri* occupies the distal quarter of the primary lamellae of *Trachurus trachurus*, while *Gastrocotyle trachuri* the proximal three-quarters of the lamellae (Llewellyn, 1962). Secondly, with increasing body size, the host may provide a relatively larger gill surface area, on which more and more monogenean species may settle. For example, with increasing body length from 10 to 60 cm of *Labeo coubie*, the number of monogenean species also increased up to 15 species (Fig. 3.5A). This is also true of *Pimelodus maculatus* infected by *Demidospermus* spp. Thirdly, *Oreochromis niloticus* and *Clarias gariepinus* are infected by both *Cichlidogyrus* sp and *Monobothrioides congolensis*. But the former is more susceptible to *Cichlidogyrus* sp and the latter to *M. congolensis* (Akoll et al., 2011). Hence, differences in the susceptibility level may allow coexistence. Fourthly, temporal separation may eliminate in the specific competition. The infection by ancyrocephalid *Urocleidus ferox* peaks on the Californian bluegill in August, January and April, but *Actinocleidus fergusoni*

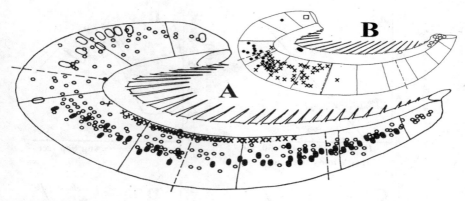

## FIGURE 3.4

The Distribution of monogenean parasites on a single gill of two carangid species. The symbols indicate the exact microhabitat of the monogenean species on the first gill; interrupted lines indicate longitudinal quarters of gills. (A) Outer: *Chorinemus tol*: ○ - *Allodiscocotyle* sp, ● - *Heterapta* sp, ✗ - *Vallisia* sp, ⊙ - Copepod sp; (B) Inner: *Caranx melampygus*: ✗ - *Cemocotylella* sp, ☐ - Monogenea sp, ● - *Protomicrocotyle* sp, ○ - Copepod *Caligus* sp, ● - Isopoda sp (modified and free hand drawing from Rohde, 1979).

in July, January and May (see Rohde, 1979). From field and experimental observations, Bagge and Velatonen (1999) reported that the infection level in *Rutilus rutilus* by 4 species of *Dactylogyrus* decreases in the following order: *D. nanus* (75%, 1.75/host) > *D. crucifer* (35%, 0.45/host) > *D. suecicus* (25%, 0.4/host) > *D. homoion* (10%, 0.1/host). The different durations, during which the dactylogyrids appear, peak and decay, minimize the interspecific competition (Fig. 3.5B).

The ectoparasitic taxa that coexist with monogeneans are ciliates, trematodes, hirudineans and arthropods including branchiurans, copepods and isopods (Table 3.7). While prevalence (Fig. 3.5C) and intensity (Fig. 3.5D) of *Caranx kalla* on *Gastrocotyle indica* is limited up to 16 cm body size, those of copepods is extended up to 24 cm size of *C. kalla*. The relatively larger isopods may occupy a larger area of the branchial chamber and thereby reduce the space available for the occupancy of monogeneans and copepods. In fact, the small sized copepods occupy the peripheral border of the inner side of the first arch, leaving a relatively wider area for the monogeneans on the gill surface of *Chorinemus tol* (Fig. 3.5A). Hence, the spatial restriction is the principal factor, which facilitates the coexistence of ectoparasitic taxa including monogeneans on the surface of resourceful gills of fishes.

Conversely, a parasite species can also be antagonistic to another. Antagonism between different parasite species is an interesting area of research. On *Gadus merlangus*, antagonism between the monogenean *Octobothrium merlangi* and copepod *Clavella devastatrix* has been reported. Similarly, *Dactylogyrus extensus* and *D. vastator* are antagonistic to the ciliate

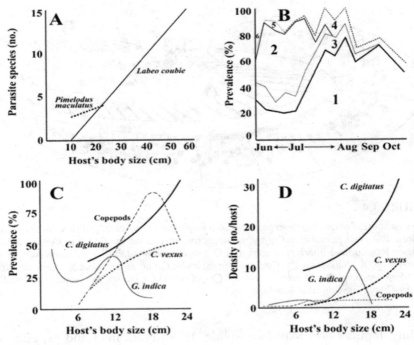

**FIGURE 3.5**

(A) Relationships between monogenean species number and host's body size (compiled from Guegan and Hugueny, 1994, Gutierrez and Martorelli, 1999). (B) Co-existence of 1. *Dactylogyrus nanus*, 2. *Gyrodactylus* sp, 3. *Dactylogyrus* juveniles, 4. *D. crucifer*, 5. *D. microcanthus* and 6. *D. suecicus* in *Rutilus rutilus* during different calendar months (modified and redrawn from Bagee and Valtonen, 1999). (C) Effect of body size on prevalence of *Gastrocotyle indica* and copepods in *Caranx kalla* as well as *Cichlidogyrus digitatus* and *C. vexus* in *Tilapia zillii*. (D) The same on intensity of infection (compiled and redrawn from Radha, 1971, Blahoua et al., 2018).

**TABLE 3.7**

Competing taxa for space and food on the gill and body surface of fishes

| Taxa | Species (No.) | Reference |
|---|---|---|
| Ciliates | < 100 | see Rohde (1987) |
| Monogeneans | 1300 | Yamaguti (1963) |
| Trematodes | *Syncoelium, Accacoelium* | see Rohde (1987) |
| Hirudineans | < 250 | see Rohde (1987) |
| Branchiura | 175 | see Pandian (2016) |
| Copepods | 1500 | Yamaguti (1963) |
| Isopods | 430 | see Pandian (2016) |
| Cymothoids | < 1250 | see Pandian (2016) |

*Trichodina carassii* (see Paperna, 1964). A field survey of carp fry infected by these gyrodactylids hinted that they can be antagonistic against each other. Undertaking an experimental study, Paperna (1964) demonstrated that (i) the dominance of *D. vastator* was gradually replaced by *D. extensus* as the age of carp advances to 60 days. Both *D. vastator* and *D. extensus* were almost completely eliminated by *D. onchoratus*, when the host attained a size of 75 mm.

## 3.2 Life Cycle and Characteristics

Notably, most reports on monogenean fecundity, incubation period, hatching success and so on are based on laboratory experiments. For example, Bondad-Reantaso et al. (1995a) collected *Neobenedenia girellae* from *Seriola dumerili* skin and allowed the parasites to lay eggs on a petri dish. Using *Diplectanum aequans* eggs collected from *Dicentrarchus labrax*, Cecchini et al. (1998) estimated the effects of temperature on hatching success and development of the fluke. Others like Mooney et al. (2006) estimated oviposition by *Zeuxapta seriolae*, a parasite on *Seriola lalandi* as a function of time in the laboratory. Admirably, Chambers and Ernst (2005) conducted field experiments to investigate the effect of tidal currents on dispersal of *Benedenia seriolae* in a sea-cage farm of *Seriola* spp. *B. seriolae* eggs were collected by spreading a nylon mesh with an eye diameter of 75 µm in the cage and infection on naïve fish in cages kept at different distance was estimated.

### 3.2.1 Life Cycle

The water-based life cycle of monogeneans is indirect and involves a single larval stage the oncomiracidium but not an intermediate host. Typically, the eggs, encapsulated with yolk cells, are released, sedimented through the water column and embryonated at the bottom until hatching. Following hatching, the oncomiracidium may settle on the skin (Fig. 3.6A) or enter the mouth through respiratory current to settle on the gills of an appropriate host fish. In anuran hosts, the entry may be through the mouth in a young tadpole or cloaca of an older tadpole/adult (Fig. 3.6A). The oncomiracidium finally settles on the urinary bladder, although some neotenic polystomatids may complete their entire life cycle on the gills of a tadpole prior to metamorphosis of the tadpole.

The shape of monogenean eggs can be spherical, ovoid, fusiform or flattened tetrahedra (Kearn, 1986b). Egg size ranges from a few micrometers to 600 µm length in *Pseudodiplorchis americanus* (for more details see Kearn 1986b). Unlike the eggs of digeneans and cestodes, many monogenean eggs have filamentous apical extensions of the egg shell material; these extensions are called filaments or appendages. A function of these appendages is to

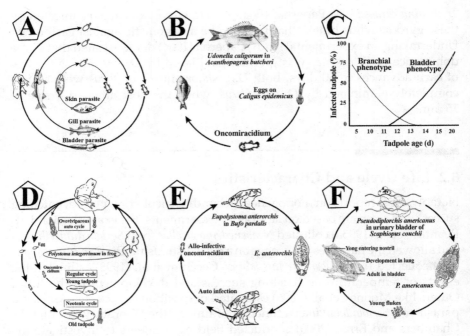

## FIGURE 3.6

(A) Simple but indirect life cycle involving oncomiracidia infecting the skin or gills of fishes or urinary bladder of amphibians. (B) *Udonella caligorum* attaching eggs on the carapace of *Caligus epidemicus* to disperse oncomiracidia and infect *Acanthopagrus butcheri* (Grobler et al., 2003). (C) *Polystoma gallieni*: effect of tadpole age of < 13 d old or > 13 d on development into branchial and bladder type in *Hyla meridionalis* (simplified and redrawn from Badets et al., 2010), respectively. (D) The complicated life cycle of *Polystoma integerrimum* involving infection of (i) young tadpole in regular cycle and (ii) older ones in neotenic cycle and (iii) ovoviviparously developed young ones from the bladder reinfect the bladder without being released to the exterior through urine (modified from Kearn, 1986b). (E) *Eupolystoma anterorchis*: the cycle involves no tadpole as its oncomiracidium hatched immediately on contact with water, directly infects the same or other *Bufo pardalis*. Note: the fully embryonated eggs in the worm. (F) *Pseudodiplorchis americanus* on the desert toad *Scaphiopus couchii*: the fully developed young ones directly infects more terrestrially through the nostril → lung → bladder (source: Tinsley and Earle, 1983).

reduce sedimentation rate of the eggs in the column of water. For example, the descent rate is 0.854, 1.26 and 3.24 cm/min for the appendaged large egg of *Hexabothrium appendiculatum*, the appendaged smaller egg of *Leptocotyle minor* and the non-appendaged egg of *Rajonchocotyle emarginata*, respectively (Whittington, 1987).

Instead of releasing the eggs, *Udonella caligorum* attaches them to the carapace of a copepod *Caligus epidemicus* (Fig. 3.6B), which serves to disperse them. *Nitzschia sturionis* and *Dionchus* spp bundle their eggs and attach them to the gills of cobias and remoras by a loop of egg shell material encircling individual primary gill lamellae. For example, *D. remorae* attaches some

Monogenea  115

25 bundles each on the first and second gill arches and 2–6 bundles on the third and fourth arches of *Echeneis naucrates* (Whittington, 1990). The remora host may improve oxygenation, afford protection to the attached eggs and may also disperse them. However, it is not clear whether the hatched oncomiracidia may infect the same host only.

A vast majority of monogeneans are ectoparasites of fishes. However, the endoparasitic Polystomatidae has radiated into amphibious and terrestrial vertebrates. It must, however, be noted that all the polystomatids are inhabitants of freshwater. Seasonal rainfall, inclusive of evaporation rate, drainage patterns and so on, has a profound implicit influence on parasite transmission. Nevertheless, the polystomatids have colonized a very wide range of hosts ranging from the lungfish *Neoceratodus forsteri* to sarcopterygians; in the latter ranges from more aquatic frogs *Rana rugosa* to the desert toad *Pseudodiplorchis americanus* as well as sporadic incidence on musk turtle *Stenotherus odoratus* (Oglesby, 1961) and African *Oculotrema hippopotami* (Tinsley and Earle, 1983). Their microhabitats are also spread over from the skin and gills in *Concinnocotylea australensis* to the pharyngeal cavity in *Polystomoides asiaticus*, palpebral cavity in *Neopolystoma palpebrae* and urinary bladder in *Pseudopolystoma dendriticum* (Badets and Verneau, 2009).

The taxonomy of 150-speciose Polystomatidae is based mainly on the attached organs of digestive and reproductive systems. Using the presence of a vaginal apparatus and size of the uterus, Badets and Verneau (2009) distinguished the bladder from branchial phenotype. According to Badets and Verneau (2009), the branchial type is characterized by the absence of a vagina apparatus and the presence of a short uterus. The type includes *Concinnocotyle, Protopolystoma, Pseudopolystoma, Sphyranura* and the branchial forms of *Polystoma*. These polystomatids infect hosts of aquatic ecology with wider durations of transmission during a year. In them, the transmission is usually through the tadpole. It is not clear whether this grouping holds true for all species belonging to a genus. For example, *Protopolystoma xenopodis* is a bladder parasite but *Protopolystoma* is classified under branchial type (Badets and Verneau, 2009). Interestingly, *Sphyranura oligorchis* is reported to infect the neotenic host *Necturus maculosus* (Sinnappah et al., 2001). More interestingly, both branchial and bladder types are exhibited by *Polystoma gallieni* and *P. integerrimum*, which infect *Hyla meridionalis* and *Rana temporaria*, respectively. Depending on the timing of cloacal opening to the exterior (and not by physiological conditions but by developmental stage) in *H. meridionalis*, the tadpoles of < 13 days old can be infected by oncomiracidium of *P. gallieni* through the gills alone but after the cloacal opening in the older (> 13 days) tadpole, more and more oncomiracidia prefer to enter through the cloacal opening (Fig. 3.6C, D). Briefly, the transmission frequency of branchial phenotype is negatively correlated with the age of the tadpole, whereas that of bladder phenotype is positively correlated with the age of the tadpole (Fig. 3.6C). As a result, those entering through the gills and cloaca

116　*Reproduction and Development in Platyhelminthes*

develop into branchial and bladder type, respectively. Those settled on the gills grow rapidly, become neotenic adults and produce eggs at the rate of 15/worm/day until the day, when the infected tadpole metamorphoses (Badets et al., 2010). On the other hand, those which settle on the urinary bladder grow slowly and require 3 months (e.g. *Protopolystoma xenopodis*, Tinsley and Owen, 1975) to 3 years (e.g. *P. integerrimum*, see Tinsley and Jackson, 2002) for sexual maturation and oviposition.

The bladder type, characterized by the presence of a vaginal apparatus and a long uterus, includes 7 + 1 genera, *Diplorchis, Eupolystoma, Neodiplorchis, Neopolystoma, Parapolystoma, Polystomoides, Pseudodiplorchis* and the bladder form of *Polystoma*. All of them infect hosts of terrestrial ecology with limited duration (one of three nights for the desert toads, Tinsley and Earle, 1983) of transmission during a year. In them, the transmission may or may not involve the tadpole; their oncomiracidia can infect only older tadpoles and develop slowly. In fact, experimental attempts to infect gill bearing young tadpoles of *Xenopus laevis* have failed. Only older tadpoles are susceptible at the post stage (57), when their urinary system is opened to the exterior and infection is through the cloaca (Tinsley and Owen, 1975). When the infected tadpole metamorphoses, the successful oncomiracidium migrates to the urinary bladder and attains maturity at the time the host goes to water for the first spawning. Among the bladder type, the transmission duration shrinks from the more aquatic hyalid *Litoria gracilenta* parasitized by *Parapolystoma bulliense* (Badets and Verneau, 2009) to the more terrestrial desert toad *Scapiopus couchii* infected by *Pseudodiplorchis americanus* (Tocque and Tinsely, 1994). The life cycle of the polystomatids is more complicated with intervening aestivation (e.g. *Xenopus laevis*, see Tinsley and Jackson, 2002) and hibernation (e.g. *P. americanus*, Tocque and Tinsley, 1994) of anuran hosts. Notably, the Arizona desert toad *S. couchii*, the host of *P. americanus* undergoes hibernation up to a period of 10 months in a year. As the toad does not feed during hibernation, its condition is measured by its abdominal fat content and Packed Cell Volume (PCV). Intensity of *P. americanus* infection increases with decreasing levels of fat body and PCV (from 0/host at 30% to 50/host at ~ 20%, Tocque and Tinsley, 1994). Encountering more and more ecological restriction for transmission, the bladder type polystomatids have developed many structural, developmental and behavioral strategies. These strategies may be grouped into ovoviviparity, as in *Polystomoidella oblonga* and *en masse* oviposition, as in *Eupolystoma anterorchis*. The South African toad *Bufo pardalis*, which hosts the bladder type of polystomatid fluke *E. anterorchis* has developed the following strategies: (i) extensive uterus capable of storing up to 300 eggs, (ii) *en masse* oviposition of completely embryonated eggs ready to hatch on exposure to water, (iii) cloacal approximation by sitting one over the other mating partner and gregarious spawning and (iv) oncomiracidim directly infecting the same or the other nearby toad by entering the cloaca and thereby avoiding the transmission through the tadpole (Fig. 3.6E). In fact, the transmission of *E. anterorchis* to

*Monogenea* 117

*B. pardalis* and *E. alluaudi* to *B. regularis* may be regarded more as a venereal infection (Tinsley, 1978). In *P. oblonga*, the oncomiracidium is hatched and developed into 6-sucker stage while still inside the parent worm (Oglesby, 1961). With transmission restricted to 12 hours on one of the three nights in the desert toad *S. couchii*, the ovoviviparous *Pseudodiplorchis americanus* has developed the following strategies: (i) elimination of miracidium and release of completely developed young ones, (ii) elimination of the tadpole and direct infection and (iii) migration over the exposed skin, entry through the nostrils and ultimate settlement on the urinary bladder through lungs (Fig. 3.6F). Hence, *P. xenopodis* and *P. americanus* are at the two extremes of transmission. In between these two extremes, there is *P. integerrimum* hosted by the mesics like *Rana temporaria*. However, transmission of *P. americanus* occurs once in a year within a short span of time; it occurs once in a year but for a prolonged few weeks in *P. integerrimum*. The characteristics of these three groups are summarized in Table 3.8.

## TABLE 3.8

Comparative account on life history characteristics of polystomid parasitic on aquatic, semi-aquatic and terrestrial hosts (condensed from Tinsley and Jackson, 2002)

| Characteristics | *Protopolystoma xenopodis* on *Xenopus laevis* | *Polystoma integerrimum* on *Rana temporaria* | *Pseudodiplorchis americanus* on *Scaphiopus couchii* |
|---|---|---|---|
| Habitat | Aquatic | Semi-aquatic | Terrestrial |
| Pre-reproductive period | 3 months | 3 years | 1 year |
| Maximum life span (y) | 2.5 | 6 | 4 |
| Transmission within host population | Continuous | Once/y (some weeks) | Once/y (< 24 hours) |
| Maximum fecundity (no./w/y) | 3650 | 2700 | 350 |
| Prevalence (%) | 40 | ~ 20 | 50 |
| Intensity (no./host) | 1–2 | 1–2 | 5–6 |

## 3.2.2 Egg Assemblage and Fecundity

In monogeneans, egg assemblage includes the duration from the release of an oocyte from generarium and its maturation, and provision of yolk cells to capsule formation around the oocyte and yolk cells. The shell of the elongate egg is provided with a lid and usually one or two threads/appendages, which function to attach the egg to the host. In *Entobdella soleae*, the assemblage is described in greater details by Kearn (1985). Following sexual maturity between the 11th and 15th day post-infection day (pid) at 2.1 mm body length in *Neobenedenia girellae*, the durations required for the release and maturation, and capsule formation are 42 and 21 seconds (s),

118    *Reproduction and Development in Platyhelminthes*

i.e. an individual egg is formed once in 63 seconds. The assembled eggs are continuously released at the rate of 12.2 and 35.4 eggs/hour by small (2 mm) and larger (> 4 mm) flukes, respectively (Bondad-Reantaso et al., 1995a). This is one of the fastest rates of egg assemblage. This period lasts for 2 and 7 minutes in *N. melleni* and *E. soleae*, respectively (Bondad-Reantaso et al., 1995a). Not all the monogeneans release the eggs continuously. In species, which exhibit egg-laying rhythm (e.g. *Zeuxapta seriola*, Mooney et al., 2006), as many as 321 eggs may be retained *in utero* up to 24 hours for some time. In ovoviviparous *Eupolystoma anterorchis*, up to 300 eggs are retained *in utero* and are expelled *en masse* (Tinsley, 1978a). In *Z. seriolae*, a pronounced peak occurs in oviposition immediately following 3 hours after the dusk, i.e. 18–21 hours (Mooney et al., 2006). In the diurnal *Neobenedenia* sp, the peak of oviposition occurs between 6 am and 9 am (Hoai and Hutson, 2014).

In monogeneans, egg release is a continuous process and is measured in number of eggs released per day. It ranges from 1.5 eggs/f (fluke)/d in *Discocotyle sagittata* to 1,398 eggs/f/d in *Benedenia seriolae* for the dermal flukes and from 9.6 eggs/f/d in *Haliotrema spariensis* to 518 eggs/f/d in *Zeuxapta seriolae* for the branchial flukes and 12–48 eggs/f/d in bladder flukes (Table 3.9). Hence, the dermal flukes are more fecund than the branchial and bladder flukes. The process of egg release is influenced by a number of factors; some of them are described below: Firstly, feeding is perhaps the most decisive factor that determines Fecundity (F). Egg production is reduced to 46%, when *Protopolystoma xenopodis* is transferred to *Xenopus laevis* that had been starved for 6 months (see Tubbs et al., 2005). Secondly, body size is the next decisive determinant of F. In *E. soleae*, a dermal parasite on *Solea solea*, F increases significantly and linearly with increasing body length (Fig. 3.7A) and with advancing age in *Gyrodactylus bullatarudis* (Fig. 3.7A-window); the relation for *E. soleae–S. solea* combination accounts for 84% of observed variations. Though a similar linear relation is apparent (Fig. 3.7B, see also Fig. 1.10A), the reported values bulge like a balloon in *Polylabroides multispinosus* and *Anopolodiscus cirrusspiralis* (Fig. 3.7B). However, no explanation is given for the widely scattered values. In fact, F increases from 0.05 eggs/f/d in the smallest (0.3 mm) *Lamellodiscus acanthopagri* to 31.6 eggs/fluke(f)/d in the large *Allomurraytrema robustum* (1.4 mm) but the egg size decreases from 98 to 63 μm in them (Roubal, 1994). In *Polystoma gallieni*, F decreases progressively with time (Fig. 3.7C). But, the neotenic branchial phenotype releases eggs at the rate of 1.5/f/d (Badets et al., 2010). In *Polystoma australis* too, the neotenics release 1.7 eggs/f/d (Kok and DuPreez, 1987). Hence, the progressive decrease in F of bladder phenotype is compensated by the neotenic branchial phonotype. In a way, the picture in these polystomatids seems to simulate the propagatory multiplication in digenean larvae. Thirdly, F is profoundly influenced by temperature in almost all monogeneans, as they are parasites of poikilothermic fish (see Fig. 1.7F) and amphibians. This has been reported in *Benedenia seriolae, Heterobothrium okamotoi, Neoheterobothrium hirame, Zeuxapta seriolae* and *Protopolystoma xenopodis* (Table 3.9). Not only

Monogenea   119

## TABLE 3.9

Fecundity of some monogeneans

| Parasite species | Host species | Fecundity (no./f/d) | Reference |
|---|---|---|---|
| **Dermal flukes** | | | |
| *Discocotyle sagittata* | *Oncorhynchus mykiss* | 1.5 @ 12°C; 12 @ 18°C | Gannicott and Tinsley (1998) |
| *Acanthocotyle greeni* | *Raja clevata* | 3 | MacDonald and Llewellyn (1980) |
| *Anoplosdiscus cirrusspiralis* | *Paguras auratus* | 17.6 | West and Roubal (1998) |
| *Allomurraytrema robustum* | *Acanthopagrus australis* | 31.6 | Roubal (1994) |
| *Entobdella soleae* | *Solea solea* | 40–60 @ 12°C | Kearn (1974) |
| *Benedenia seriolae* | *Seriola lalandi* | 40 @ 21°C; 51 @ 17.5°C; 18 @ 13°C | Tubbs et al. (2005) |
| *B. seriolae* | *S. quinqueradiata* | 1398 @ 24°C | Kearn et al. (1992) |
| *Neobenedenia* sp | *Lates calcarifer* | 190 self, 191 cross | Hoai and Hutson (2014) |
| *Neobenedenia grillae* | *Seriola dumerili* | 292, 2.5 mm; 850, 4.6 mm | Bondad-Reantaso et al. (1995a) |
| *Heterobothrium okamotoi* | *Takifugu rupripes* | 142 @ 15°C; 307 @ 20°C; 54 @ 25°C; 301 @ 30°C | Yamabata et al. (2004) |
| *Neoheterobothrium hirame* | *Paralichthys olivaceus* | 203 @ 10°C; 578 @ 15°C; 781 @ 20°C; 658 @ 25°C | Tsutsumi et al. (2002) |
| **Branchial flukes** | | | |
| *Haliotrema spariensis* | *Acanthopagrus australis* | 9.6 | Roubal (1994) |
| *Pseudorhabdosynochus lantanensis* | *Epinephelus coioides* | 10–22 | Erazo-Pagador and Cruz-Lacierda (2010) |
| *Polylabroides multispinosus* | *A. australis* | 14.5 | Diggles et al. (1993) |
| *Lamellodiscus squamosus* | *Acanthopagrus australis* | 31.6 | Roubal (1994) |
| *L. major* | *Acanthopagrus australis* | 31.7 | Roubal (1994) |
| *Heteraxine heterocerca* | *Seriola quinqueradiata* | 404 @ 24°C | Mooney et al. (2008) |
| *Zeuxapta seriolae* | *Seriola lalandi* | 208 @ 21°C; 309 @ 17.5°C; 207 @ 13°C; 518 @ 18°C, 40‰ S | Tubbs et al. (2005) Mooney et al. (2006) |
| **Bladder flukes** | | | |
| *Protopolystoma xenopodis* | *Xenopodis laevis* | 12 @ 20°C; 0.9 @ 8°C; 0.6 @ 6°C | Tinsley and Jackson (2002) |
| *Pseudodiplorchis americanus* | *Scaphiopus couchii* | 48 @ 26°C | Tinsley and Earle (1983) |

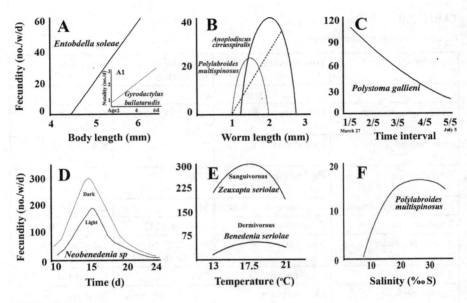

### FIGURE 3.7

Fecundity of monogeneans. (A) Fecundity (F) as function of body length in *Entobdella soleae* (modified from Kearn, 1985). A1 window: Natality as function of age in viviparous *Gyrodactylus bullatarudis* (modified from Scott, 1982). (B) F as function of body length in *Polylabroides multispinosus* and *Anoplodiscus cirrusspiralis* (compiled from Diggles et al., 1993, West and Roubal, 1998). (C) F as function of time interval, each lasting for ~ 24 days from March 27th to July 5th in *Polystoma gallieni*. (D) Effect of photoperiod on fecundity of *Neobenedenia* sp (modified from Hoai and Hutson, 2014). (E) Effect of temperature on fecundity of dermivorous *Benedenia seriolae* and sanguivorous *Zeuxapta seriolae*, parasites on *Seriola lalandi* (compiled from Tubbs et al., 2005). (F) Effect of salinity on f of *P. multispinosus* (modified from Diggles et al., 1993).

the daily egg output but also the lifetime F is also significantly influenced by temperature. For example the estimated lifetime F values are 18,000, 24,000 and 40,000 eggs/f for *N. hirame* on *Paralichthys olivaceus* at 25°, 20° and 15°C, respectively (Tsutsumi et al., 2002, 2003). Fourthly, solid and fluid food types seem to play an important role on F. At 21°C, F is 40 and 208 eggs/f/d for the dermativorous *Benedenia seriola* and sanguivorous *Zeuxapta seriolae*, respectively, both being parasites on *Seriola lalandi* (Table 3.9). Halton (1997) suggested that feeding on readily digestible fluid, the hematophagous parasites are able to minimize the digestion cost and divert the cost for egg production. However, it must be noted that on an average, the dermal flukes are more fecund than branchial flukes; other values (1,398 eggs/f/d) reported for *B. seriolae* on *S. quinqueradiata* by Kearn et al. (1992) are far higher than that reported for *Z. seriolae*. Fifthly, intra-specific competition among parasitic individuals affects fecundity more than body size. Rearing *Sebastes melanops* infected with *Microcotyle sebastis*, Thoney (1986a) found that super intensity of 760 flukes/host does not stunt the growth of the worm but its

Monogenea 121

fecundity is significantly reduced from 13.5 eggs/f d to 3.2 eggs/f/d. Other environmental factors like photoperiod (Fig. 3.7D) and salinity (Fig. 3.7F) also influence F.

### 3.2.3 Incubation and Hatching

In monogeneans, egg size ranges from 63 µm in *Allomurraytrema robustum* (Roubal, 1994) to 600 µm in *Pseudodiplorchis americanus* (Tinsley and Jackson, 2002). It may play the most important role on the duration of the incubation period. *In vitro* incubation studies have shown that (1) The dermal fluke *Benedenia seriolae* requires double the time (8 days at 20°C) than the branchial fluke *Diplectanum aequans* (4 days at 20°C) (Table 3.10). This holds true for those incubated at 28°C also. Figure 3.8 A illustrates the effect of temperature on cumulative *in vitro* egg hatching of *Zeuxapta seriolae* and *B. seriolae*. Similarly, Fig. 3.8B shows the effects of temperature and salinity on cumulative hatching of *B. seriolae*. The viviparous *Gyrodactylus bullatarudis* produces four successive generations, each of which generates 25 to 50 offspring within a period of 12 days (Fig. 3.8C). Tubbs et al. (2005) reported that the branchial fluke *Zeuxapta seriolae* releases 20% more number of hatchlings than *B. seriolae* within 6 days (instead of 9 days) at 21°C, both being parasites on *Seriola lalandi* (Table 3.10). It is not clear whether *Z. seriolae* releases more number of smaller hatchlings. (2) With increasing temperature from 14°C to 30°C, the incubation duration progressively decreases both in the dermal and branchial flukes. This may be one reason for the abundance of monogeneans in temperate waters (see Rohde, 1982).

The process of hatching of monogenean eggs, especially due to pressure, has been elaborated by Kearn (1986b). In *Branchotenthes octohamatus*, eggs are hatched, only when disturbed. But *Enoplocotyle kidakoi* eggs hatch, only after physical contact between them and their host moray eel (Whittington and Kearn, 2011). In others, a number of chemicals arising from the host's skin tissue, skin mucus, urea and ammonium chloride have been found to induce hatching (Table 3.11). In some others, embryos within eggs seem to undergo development at different rates. For example, *Acanthocotyle labianchi* eggs hatch, only when induced by the mucus of its host *Raja* spp. However, hatching success increases from 32% in ~ 18 days old eggs to 92% in 23–80 days old eggs (MacDonald, 1974). Still others display a hatching rhythm that may facilitate better chances for infection of an appropriate host. For example, 40% of *Entobdella soleae* eggs hatch between 9 am and 12 noon, 60% between 22 and 24 hours in *Diplozoon homoion gracile*, 95% between 23 and 24 hours in *E. hippoglossi* (see Kearn, 1986b) and 81%, 3 hours following light in *Neobenedenia* sp (Hoai and Hutson, 2014).

Egg viability is more important than hatching success. In *Protopolystoma xenopodis*, an egg hatches on the 20th day following oviposition at 26°C, when many yolk cells remain unutilized. However, the remaining unutilized yolk cells can sustain egg viability up to 32 days after oviposition. Subsequently,

## TABLE 3.10
Incubation period and hatching success of some monogeneans

| Species | Duration | Reference |
|---|---|---|
| **Incubation period** Dermal flukes |||
| *Neobenedenia girellae* | 3 days @ ~ 28°C; 5.5 days @ 20°C; 7.5 days @ 18°C | Bondad-Reantaso et al. (1995a) |
| *Benedenia seriolae* | 17 days @ 14°C; 8 days @ 20°C; 7 days @ 28°C; 15 days @ 15‰ S; 9 days @ 35‰ S; 13 days @ 50‰ S | Ernst et al. (2005) |
| Branchial flukes |||
| *Diplectanu aequans†* | 9 days @ 15°C; 4 days @ 20°C | Cecchini et al. (1998) |
| *Pseudorhabdosynochus lantanensis* | 4 days @ 30°C | Erazo-Pagador and Cruz-Lacierda (2010) |
| *Dawestrema cycloancistrium* | 72–96 hours @ 29°C | Maciel et al. (2017) |
| Bladder fluke |||
| *Eupolystoma anterorchis* | Hatches immediately | Tinsley (1978) |
| **Hatching success** Dermal flukes |||
| *Benedenia seriolae* | 70% on 9 days @ 21°C; 75% on 11 days @ 17.5°C; 76% on 20 days @ 13°C | Tubbs et al. (2005) |
| Branchial flukes |||
| *Polylabroides multispinosus* | 80% @ 30–35‰ S | Diggles et al. (1993) |
| *Zeuxapta seriolae* | 90% on 6 days @ 21°C; 90% on 8 days @ 17.5°C; 75% on 14 days @ 13°C | Tubbs et al. (2005) |

† = time required for 50% hatching

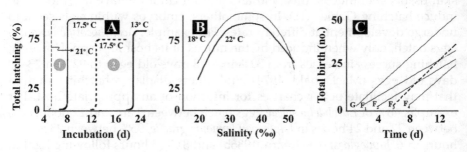

### FIGURE 3.8
(A) Cumulative *in vitro* egg hatching of ①*Zeuxapta seriolae* and ②*Benedenia seriolae* as function of incubation period and temperature (compiled from Tubbs et al., 2005). (B) Effect of salinity and temperature on hatching of *Benedenia seriolae* eggs (modified and redrawn from Ernst et al., 2005). (C) Cumulative birth of *Gyrodactylus bullatarudis* through successive generations (source: Scott, 1982).

## TABLE 3.11

Chemical cues attracting oncomiracidium to settle on *Leptocotyle minor* skin (condensed from Whittington, 1987, see also Kearn, 1986a)

| Egg age (d) | Chemical cue | Hatching (%) |
|---|---|---|
| 20 | Host skin tissue | 71 |
| 22 | Sole mucus | 67 |
| 22 | Ray mucus | 60 |
| 44 | 1 µM urea/ml | 56 |
| 22 | 100 µM thiourea/ml | 33 |
| 37 | 1 µM $NH_4$ cl/ml | 48 |

sole = *Solea solea*, ray = *Rajonchocotyle emarginata*

the embryo dies because of starvation (Tinsley and Owen, 1975). Conversely, the duration of egg viability of the branchial fluke *Polylabroides multispinosus* is so short that in < 24 hours following oviposition, the viability is reduced to 40% (Diggles et al., 1993). As indicated by hatching success, it is ~ 90% in another branchial fluke *Zeuxapta seriolae* at 17.5 and 21°C but ~ 70–75% in the dermal fluke *Benedenia seriolae* at the same temperatures (Fig. 3.8A). With regard to fecundity, duration of incubation and hatching success, there are arguable differences between dermal and branchial monogenean flukes and further study on this is suggested. Self fertilization occurs in some monogeneans (e.g. *Neobenedenia* sp, Hoai and Hutson, 2014; *Eupolystoma anterorchis*, Tinsley, 1978a). In *Neobenedenia*, hatching success increases from 78% for $G_o$ selfers to 86% for $F_1$ selfers (Hoai and Hutson, 2014).

## 3.2.4 Oncomiracidium and Dispersal

Information on ciliated and unciliated oncomiracidia is indeed very limited. Of 8 species listed by Kearn (1986a), only 2 are ciliated in each of the skin and gill parasites. The eggs of unciliated oncomiracidia may require a chemical cue from the appropriate host to enable their hatching or they may be at the mercy of waves and currents for dispersal. Similarly, only limited information is available on the life span of oncomiracidium (Table 3.12). Interestingly, even the limited information seems to hint at the life span being limited to < 24 hours, 24–36 hours and > 48 hours for the branchial, dermal and bladder flukes. Immediately following the respective fixed life span, the oncomiracidia suffer heavy mortality. The field experiments suggest that egg density is an important factor on the prevalence of *Benedenia seriolae* infection on *Seriola* spp in cages kept at different distances of 3.7 and 0.7% at 8 and 18 km from the source (Chambers and Ernst, 2005). Employing transparent *Neobenedenia* sp, Trujillo-Gonzalez et al. (2015) estimated the progressive increase in infection on the head, fins and body of *Lates calcarifer* as a function

124　*Reproduction and Development in Platyhelminthes*

## TABLE 3.12

Life span of oncomiracidium of some monogeneans

| Species | Life span (h) | Reference |
|---|---|---|
| **Dermal flukes** | | |
| *Entobdella soleae* | 24 | Kearn (1967) |
| *Polylabroides multispinosus* | 36 | Diggles et al. (1993) |
| *Neobenedenia* sp | 37, 25°C, 35‰ S | Trujillo-Gonzalez et al. (2015) |
| **Branchial flukes** | | |
| *Pseudorhabdosynochus lantanensis* | 4–8 | Erazo-Pagador and Cruz-Lacierda (2010) |
| *Microcotyle sebastis* | 12–24 | Thoney (1986b) |
| **Bladder flukes** | | |
| *Protopolystoma xenopodis* | 17–24 | Tinsley and Owen (1975), Theunissen et al. (2014) |
| *Pseudodiplorchis americanus* | 48 | Tinsley and Earle (1983) |

Dermal flukes = 24–36 hours, Branchial flukes = < 24 hours, bladder flukes = 17–48 hours

of time. The duration required for establishment of oncomiracidium is increased up to 96 hours, 8 days and 12 days on the body, fins and head, respectively and is subsequently decreased. Apparently, establishment and colonization following infection require a shorter duration on the soft body but longer durations on the relatively harder fin and head. Of ~ 63 successful larvae, 31, 24 and 8 of them have infected and settled on the body, fin and head, respectively. Infection success is 20 and 98% for branchial and bladder phenotypes of *Polystoma gallieni* (Badets et al., 2010). It is 38% for *P. australis* on the gills of *Kassina senegalensis* tadpole. Incidentally, the foregone account on life history traits like fecundity, incubation and hatching suggests that Monogenea may have to be viewed more from the angle of dermal, branchial and bladder types.

### 3.2.5 Growth and Maturation

Growth in body length of the flukes is profoundly influenced by host body size (Guegan et al., 1992) and temperature. Figure 3.9 A shows the relationship between the body length of 5 fluke species and that of *Scomberomorus commersoni*. Growth and maximum body size of each fluke species depend on the body size of the host *S. commersoni*. The range of *S. commersoni* body size that supports growth and maximum size attained differs for each fluke species. Between the host size ranging from 1 to 6 mm body length, *Pricea multae*, *Pseudothoracocotyle gigantica* and *Gotocotyle secunda* coexist up to the host size of 6–7 mm; similarly, *G. bivaginalis* and *Ps. gigantica* are accommodated in the host size between 7 and 15 mm. Hence, coexistence is limited to a maximum of 2–3 fluke species at a given body

Monogenea 125

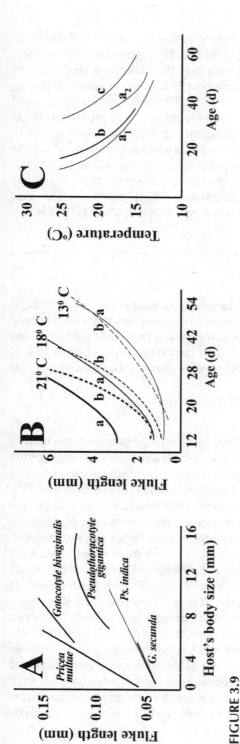

**FIGURE 3.9**

(A) Growth and maximum size attained by five buccal sucking fluke species in relation to host body size of the fish *Scomberomorus commersoni* (redrawn from Rohde, 1979). (B) Growth of *Benedenia seriolae* (a, continuous lines) and *Zeuxapta seriolae* (b, dotted lines) on *Seriola lalandi* in relation to age and temperature (source: Tubbs et al., 2005). (C) Maturity as function of age and temperature (a₁) *B. seriolae* on *S. lalandi* (source: Tubbs et al., 2005, Lackenby et al., 2007), (a₂) *Z. seriolae* on *S. lalandi* (source: Tubbs et al., 2005). (b) *Diplectanum aequans* on *Dicentrarchus labrax* (source: Cecchini et al., 1998), (c) *Neoheterobothrium hirame* on *Paralichthys olivaceus* (source: Tsutsumi et al., 2003).

126 *Reproduction and Development in Platyhelminthes*

size of the host fish. The relationships between size and age of *Seriola lalandi* and *Zauxapta seriolae* are shown in Fig. 3.9B-a-b. The figure also shows effect of temperature range from 13 to 21°C. To attain the maximum size of 6 mm *Benedenia seriolae* requires a period of 22 days at 21°C but even on the 54th day, the fluke attains the size of only 5 mm at 13°C. This observation holds good for *A. cirrusspiralis*. Similarly, both age and temperature have profound effects on sexual maturity. For example, the duration required for sexual maturity is 25 and 45 days for the dermitivorous *B. seriolae* and sanguivorous *Zeuxapta seriolae*, respectively, both flukes being parasites of *Seriola lalandi* (Fig. 3.9C-a$_1$, a$_2$). Both *B. seriolae* and *Diplectanum aequans* attain sexual maturity at the age of 30 days at 15°C but require 15–20 days at 25°C (Fig. 3.9C-b). This also holds true for *Neoheterobothrium hirame* on *Paralichthys olivaceus* (Fig. 3.9C-c).

## 3.3 Host Specificity

In general, monogeneans are viewed as highly host specific but a detailed analysis suggests that the specificity ranges from oioxenics that can live only on a single host species to stenoxenics that infects a few closely related host species (species of the same genus) and to euryxenic fluke species that is found on several distantly related host species sharing the same ecological niche (Lambert and Gharbi, 1995).

### 3.3.1 Specificity and Range

Most monogeneans are known for their distinct 'phylogenetic host specificity'. Their specificity may be restricted to a single species (oioxenics), a single genus or a single family of host (stenoxenics). Examining 2,097 specimens collected from 122 host species belonging to 42 families, Rohde (1979) found that 71% of them are restricted to one host species, i.e. oioxenics, 85% to one or two host species, e.g. stenoxenics, and 8% on one family of hosts, i.e. euryxenics. Only 2% of 435 species infect more than one order of fishes (i.e. euryxenics). Capsalids parasitize the skin, fins and gills of ~ 200 marine fish species in 44–46 genera and 9 subfamilies. The euryxenic *Neobenedenia melleni* has broad host specificity with > 100 host species from 30 families in teleostean orders (Whittington, 2004). In another analysis, Rohde (1979) recorded that many species belonging to the genus *Benedenia* are limited to one host. However, the 'phylogenetic host specificity' is replaced by 'ecological host specificity' in *B. hawaiiensis*, which infects as many as 24 hosts belonging to 11 families (see p 15–16). But all these host species are associated with Hawaiian coral reefs. A few monogeneans exhibit 'geographical host specificity', i.e. they have different hosts in different geographical areas. *Octodactylus minor* infects *Micromesistus poutassou* (Gadidae) in the Barents Sea but it is reported to occur on *Gadus merlangus* (Gadidae) on the coast of Norway. The same holds true for *Diclidophora denticulata*, which occurs in *Pollachius virens* (Gadidae) in the Barents Sea but it is found also on *Merluccius merluccius* (Meruluccidae) and *G. minutus* (Gadidae) in the European and American Atlantic (see Rohde, 1979). It is likely that almost all branchial flukes are oioxenics or stenoxenics but dermal flukes are euryxenics.

Reexamination of the list of monogeneans from the Chinese marine host species provides another opportunity to identify the host families that are responsible for smaller or large modulations on host specificity. Of 33 fluke families, 13 are oioxenics and strictly host specific (Table 3.13). The second group is oioxenics and comprises 223 fluke species in 12 families. In them, only 0.5–0.95 host species is available for each fluke species. The third group includes 66 parasite species in eight families found on 46 host species, i.e. for each fluke species, 1.2–3.0 host species are available. An exception can be 104 ancyrocephalid species, for which only 1.04 host species are available for each fluke species. From this analysis, the following inferences can be made: (1) Of 337 Chinese marine monogenean species, 14.2% are strictly host specific. In another 223 species or 66.2% of fluke species can also inhabit only on a single host species. Hence, 80.4% of them can be considered host specific, i.e. 25 out of 33 (76%) parasite families are host specific. (2) Some 19.6% of the parasites are stenoxenics or euryxenics, i.e. each fluke species in this group can be hosted by > 1 but < 3 host species. (3) Strikingly, Carangidae and Terapontidae modulate host specificity both towards oioxenicity and stenoxenicity. (4) Flukes parasitizing elasmobranchs are all oioxenics, as there are 17 elasmobranch species for 17 fluke species.

Considering fluke size (0.25–3.0 mm), mean prevalence (5–100%) and number of known host species (1–5) in relation to phylogeny, Sasal et al. (1999) made a comparative analysis of 74 oioxenic (specialists, host specifics) and euryxenics (generalists) monogenean parasite species on 2,547 specimens from 48 Mediterranean host species. Their findings are listed below: (1) The size of euryxenics ranges from 1 to 2.75 in log cm peaking at 1.66 in log cm and the oioxenics also from 1 to 2.75 in log cm but with the peak at 1.85 in log cm size. Hence, oioxenics parasitize larger hosts than euryxenics (Simkova et al., 2000, 2006). (2) A negative correlation between parasite body size and prevalence is apparent for euryxenics. (3) A significant positive correlation is found between host body size and its health status (condition factor, see p 101 also). (4) A significant positive correlation is also apparent between body size and parasite richness in oioxenics. Some of their findings confirm the observations reported by other authors.

### 3.3.2 The Gyrodactylids

As parasites, the gyrodactylids have attracted much attention for following reasons: (1) Direct life cycle (e.g. viviparous *Gyrodactylus gallieni*, Harris and Tinsley, 1987) involving no specialized transmission through oncoromiracidial

128  *Reproduction and Development in Platyhelminthes*

## TABLE 3.13

Host-parasite relationship in the Chinese marine fishes infected by monogeneans (basic data from Table 3.1)

| Family name | Fluke species | Host species | Host: fluke ratio | Fluke: host ratio | Host family |
|---|---|---|---|---|---|
| **Group I–No. of host species/ fluke species** | | | | | |
| Monocotylidae | 14 | 14 | 1:1 | 1:1 | Elasmobranchs |
| Capsalidae | 13 | 13 | 1:1 | 1:1 | *Terapontidae* |
| Gyrodactylidae | 2 | 2 | 1:1 | 1:1 | Mugilidae |
| Neocalceostomatidae | 1 | 1 | 1:1 | 1:1 | Ariidae |
| Calceostomatidae | 3 | 3 | 1:1 | 1:1 | Serranidae |
| Hexabothridae | 3 | 3 | 1:1 | 1:1 | Elasmobranchs |
| Neothoracocotylidae | 2 | 2 | 1:1 | 1:1 | Scombridae |
| Gotocotylidae | 4 | 4 | 1:1 | 1:1 | Scombridae |
| Hexostomatidae | 2 | 2 | 1:1 | 1:1 | Scombridae |
| Paramonaxinidae | 1 | 1 | 1:1 | 1:1 | *Gerreidae* |
| Pterinotrematidae | 1 | 1 | 1:1 | 1:1 | Alubulidae |
| Anchorophoridae | 1 | 1 | 1:1 | 1:1 | Cynoglossidae |
| Megamicrocotylidae | 1 | 1 | 1:1 | 1:1 | Chirocentridae |
| **Subtotal** | **48** | **48** | | | |
| **Group II–No. of host species/ fluke species** | | | | | |
| Tetraonchoididae | 7 | 9 | 0.78:1 | 1.29 | Labroidae |
| Amphibdelliatidae | 2 | 3 | 0.67:1 | 1.50 | Nakidae |
| Ancyrocephalidae | 104 | 109 | 0.95:1 | 1.04 | *Gerreidae* |
| Diplectanidae | 31 | 35 | 0.88:1 | 1.13 | Sparidae |
| Plectanocotylidae | 3 | 4 | 0.75:1 | 1.25 | Trichiuridae |
| Pseudodiclidophoridae | 1 | 2 | 0.50:1 | 2.00 | **Carangidae** |
| Allodiscocotylidae | 4 | 7 | 0.57:1 | 1.75 | **Carangidae** |
| Gastrocotylidae | 14 | 20 | 0.70:1 | 1.42 | **Carangidae** |
| Axinidae | 6 | 8 | 0.75:1 | 1.33 | Hemiramphidae |
| Heteraxinidae | 9 | 11 | 0.82:1 | 1.33 | **Carangidae** |
| Microcotylidae | 26 | 38 | 0.68:1 | 1.46 | **Carangidae** |
| Diclidophoridae | 16 | 20 | 0.80:1 | 1.25 | Engrualidae |
| **Subtotal** | **223** | **266** | | | |

*Table 3.13 contd. ...*

*Monogenea* 129

*... Table 3.13 contd.*

| Family name | Fluke species | Host species | Host: fluke ratio | Fluke: host ratio | Host family |
|---|---|---|---|---|---|
| Group III–No. of fluke species/ host species | | | | | |
| Dionchidae | 6 | 4 | 1.50:1 | 0.67 | Rachycentridae |
| Dactylogridae | 3 | 1 | 3.00:1 | 0.33 | Lateolabricidae |
| Protogyrodactylidae | 13 | 7 | 1.85:1 | 0.54 | *Terapontidae* |
| Mazocraeidae | 22 | 18 | 1.22:1 | 0.82 | **Carangidae** |
| Protomicrocotylidae | 8 | 5 | 1.60:1 | 0.63 | **Carangidae** |
| Chauhaneidae | 4 | 3 | 1.33:1 | 0.75 | Sphyraenidae |
| Bychowskicotylidae | 3 | 2 | 1.50:1 | 0.67 | Haemulidae |
| Heteromicrocotylidae | 7 | 6 | 1.17:1 | 0.86 | **Carangidae** |
| **Subtotal** | **66** | **46** | | | |
| **Grand Total** | **337** | **360** | | | |

stage, (2) Amenability to laboratory culture and (3) Shortest generation time and life span facilitating the estimates on growth and reproduction. As hosts, the susceptible salmonids (e.g. *Salmo salar*, Harris et al., 2000) attracted equal attention in view of their economic importance and aquaculture potency. The intense investigations have led to the identification of at least a dozen strains in the Atlantic salmon *Salmo salar*, three in *Salvelinus fontinalis* and two each in Arctic charr *S. alpinus* and grayling *Thymallus thymallus* (Bakke et al., 2002). Gyrodactylids include some extremely wide range supertramp species with highly efficient dispersal but poor competitive ability leading to the relatively rapid exclusion (e.g. *G. salaris*). The highly migratory Atlantic salmon shows more local variations in life history strategies and reproductive biology, and each of the local variance is a distinct stock/strain (see Bakke et al., 2002). Hence, the directly transmitted viviparous speciose gyrodactylids constitute a guild of species ideal for studies on the evolutionary significance of host specificity (Bakke et al., 2002). The following are the characteristic features of the guild, some of which are unique to gyrodactylids: (1) With 402 described species (out of a potential of 20,000 species), the guild has the widest host range for any monogenean family, infecting 20 of the 45 orders of fish, and cephalopods (e.g. *Isancistrum*) and aquatic anurans (e.g. *Gyrodactylus*). (2) Its host specificity ranges from 71% of (valid 402 species) oioxenics (specialists) to extremely euryxenics (generalists); for example, *Gyrodactylus alvica* infects 16 host species (Dmitrieva and Gerasov, 2000) and interestingly, *Phoxinus phoxinus* can serve as host for 14 gyrodactylid species (Bakke et al., 2002).

130   *Reproduction and Development in Platyhelminthes*

Of 27 Canadian gyrodactylids, 56% infect a single host, while the average number of fluke species infected is ~ two hosts (Poulin, 1992). Bakke et al. (1992a) found that of 319 species, 74% infect a single host but 4% of them infect four host taxa. Of 51 gyrodactylid species parasitic on cyprinids, only 49% are oioxenics or specific to a single host species; another 37% of them are stenoxenics, infecting hosts belonging to the same genus, clade or subfamily and the remaining 12% are euryxenics. Hence, some gyrodactylids are less oioxenic. In view of their economic importance, a larger scale analysis of the most speciose genus *Gyrodactylus* was made by Bakke et al. (2002). Irrespective of the existence of 1,496 host genera, the incidence is limited to only 2.6% of perciformean genera, i.e. only 59 perciformean species are susceptible to *Gyrodactylus*. However, of only 11 salmoniferan genera, 64% of them are susceptible to *Gyrodactylus* (Table 3.14). The number of species susceptible to *Gyrodactylus* increases in the following ascending order: > Gadiformes (13) Siluriformes (14 species) > Salmoniformes (19) > Scorpiaenoformes (20) > Cyprinodontiformes (22) > Perciformes (59) > Cypriformes (191). Table 3.14 also lists the number of oioxenic host specificity in the selected host families. The values range from 47% for Salmoniformes to 78% for Perciformes. Poulin (1998) predicted that speciose host taxa are less host specific than species-poor host taxa. However, the most speciose Perciformes comprising 1,496 genera includes only 78% of oioxenic host specificity in 59 species. (3) To some extent gyrodactylids have radiated with their hosts. The restriction of *G. pleuronecti* to pleuronectiform fishes alone and *G. nemacheili* to balitorid loaches alone suggests host-parasite co-evolution. However, host switching has been the predominant mode of radiation, evolution and speciation (Harris, 1993). (4) Within gyrodactylids, morphological diversity is minimal. Being eutelic (a constant number of nuclei usually established prior to birth), they contain ~ 1,000 syncytial cells (e.g. *G. gasterostei*) that facilitate material transport within syncytial layers (Bakke et al., 2007). The progenetic

## TABLE 3.14

Distribution of *Gyrodactylus* in finfish (condensed from Bakke et al., 2002)

| Available host | | Infected host | | Host species (no.) | | | | Oioxenics (%) |
|---|---|---|---|---|---|---|---|---|
| Order | Genera (no.) | Genera | | Species | 1 | 2 | 8 | 14 | |
| | | (no.) | % | (no.) | | | | | |
| Perciformes | 1,496 | 39 | 2.6 | 59 | 46 | 6 | 3 | 1 | 78 |
| Siluriformes | 412 | 13 | 3.2 | 14 | 9 | 5 | – | – | 64 |
| Cypriformes | 279 | 50 | 17.8 | 191 | 141 | 28 | – | – | 74 |
| Scorpaeniformes | 266 | 12 | 4.5 | 20 | 14 | 3 | 1 | – | 70 |
| Cyprinodontiformes | 88 | 8 | 9.1 | 22 | 14 | 5 | 1 | 1 | 64 |
| Gadiformes | 85 | 10 | 11.8 | 13 | 13 | 2 | – | 1 | 76 |
| Salmoniformes | 11 | 7 | 63.6 | 19 | 9 | 6 | – | – | 47 |

Monogenea 131

gyrodactylids are highly modified for viviparity: oocyte maturation and sperm storage occur in a single chamber, and a mature oocyte passes into the uterus after the birth of the successive $F_1$ fluke. Embryos receive nutrients via the uterus but not from yolk cells (Cable and Harris, 2002). (5) Gyrodactylids are viviparous and progenetic. The exceptional oviparous gyrodactylids are restricted to catfishes within the Amazon basin (e.g. *Oegyrodactylus farlowellae*, *Phanerothecium caballeroi*, Harris, 1983). Being highly progenetic, the adult closely resembles the larva and young male of *O. farlowellae*. Consequent to precocious maturation, the viviparous gyrodactylids give birth to live young ones, which are already pregnant, when born, a phenomenon termed hyperviviparity (Malmberg, 1970). Hence, the generation time is considerably abbreviated. Viviparous gyrodactylids are protogynous hermaphrodites, whereas all other monogeneans including the oviparous gyrodactylids are protandrous. (6) Despite the twin constraints of progenesis (paedomorphism) and viviparity (low fecundity), the gyrodactylids can rapidly colonize an entire aquatic system (Bakke et al., 1992b), indicating that they are extremely efficient in dispersal and finding naive hosts (Whittington, 1997).

7. *Sexuality*: Monitoring reproduction in *G. wageneri* for several generations on individually maintained goldfish, Braun (1996) found that the fluke continued to parturiate at least for 20 generations, when mother and daughter flukes were separated at birth. The isolated newborn flukes continued reproduction unhindered, but the mothers, at the time of separation from their first-born daughters did not possess a functional male reproductive system (Harris, 1985). Newborn flukes also did not yet have a functional male reproductive system. Hence the first-born daughter could have arisen by clonal multiplication. Confirming Braun's findings, Harris (1998) reported that the isolated lineages of *G. gasterostei*, a parasite on *Gasterosteus aculeatus*, too continued reproduction up to 19 generations, yielding 224 births. The application of quantitative genetic techniques to estimate the variance in dimensions of hamulus and marginal hooks, after neutralizing the effects of body size and temperature, led Harris (1998) to conclude that reproduction in isolated *G. gasterostei* neither involved sex nor relied on regular insemination. Sexual reproduction in this fluke may therefore be rare, suggesting the possible occurrence of clonal reproduction. Incidentally, two observations of Braun have to be noted: (1) The presence of meiotic chromosomes deep in the embryo cluster suggesting the extreme sexual paedomorphosis. (2) The first-born daughter and the subsequent daughters arise from oocytes. Clearly, the first-born daughter is derived clonally from the mitotic division of a single quiescent macromere, when the mother is an embryo *in utero*. However, Cable and Harris (2002) confirmed only the mitotic division in the intra-embryonic generation and the clonal origin of the first-born daughter. The latter may involve pre-meiotic doubling of chromosomes. However, the precise mechanism and the relative proportion of sexual and parthenogenic offspring are not yet known (Cable and Harris, 2002). Evidence for the

132  *Reproduction and Development in Platyhelminthes*

occurrence of sexual reproduction was brought by Schelkle et al. (2012) in *G. turnbulli*, a parasite on the guppy *Poecilia reticulata*. Hybrids were generated between three strains of the guppy that had been isolated and inbred for 1, 8 and 12 years after each of the strain had passed through 2 x $10^2$ to 2 x $10^8$ generations. Using microsatellite marker, sexually reproduced hybrids were identified in 3.7–10.9% of the offspring. It is not clear whether the remaining offspring are products of parthenogenesis. Being capable of clonal, parthenogenic and sexual reproduction, the gyrodactylids are unique among other sexually reproducing monogeneans.

8. *Transmission*: Combination with progenesis and hyperviviparity facilitates the gyrodactylids to generate offspring within 24 hours resulting in explosive population growth, which is usually known as epidemic. With the elimination of free-swimming oncomiracidium, the dactylogyrids are largely transmitted between fish by direct host-host contact with a single parasite being adequate to seed an entire population (Cable and Harris, 2002). Bakke et al. (1992b) describe the following transmission pathways: (i) detached parasite in the drift (e.g. *G. salaris*), (ii) from dead fishes (e.g. *G. bullatarudis*) and (iii) transmission between live host species. The behavioral mechanism of transmission is perhaps the most adopted one. In the protogynous *G. sphinx*, known for its host specificity, 72% of the transmitted flukes possess a functional male reproductive system, suggesting that the development and completion of male reproductive system trigger the migratory behavior. The mature flukes already possess 40% of post-second and post-third daughters (Dmitrieva, 2003). Incidentally, the migration of fully mature flukes (with post-second and post-third offspring) to other hosts may also maintain heterozygosis in the population and increase genetic diversity. With density-dependent transmission, two modes of transmission are commonly recognized: (i) the contact rate may directly increase with population density and (ii) the relative frequency of a susceptible host in a population, when the contact rate remains constant, irrespective of differences in population density. For example, mating aggregation and/shoaling may increase the contact rate, even when population density is low. The dactylogyrids are known to aggregate prior to fertilization (Dorovskikh and Matrokhina, 1987). From an experimental study on the effect of density (3–24 infected individuals) on the prevalence of epidemic growth in *G. turnbulli* on *P. reticulata*, Johnson et al. (2011) found that the contact rate, rather than population density, is more important in transmission and the probability of inducing an epidemic growth. The female guppies are infected earlier in the epidemic than the males, as they displayed more shoaling activity than the males.

9. *Growth and reproduction*: The life span of gyrodactylids is very short. The shortest generation time (~ 1 day) and progenesis have facilitated the so called epidemic multiplication to abound the fluke population. *G. ehrhardti* and *G. schwarzi* are natural parasites of *Corydorus paleatus* and *C. ehrhardti*. However, *G. anisopharynx* can also successfully infect and abound itself on

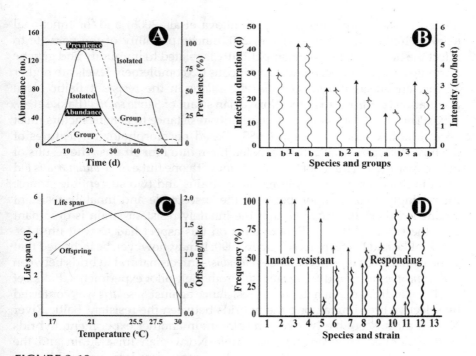

**FIGURE 3.10**

(A) Temporal changes in abundance and prevalence of isolated (thick continuous lines) and grouped (thin broken lines) *Gyrodactylus anisopharynx* on *Corydoras palaetus*. (B) Infection duration (straight vertical lines) and intensity (wavy vertical line) of *G. anaisopharynx* on *Corydoras* (1) *paleatus,* (2) *ehrhardti,* (3) *schwartzi.* 'a' represents isolated presence and 'b' grouped presence of parasite (compiled from Boeger et al., 2005). (C) Life span and offspring number of *G. bullataridis* on *Poecilia recitulata* (compiled from Scott and Nokes, 1984). (D) Frequency of innate resistant and responding species and strains in individually isolated fish after exposure to highly infected *Salmo salar* with *Gyrodactylus salaris* Lierelva strain. 1. *Salmo trutta* Lierelva, 2. *Salmo trutta* Fossbekk, 3. *Salvelinus alpinus* Korssjoen, 4. *Salmo trutta* Lake Tunhovd, 5. *Coregonus lavaretus*, 6. *Thymallus thymallus* 1+, 7. *Salvelinus namaycush,* 8. *Oncorynchus mykiss,* 9. *Salvelinus alpinus* Hammerfest, 10. *Thymalus thymalus* 0+, 11. *Salvelinus fontinalis,* 12. *Salmo salar* Neva, 13. *Salmo salar* Mean Norwegian stocks (modified and redrawn from Bakke et al., 2002).

*C. paleatus* (Fig. 3.10A). Its prevalence and abundance are profoundly influenced by its presence in isolation or a group. Host species considerably influence the duration and intensity of infection (Fig. 3.10B). Temperature is another important factor that has a dominant effect on the life span and offspring production, as in *G. bullatarudis* (Fig. 3.10C).

### 3.3.3 Susceptibility and Resistence

Susceptibility indicates the inability of a population/species to resist infection by a parasite species. In fact, susceptibility and resistance to an infection represent the spectral ends of host-parasite interactions. Resistance may be

134 *Reproduction and Development in Platyhelminthes*

related to the morphological (e.g. Simkova et al., 2006) and/or functional (e.g. Buchmann and Lindenstrom, 2002) incompatibility of the parasite to infect the host. Functional incompatibility is related to immunity and genetic attributes. For example, blood and mucous C3 complement levels are higher against an infection with *Gyrodactylus salaris* in the resistant Baltic Neva population (strain) than in the Norwegian strain of *Salmo salaris* (Bakke et al., 2000). The credit for relating susceptibility-resistance to genetic factors must go to Madhavi and Anderson (1985). Based on morphological features of the host *Poecilia reticulata*, they divided them into four strains. The results of their experimental infection of each strain with one fluke of *G. bullatarudis* led them to discover two innately resistant strains and two susceptible strains. Interestingly, hybridization between the susceptible and innately resistant strains led them to recognize that the innately resistant strain is dominant over the susceptible one. Experiments on this aspect had to wait until Dr. T.A. Blakke and his team took it up in 1990. It may, however, be indicated that Cone et al. (1983) reported that *G. colemansis* infects natural hybrids between *Salvelinus fontinalis* and *S. namaycush*. From their pilot experiments, Bakke et al. (1990) indicated that a degree of resistance against *S. salaris* was conferred by selective breeding among the hybrids between the resistant Baltic River Neva x the susceptible Norwegian salmon. In another experiment, hybrids were generated between the susceptible Norwegian Imsa strain and the resistant Baltic Neva strain. The hybrid progeny were less susceptible than the pure-bred Imsa strain but more susceptible than the pure-bred Neva strain. These experiments confirm that susceptibility/resistance to *G. salaris* is a heritable trait and the resistance is partially dominant to susceptibility (see Bakke et al., 2002). In an another, more elaborate experiment, Bakke et al. (1999) generated hybrids between the susceptible Alta strain of *S. salar* x *S. trutta*, using *S. salar* in either direction as sires and dams. Both hybrid and pure-bred progeny were infected with *D. derjavani*, a trout parasite and *G. salaris*, a parasite of Atlantic salmon. The growth rate of these gyrodactylids was slower in the hybrids than in the pure-bred natural hosts. But the pattern of resistance was maternally inherited, i.e. when the dam was *S. trutta*, the susceptibility of the hybrids to *S. salaris* was similar to that of pure *S. trutta*. However, a spectrum of responses from moderate susceptibility to innate resistance was noted, when the dam was *S. salar*. Briefly, Bakke and his team demonstrated that (i) selective breeding can lead to resistance, (ii) maternal inheritance of resistance and (iii) probable sex-linked transmission of susceptibility.

Considering gyrodactylid infections, Blakke et al. (2002) divided salmonids into three categories: (1) 'Susceptible', in which *G. salaris*, for example, increases to several thousands in numbers until the host, the Norwegian *S. salar* dies. (2) 'Responding', in which the partially susceptible *S. trutta* is capable of mounting an immune response and self curing following an initial growth of *G. salaris*. (3) 'Innately resistant', in which *G. salaris* fails to grow on *Salvelinus alpinus*. Experimental infection with *S. salaris* have also failed to establish infection in the innately resistant (to *G. salaris*) cyprinids *Rutilus rutilus*, *Phoxinus phoxinus*, percid, *Perca fluviatilis*, gasterosteids, *Gasterosteus aculeatus*, *Pungitus pungitus*, anguillid, *Anguilla anguilla* and pleuronectid, *Platichthys flesus*, despite, initial 100% prevalence and intensity up to 67 flukes/host (see Bakke et al., 2002).

Regarding infection by *G. salaris*, no taxonomic pattern is apparent within salmonids, suggesting that gyrodactylids evolve by host switching rather than host speciation (Desdevises et al., 2002, Zietara and Lumme, 2002). For example, *S. trutta*, a sister taxon to *S. salar*, is almost entirely resistant to infection with *G. salaris* (Jansen and Bakke, 1995). But the less closely related anadromous strain of *S. alpinus* includes both innately resistant and responding hosts, in which all the infections are sooner or later eliminated. The relative frequency of responding and innate resistant individuals of different salmonid species and some of their strains are shown in Fig. 3.10D. Liereva and Fossbekk strains of *S. trutta* consists of only innately resistant individuals to *G. salaris*. At the other end of the spectrum, all individuals of *S. salar* Neva and *S. fontinalis* respond to *G. salaris*. The remaining salmonids and some of their strains lie between these spectral limits.

Progressive narrowing of host specificity involves (i) increasing structural specialization and (ii) functional interaction between the host and parasite. Considering the morphological attributes of attachment organs in 51 *Dactylogyrus* species on cyprinids, Simkova et al. (2006) compared the likely processes connected with evolution of host specificity. Two major findings have emerged from their analyses: (1) The degree of host specificity is correlated with morphological attributes, i.e. higher specificity or narrower host range imposes selection on the development of anchors and associated structures that are important for attaching the fluke on the host. (2) Narrowing host specificity is more common among larger host species than on smaller fish that host more euryxenics (see p 127).

Regarding functional compatibility, Buchmann and Lindenstrom (2002) elucidated a range of dynamic interactions between monogeneans and the host fish. These dynamic reactions are responsible for (i) host finding, (ii) host specificity and (iii) host immunity. Diffusion of soluble molecules emanating from the host are attractants for the parasites. Mucus extract of the host skin is reported to attract *Entobdella soleae* oncomiracidium to its natural host *Solea solea* (Kearn, 1967). The corneal epithelium, devoid of mucous cells, does not entice oncomiracidia, suggesting that the attractant is associated with mucous cells. The mucus extract from *Oncorynchus mykiss* contain a number of proteins, polypeptides, amino acids and carbohydrates (Buchmann and Bresciani, 1999). The mucous cells also release a number of polysaccharides carrying galactose, mannose, lactose and fucose epitopes (Buchmann, 1998b). Lectins are also found in fish serum and mucus (e.g. *Anguilla anguilla*, Gercken and Renwrantz, 1994). Interestingly, fish immunity is associated with two mannose-rich glycoproteins (Sigh and Buchmann, 2000). Such molecules are considered as important antigens in fish parasites (Woo, 1996).

The glands located in the anterior and posterior attachment areas of monogeneans, their secretions, especially the amino acid composition of the secreted proteins used for adhesion play a major role in host specificity. Monopisthocotyleans feeding on the host's epithelia may have to neutralize the hostile substances present in the epidermis. For example, the parasite tegument is perforated by the complement factor and macrophage activity in the skin of the imminent host (Buchmann and Bresciani, 1999). Polyopisthocotyleans, ingesting the host's blood and thereby exposing their gut, may have to encounter the antibodies, complement factors and immunologically competent cells (Buchmann and Lindenstrom, 2002). In salmonids, the detected complement factors are proteins (Buchmann, 1998b). In them, a number of isomers occur in teleosts (Sunyer et al., 1997). With different affinities to bind the parasite epitopes, a specific isomer may trigger either the destructive pathway and expel the parasite, or induce its settlement pathway. Investigations on search for specific fish antibodies against monogenean infection have reported contrasting results. Some of them are listed below:

| Positive results | Negative results |
| --- | --- |
| 1. Antibodies from carp bound to *Dactylogyrus vastator* antigens were adsorbed to heterologous erythrocytes (Vladimirov, 1971). | 1. No antibodies were found from *Oncorynchus mykiss* to recognize the infection by *Gyrodactylus derjavini* (Buchmann, 1998a). |
| 2. Weak binding of serum antibodies from the infected eels to antigens of *Pseudodactylogyrus bini* (Buchmann, 1997a). | 2. No rise in antibody titres in *Paralichthys olivaceus* infected with *Neobenedenia girellae* (Bondad-Reantaso et al., 1995b). |
| 3. (a) Specific immunoglobulin from *Takifugu rubripes* reacting with *Heterobothrium okamotoi* (Wang et al., 1997). (b) Specific immunoglobulin in *Anguilla anguilla* against *Pseudodactylogyrus bini* (Mazzanti et al., 1999). | 3. No detectable binding of immunoglobulin from the fish *Leiostomus xanthurus* to homogenates of its parasite *Heteraxinoides xanthophilis* (Thoney and Burreson, 1988). |

The fact that the carp evades dactylogyrid infection and rockfish partly evades microcotylid infection following vaccinations with parasite homogenates (see Buchmann and Lindenstrom, 2002) suggests that at least some fishes have developed immunity against a few monogeneans. The contrasting observations on the presence or absence of immunity in some of these infected fish can be explained in the context of immune-suppressants. Induced by stress, cortisol is known to be a stress hormone in fish (see Pandian, 2013). Interestingly, cortisol treatment rendered three salmonid species, which consisted of a high proportion of innately resistant individuals more susceptible to *Gyrodactylus salaris* than control (Harris et al., 2000). Hence, an important feature of immune response is its susceptibility to immune-suppressants like dexamethasone (Lindenstrom and Buckmann, 1998) or hydrocortisol (Harris et al., 2000). In unstressed salmonid fish, plasma cortisol level ranges from 0 to 5 ng/ml; when stressed, the level can go to 40–200 ng/ml. Chronic stresses like prolonged confinement or crowding result in an elevation of plasma cortisol level up to 10 ng/ml (Pickering and Pottinger, 1989). The cortisol axis has been reported as vital in modulating the response against gyrocotylids (Lester and Adams, 1974). Aqueous aluminum and low pH may also act as immune-suppressants in infected fish. For example, *G. salaris* is susceptible to aluminum ions and is eliminated (by suppressing immunity) after 4 days at 202 µg Al/l at pH 5.9. Notably, lower pH eliminates the fluke in aluminum poor waters (Soleng et al., 1999). Hence, immunity levels have to be estimated in infected fish that are not suppressed by immune-suppressants like cortisol, low pH and aluminum pollutants.

# 4

## Digenea

### Introduction

The class Trematoda comprises three orders namely Monogenea, Aspidocotylea and Malacocotylea or Digenea. The aspidocotylids include only four families, 13 genera and 61 species (Rohde, 2001). Like monogeneans, the digeneans are also covered externally by a non-ciliated syncytial tegument resting directly on the mesenchyme (Hyman, 1951). However, many features differ from those of monogeneans (Table 4.1). The digenean life cycle is inextricably linked with alternation of generations and alternation of hosts. The cycle involves typically four larval forms and one (e.g. *Lymnaea truncatula* infected with *Fasciola hepatica*) to four intermediate hosts; the site (tongue) fedelitic hemiurid *Halipegus accidualis* passes through the snail *Helisoma anceps*, an ostracod, a dragonfly nymph and *Rana clamitans* (Esch et al., 2002).

### 4.1 Taxonomy: Parasites and Hosts

Trematoda and Cestoda are considered as the most successful among metazoan parasites by their numerical abundance, geographical reach and host (habitat) diversity (Poulin and Morand, 2000). The life cycle of the latter is completed entirely through trophic transmission, whereas that of the former may involve at least one developmental stage that penetrates its next host (see Table 4.6). Vertebrates serve as definitive hosts for both trematodes and cestodes. However, the involvement of a mollusc is obligate for the former and an arthropod for the latter. Considered *en masse*, adaptability, flexibility and plasticity are the hallmarks of the endoparasitic flatworms. Digenean phylogeny is constructed on the basis of 56 morphological characters; only two are concerned with miracidium and six with sporocyst and redia. It

## Digenea 139

## TABLE 4.1

Contrasting features of monogeneans and digeneans (condensed from Cribb et al., 2002, added from Littlewood et al., 2015)

| Feature | Monogeneans | Digeneans |
|---|---|---|
| Parasitism | Ectoparasites, except for a few polystomids | Endoparasites except for a few on fish skin |
| Habitat | Aquatic fishes, except a few on anurans | Mostly aquatic but many in terrestrial tetrapods |
| Life cycle | Simple, indirect involving a single penetrating larva | Complicated, indirect involving 2–6 larval stages in 1–4 intermediate hosts |
| Life span | Short, mean 631 days (see Table 1.5) | Long, mean 2,102 days (see Table 1.5) |
| Reproduction | Sexual only | Sexual in vertebrate host. Parthenogenic/polyembryonic in invertebrate host |
| Taxonomy | 53 families including 3 megadiverse genera *Dactylogyrus* (900 nominal species), *Gyrodactylus* (402 species), *Heliotrema* (> 156 species) | > 150 families, 1,777 genera and 12,012 species. No genus contains 2% of all digenean species. *Phyllodistomum* (116 species), *Stephanostomum* (102 species) |
| Distribution | > 99% of them are parasites on fishes. < 1% have radiated into anurans and rarely into tortoise and hippopotamus | Of 150 families, 46 are primarily fish parasites. The remaining (> 30%) have radiated into terrestrial tetrapods |
| Specificity | 78% of them are specific to a single host genus but with mean of 1.45–2.85 host genera/parasite species | Tending toward euryxenic with regard to both definitive and intermediate hosts |

comprises a bewildering array of two subclasses, 24 superfamilies, 150 families, 1,777 genera and 12,012 species (Table 4.2). With a rare exception of transversotrematoids, they are all endoparasites in hosts of marine, freshwater and terrestrial inhabitants. The relatively less speciose subclass Diplostomida, comprising three superfamilies and 19 families, is restricted to tetrapods as definitive hosts. But the relatively more speciose subclass Plagiorchiida includes 21 superfamilies and 131 families and is exclusively parasites of teleosts. Rough estimate indicates that 52% (46% genera) and 48% (54% genera) of species are hosted by tetrapods and fish, respectively (Littlewood et al., 2015). The digeneans differ in size and shape, sucker size, intestinal length and details of the reproductive system. Individuals may be host specific and even site-specific. However, 78% of the oioxenic monogeneans are specific to a single host species but the digeneans tend toward euryxenicity with regard to both definitive and intermediate hosts. Remarkably, none of the digenean genera contains > 2% of all digenean species, whereas *Dactylogyrus* (900 nominal species), *Gyrodactylus* (402 species) and *Heliotrema* (> 150 species) are megabiodiverse monogenean genera. Most digenean genera (79%) have relatively few species (1 to 5). Even

140   *Reproduction and Development in Platyhelminthes*

**TABLE 4.2**

Phylogeny of major digenean groups indicating numerical diversity of families, genera and species. Molecular phylogenies for trematodes are derived mostly from *ssr*DNA (18S) and *lsr*DNA (28S) ribosomal RNA genes (condensed from Littlewood et al., 2015)

| Phylogeny | Taxonomic group | Family (no.) | Genus (no.) | Species (no.) |
|---|---|---|---|---|
| | Brachylaimoidea | 7 | 29 | 227 |
| | Diplostomoidea | 6 | 97 | 797 |
| | Schistosomatoidea | 6 | 84 | 453 |
| | Bivesiculoidea | 1 | 5 | 28 |
| | Transversotrematoidea | 1 | 4 | 30 |
| | Azygioidea | 1 | 4 | 40 |
| | Hemiuroidea | 13 | 212 | 1334 |
| | Heronimoidea | 1 | 1 | 1 |
| | Bucephaloidea | 2 | 29 | 416 |
| | Gymnophalloidea | 4 | 44 | 231 |
| | Paramphistomoidea | 11 | 135 | 431 |
| | Pronocephaloidea | 6 | 49 | 293 |
| | Haplosplanchnoidea | 1 | 9 | 50 |
| | Echinostomatoidea | 10 | 118 | 1098 |
| | Opisthorchioidea | 3 | 160 | 839 |
| | Apocreadioidea | 1 | 21 | 111 |
| | Lepocreadioidea | 10 | 108 | 542 |
| | Monorchioidea | 2 | 56 | 336 |
| | Gorgoderoidea | 12 | 123 | 1084 |
| | Haploporoidea | 2 | 40 | 170 |
| | Opecoeloidea | 3 | 96 | 932 |
| | Brachycladioidea | 2 | 27 | 181 |
| | Plagiorchioidea | 26 | 129 | 953 |
| | Microphalloidea | 19 | 197 | 1335 |
| | Total | 150 | 1777 | 12012 |

the most diverse 25 genera are relatively less speciose (20–78 species) and account for only 21% of the total digenean genera (Cribb et al., 2002). The incidence of trematode species decreases in the following descending order: Warm temperate Northeast Pacific (529 species) > West and South Indian Shelf (391 species) > Bay of Bengal (348 species) > Northeast Australian Shelf (340 species) > Marshall, Gilbert and Elis Islands (328 species) > South China Sea (319 species) > Cold temperate Northwest Pacific (249 species) > Red Sea and Gulf of Aden (200 species) (Cribb et al., 2016).

## 4.2 Larval Forms

In digeneans, the life cycle is highly complicated involving typically four larval forms, Miracidium (M), Sporocyst (S), Redia (R) and Cercaria (C), and two (an obligate vertebrate Definitive Host, DH and an obligate first intermediate molluscan host, e.g. schistosomes), three (snail–tadpole–snake, e.g. *Strigea*, see Shoop, 1988) and rarely four (snail–frog–small rodents–cats/ dogs, e.g. strigeid *Alaria*) intermediate hosts. In a classic, A.P. Thomas was the first to describe the life cycle of liver fluke in sheep. In digeneans, the First Intermediate Host (FIH) is usually a gastropod, less often a bivalve (e.g. *Turtonia minuta* for the gymnophallid *Parvatrema margaritense*, Galaktionov et al., 2006), rarely a scaphopod (e.g. *Ptychogoniomus megastoma* in *Dentalium* spp, *Lecithophyllum botryophorum* in *Antalis entails*, [= *Dentalium entale*], see Koie et al., 2010) and very rarely an annelid (e.g. *Sanguinicola inermis* in polychaete *Artacana proboscidae*, see Pandian, 2017). However, DH may belong to any of the five classes of vertebrates. As adults, they occur primarily in the digestive track, especially the intestine (e.g. Lecithasteridae, Cribb et al., 2002); some of them occur on the gills, operculum, buccal cavity or pharynx (e.g. didymozoonids, clinostomatids, Hyman, 1951), lungs (e.g. saphaedrine plagiorchids, *Pneumobites, Crepidostomum cooperi*), liver (e.g. *Fasicola hepatica*, Thomas, 1883), pyloric caeca (e.g. *Echinostoma revolutum*), gall bladder (e.g. *F. gigantica*, Phalee et al., 2015), bile passage (e.g. dicrocoeliid *Dicrocoelum dentrictum*), urinogenital tract of reptiles and birds of plagiorchids (Hyman, 1951), eye lens (e.g. *Diplostomum*, Ubels et al., 2018) or gonads (e.g. *Maritrema novaezealandensis*, Keeney et al., 2007); for more information, Mehlhorn (2016) may be consulted. Some families represented by the genera *Sanguinicola*, *Spirorchis* and *Bilharzia* (= *Schistosoma*), the flukes with elongated body inhabit the blood of fish, turtles, birds and mammals including human (Hyman, 1951).

*Miracidium*: In most digeneans, eggs (Fig. 4.1A) are voided with excreta of the definitive host. Miracidium (Fig. 4.1B) is the first larval form in the digenean life cycle. Its entry into an appropriate host is achieved by free-swimming active penetrative miracidium (e.g. *Fasciola hepatica*) or by passive trophic ingestion of an egg (e.g. *M. novaezealandensis* eaten by *Zeacumantus subcarinatus*, Keeney et al., 2007) or rarely miracidium (e.g. *Mesostephanus haliasturis*, Barkar and Cribb, 1993). In Fasciolidae, Echinostomatidae, Philophthalamidae, Schistosomatidae and others, the miracidium is large

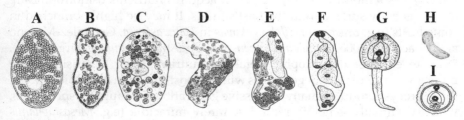

**FIGURE 4.1**

Digenean larval forms: (A) egg, (B) miracidium, (C) mother sporocyst, (D) daughter sporocyst, (E) mother redia, (F) daughter redia, (G) cercaria, (H) mesocercaria and (I) metacercaria (from Phalee et al., 2015).

142  *Reproduction and Development in Platyhelminthes*

in size (80–150 µm up to 340 µm in *Cyclocoelum microstomum*) and complex in organization (Galaktionov and Dabrovolskij, 2003). On hatching from the egg in water, it swims erratically until interception of a chemical signal originating from the potential host by the signal-receiving organ located on its body surface. Then, it swims along the gradient of this chemical toward the direction of the potential host, adheres into its surface and initiates penetration through the host body surface. Chemicals, usually in the form of short chain fatty acids or specific amino acids arising from the host, attract the miracidium toward the host (Esch et al., 2002). During penetration, it sheds the ciliated plates from its surface, which are replaced by the instantly forming tegument. Haas et al. (1995), however, reported that *Schistosoma japonicum* miracidia respond to the aminoglycans of 30 kDa molecular weight and are able to distinguish among signals arising from different snails and correctly select the host *Biomphalaria glabrata*. More information on the mechanism responsible for the entry is provided later.

The passive miracidia of Plagiorchiidae, Hemiuroidea, Brachylaimoidea and others are small in size (20–60 µm up to 4–10 µm in Microcephallidae, Lecithodendrioidea) and their somatic organization is more simplified (Galaktionov and Dobrovolskij, 2003). Their eggs are hatched only after ingestion by the host/predator. In the passive and active processes of infection, there are advantages and disadvantages. The disadvantage of active host invasion by free-swimming, penetrating miracidia is their short life span. However, this is compensated by their ability of intercepting a host specific signal emitted by the appropriate molluscan host in the aquatic system. The disadvantage of passive trophic ingestion is the high risk involved in being ingested by an inappropriate host and consequent mortality of the egg/miracidium. However, the advantage is the protection afforded to the embryonated eggs for a longer duration by the parent. Active and passive invasions occur in aquatic medium but passive only on land (Niewiadomska and Pojmanska, 2011).

*Sporocyst* (S): In the molluscan host, the miracidium undergoes regressive metamorphosis and develops into sac-like mother sporocyst (Fig. 4.1C), which lacks a mouth and gut. Hence, it acquires nutrients osmotrophically across its body surface from the host's fluids. It may be highly branched in some hosts (e.g. *Brachylaima ilobregatensis* in *Helix aspersa*, Gonzalez-Moreno and Gracenea, 2006, *Renylaima capensis* in *Ariostralis nebulosa*, Sirgel et al., 2012) to enhance osmotrophic uptake of nutrients. Not surprisingly, the miracidial musculature degenerates within 2 post-infection days (pid). Often, the sporocyst produces many successive generations of daughter sporocysts, which eventually generate redia but rarely miracidia (e.g. *Mesostaphanus haliasturis*, Barkar and Cribb, 1993). In fact, the presence of miracidium in sporocysts (Sewell, 1922), and simultaneous production of miracidia and daughter sporocysts in the snail *Melanoides tuberculata* (Premvati, 1955,

*Digenea* 143

Mohandas, 1975) has been earlier reported. In the sporocyst, the germinal cells become apparent by the 4th pid, when they grow to a size (252 x 74 µm) and attain an irregular shape. By the 12th pid, the large sporocyst includes a spacious central cavity and contains different stages of daughter sporocysts. Some mother sporocysts, from which majority of daughter sporocysts have already escaped, begin to collapse by 16–20 pid. But they can persist up to 154 pid (e.g. *Schistosomatium douthitti* in *Lymnaea catascopium*, Loker, 1978).

*Daughter sporocyst*: To begin with, the daughter sporocyst (Fig. 4.1D), a simple sac like the sporocyst, is small (25 x 12 µm) and becomes spherical in shape on the 12th post-birth day (pbd), when it loses the distinct epithelial boundary and grows to a size of 177 x 48 µm. It is a large cylindrical sac (395 x 48 µm) and contains intensively stained germinal cells on the 16th pbd. Its life span may last up to 470 days in *S. douthitti* (Loker, 1978). Both mother and daughter sporocysts are regarded as paedomorphic embryos, which reproduce by means of apomictic parthenogenesis.

*Redia*: In general, the mother and/or daughter sporocysts generate rediae. The sporadic occurrence of a single mother redia (Fig. 4.1E) has been reported from the cyclocoelids (Johnstone and Simpson, 1940), *Parorchis acanthus* (Rees, 1940) and paramamphistomoid *Stichorchis subtriquetrus* (see Pearson, 1972). Rediae are usually characteristic of primitive trematodes like Fasciolidae, Echinostomatidae, Paramphistomoidae, Psilostomidae, Philophthalmidae, Lepocreadiidae and other specialized groups Halipeginae, Aporocotylidae, Heterophyridae and Opisthorchiidae (Galaktionov and Dobrovorlkij, 2003). According to Galaktionov and Dobrovolskij (2003), substitution of rediae by the daughter sporocyst took place at least four times during trematode evolution. They have an elongated cylindrical body with the mouth opening terminally at anterior end. In them, a primitive type of digestive tract consisting of a muscular pharynx, esophagus and a sac-like gut is present. They emerge out of the sporocyst, move around tearing and swallowing not only the host tissues but also rediae of other species. Depending on the temperature and season, they may generate daughter rediae (Fig. 4.1F) and/ or cercariae. Production of two successive generations of rediae occurs at cooler temperatures. Experimental studies indicate different responses to temperature. In the snail *Lymnaea truncatula, F. hepatica* produces a single generation of rediae at 11–21°C but both rediae and immature cercariae at 4–5°C. In *F. gigantica*, 20°C switches from redia to cercaria production in *L. natalensis* (see Whitfield and Evans, 1983). However, the potency to produce successive generations of one or other larval forms seems to be inherent in digeneans but its expression is regulated by an unknown host substance. Following serial transplantations to naive *L. stagnalis*, the rediae of *Echinoparyphium aconiatum* and *Isthmiophora melis* continue to produce successive generations of rediae up to 27 and 40 generations and the flukes retain the multiplication potency as well as the cercarial infectivity (Donges, 1971). On transplantation

144   *Reproduction and Development in Platyhelminthes*

of *Schistosoma mansoni* sporocyst to naïve *Biomphalaria glabrata*, the grafted mother sporocyst produces six successive generations of daughter rediae for a period of one year (Jourdane and Theron, 1980). Briefly, the trematode larval forms have retained the multiplication potency and the sporocyst in either direction. This is an important research area to identify (i) whether the nutrient depletion limits the propagatory cycle, (ii) whether the hosts secrete an inhibiting substance (e.g. schistosomin, De Jong-Brink et al., 1991) not to allow further damage and loss and/or (iii) whether the cycle shall still continue, even on transplantation to an unsual host like a bivalve or leech.

*Cercaria* (Fig. 4.1G) may originate from the mother (e.g. plagiorchid *Plagiorrchis*) or daughter (e.g. *Diplostomum, Alaria*) sporocyst or redia I (e.g. *Echinostoma, Heterophyes*) or redia II (e.g. *Fasciolopsis, Paramphistomum*) or rarely from the miracidium (e.g. *Stichorchis subtriquetrus, Echinostomum margarinatum*). They are more structurally organized than miracidium, sporocyst and redia, and consist of a broader or elongated oval body and a simple or forked tail. Based on the structural features, they are broadly grouped into gasterostome, monostome, amphistome, echinostome, holostome and distome cercariae. Based on the tail features, the distome cercariae are further grouped into rhopalocercous, microcercous, cotylocercous, furcocercous, plerurolophocercous, triclocercous, cystocercous and macrocercous types (Hyman, 1951, for figures see Esch et al., 2002). Cercariae with a reduced tail encyst within the first intermediate host (e.g. *Heronimus chelydrae, Burnellus trichofurcatus*). Loker (1978) describes the development of *Schistosomatium douthitti* in *Lymnaea catascopium*. The cercarial embryo grows from 18 x 18 µm size with 20–30 cells on the 12th day to 92 x 38 µm on the 24th day. From the 24th day onwards, its body is elongated, and a tail bud is formed. Examination of the snails has indicated that 51% of all cercariae within daughter sporocysts have not yet begun to differentiate, 17% have begun to differentiate and the remaining 32% are morphologically mature.

The developed cercaria enters the invasion stage, which consists of (i) dispersion phase, when cercaria is emitted from FIH and (ii) directional host finding phase, when cercaria is reaching an appropriate host. During the dispersion phase, cercaria utilizes signals like (i) gravity of water, (ii) turbulence, (iii) light, (iv) temperature and (v) ion concentration originating from the ambient water. It responds to the signal of chemical gradients emanating from the host during the host finding phase. Low molecular weighing hydrophilic chemicals are reported to attract cercariae of *Echinostoma trivolvis* and *E. caproni* (Reddy et al., 1997). In trophically transmitted cercariae, special adaptations have been developed.

*Metacercaria* (Fig. 4.1H) arises only from cercaria. In families like Echinostomatoidea, Paramphistomoidea, Notocotyloidea, Cyclocoeloidea and Megaseridae, cercaria generates metacercaria (Pearson, 1972). On entry into Second Intermediate Host (SIH), cercaria sheds its tails and becomes metacercaria, which is a miniature of sexually immature adult but

*Digenea* 145

in some of which the reproductive system is present and functional (e.g. progenetic *Plagioporus sinitsini*). Four types of metacercariae are recognized: (i) those capable of directly infecting a definitive host (e.g. *Echinostoma trivolvis*, Pechenik and Fried, 1995), (ii) those capable of only encystation on vegetation or animals or substratum, (iii) those capable of infecting the second intermediate host and (iv) those developing into mesocercaria in the second intermediate host (Toledo et al., 2006). Those digeneans characterized by two host life cycle like the paramphistomids, fasciolids and haploporids encyst on plants; others like notocotylids, pronocephalids, philophthalmids and echinostomatids do it on animals; still others like the haplosplanchnids on plankton (Pearson, 1972). Encystation on vegetation requires only a few hours. Immediately following it, the encysted metacercariae become infective (e.g. *Fasciola hepatica*). In both the structure and chemical composition of cyst layers, extreme variations range from species to species. Fried (1994) provides a long list for *in vitro* excystation of metacercarial cyst. Briefly, metacercaria is an adaptive larval form to prolong the infective life of the cercaria and thereby enhance the chances of ingestion by definitive host.

Some gymnophallid metacercariae like *Cercaria margaretensis* (Irwin et al., 2003) and *Parvatrema margaritense* (Galaktionov et al., 2006) are also capable of parthenogenic multiplication. In the latter, the bivalve *Turtonia minuta*, the gastropod *Margarites helicinus* and the marine duck *Somateria mollissima* serve as FIH and SIH, and DH, respectively. Two generations of propagatory multiplications of metacercaria $MC_2$ and $MC_3$ occur in the extrapallial cavities of the snail. Feeding on the pallial fluid and cells, the metacercaria ($MC_1$) grows slowly to a size of 730 x 530 μm but its body cavity is filled with $MC_2$ (180 x 120 μm) and in whose brood sac, embryonic balls of $M_3$ (130 x 76 μm) are present, as the metacercarial development proceeds (Galaktionov et al., 2006, see also Galaktionov, 2006).

*Mesocercaria*: In the strigeid genus *Alaria*, the cercaria penetrates SIH, the tadpoles and frogs but does not develop further and remains as mesocercaria. Following ingestion by third intermediate hosts like small mammals, it develops into a metacercaria, which finally infects the definitive host, when the second intermediate host is devoured. Like metacercaria, the mesocercaria is also another larval form to extend the infective life of cercaria. Incidentally, an array of third intermediate hosts like mice, rats and raccoons, and definitive hosts like dogs, cats, minks, weasels and so on greatly enhances the chances of the third intermediate host and definitive host ingesting the second and third intermediate host, respectively.

Life cycles of selected digeneans are shown in Fig. 4.2. Notably, one or more larval forms are added or excluded in the cycle. With incorporation of terrestrial vertebrates as the definitive host, the transmission is switched to the inclusion of the terrestrial snail as obligate first intermediate host (e.g. brachylaimoids).

146  *Reproduction and Development in Platyhelminthes*

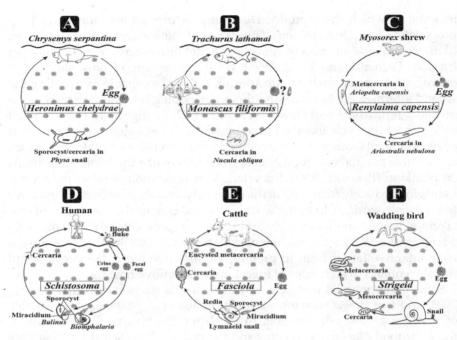

**FIGURE 4.2**

Life cycle of selected digeneans. (A) *Heronimus chelydrae* involving one intermediate host (redrawn from Shoop, 1988). (B) *Monascus filiformis* involving two intermediate hosts with choices for second intermediate host (a) *Sagitta* sp, (b) *Eucheilota ventricularis* (c) *Liriope tetraphylla*, (d) *Proboscidactyla mutabilis* and (e) *Aequorea* spp (modified and redrawn from Diaz Briz et al., 2016). (C) *Renylaima capensis* involving two intermediate hosts. (modified and redrawn from Sirgel et al., 2012). (D) *Schistosoma* sp involving one intermediate host, in which sporocyst and cercaria are developed. (E) *Fasciola* sp involving one intermediate host, in which sporocyst, redia, cercaria and encysted metacercaria are developed. (F) *Strigeid* sp involving 3 intermediate hosts, snail, tadpole and snake (redrawn from Shoop, 1988). Note the life cycles of A and B are completely aquatic, C completely terrestrial, D, E, and F are partially aquatic involving terrestrial definitive hosts.

## 4.3 Intermediate Hosts

The taxonomic distribution of monogeneans is greatly restricted to aquatic fish and amphibious anuran species, as their indirect life cycle involves no intermediate host. In contrast, the incorporation of one or more intermediate hosts by digeneans and cestodes has vastly expanded the definitive host to all five classes of vertebrates as well as the geographical reach from aquatic to terrestrial habitats. The taxonomic distribution of digenean definitive hosts has already been elaborated in Chapter 1.3. The following represents a snap shot of the digenean distribution on intermediate hosts. The new database generated by Cribb et al. (2001) indicates that information on digenean life

Digenea 147

cycle is limited to ~ 1,000 species. Description of a digenean species based on adult morphology and/or experimental elucidation of partial or complete life cycle may require a long time, especially with progressive 'extinction' of taxonomic experts and challenges encountered in rearing the larval forms. Molecular approaches offer significant methodological advantages by matching sequence data for different larval and adult forms. For example, Littlewood et al. (1999) constructed morphological- and molecular-based phylogenesis of the phylum Platyhelminthes using morphological characters, and 18S and two partial 28S rRNA gene sequences to evaluate the emergence and subsequent divergence of parasitic forms. Using 18S rDNA sequence, Cribb et al. (2001) inferred the phylogeny and distribution of ~ 51 digenean superfamilies that employ gastropods (30), bivalves (20) and scaphopod (1) as FIH as well as 56 orders of fishes that serve as DH. At species level, Huston et al. (2016) employed ITS2 rDNA sequences generated for the larval trematodes from the infected snail *Echinolittorina austrotrochoides* to describe the complete life cycle of *Gorgocephalus yaaji*. By means of D2 LSU, ITS1 and COI DNA sequence analysis, Gonchar and Galaktionov (2017) resolved the life cycle of notocotylid *Tristriata anatis*, for which periwinkles *Littorina* spp and sea duck *Somateria* serve as FIH and DH, respectively.

Cribb et al. (2001) generated a new database containing information on life cycles of > 1,000 trematodes and 5,000 host fish species from ~ 20,000 published records. An analysis of the distribution of digenean families in Mollusca reveals that (i) the number of digenean families engaging the molluscan classes as FIH is 66, 9 and 1 + 1 (see Koie, 1995, Koie et al., 2010) for gastropods, bivalves and scaphopods, respectively, (ii) bivalves, scaphopods and polychaetes began to serve as FIH due to host switching (Table 4.3), (iii) hence, the digeneans are associated primitively with gastropods alone, (iv) the number of digenean families distributed on definitive vertebrate aquatic host is 2, 4, 13, 2, 10, 5, 6 and 80 for Petromyzontiformes, Holocephali, Elasmobranchii, Dipnoi, Chondrostei, Semionotiformes, Amiiformes and Teleostei, respectively, (v) digeneans may have been primitively associated with teleosts as definitive hosts, and (vi) Engaging Petromyzontiformes and Chondrostei as a definitive host is also due to host switching.

Description of the digenean life cycle is a continuous process. For example, one of the most complex life cycles among trematodes was described for *Parvatrema margaritense* in 2006 by Galaktionov and his coauthors, subsequent to that published by Cribb et al. (2001). In bucephalids, for example, the discovery of new molluscan superfamilies serving as FIH has been ongoing from one in 1905 to 17 in 1985. Except for Bucephalidae, Sanguinicolidae and Echinostomatidae, the relationship between the number of superfamilies of molluscs and polychaetes engaged as FIH increases from 4–5 to 5–6 in ~ 50 reported life cycles and subsequently levels off. Hence, the conclusions arrived by Cribb et al. (2001) can only be fine tuned but not changed altogether. The following are the highlights of their findings:

148   *Reproduction and Development in Platyhelminthes*

**TABLE 4.3**

Families of major aquatic taxonomy groups and their first, second and definitive aquatic hosts. First intermediate hosts: ⬡ = gastropod or ◯ = bivalve or ◖ = scaphopod or ʃ = polychaete, second intermediate hosts: 🐸 = anuran, ◯ = mollusc, 🦐 = arthropod, ʃ = annelid, ✳ = echinodermate, definitive hosts: 🐟 = teleosts, 🦈 = chondrichthyans, 🐾 = tetrapods (compiled and reorganized from Cribb et al., 2003)

| Family | FIH | SIH | DH |
|---|---|---|---|
| Aspidogastrea | gastropod or bivalve | | tetrapod |
| Brachylaimoidea | gastropod | mollusc | tetrapod |
| Diplostomoidea | gastropod | tetrapod, mollusc | tetrapod |
| Clinostomidae | gastropod | tetrapod, mollusc | tetrapod |
| Sanguinicolidae | gastropod or bivalve | annelid | teleost |
| Spirorchiidae | gastropod | | tetrapod |
| Schistosomatidae | gastropod | | tetrapod |
| Bivesiculoidea | gastropod | tetrapod | teleost |
| Transversotrematoidea | gastropod | | teleost |
| Azygioidea | gastropod | tetrapod | teleost |
| Hemiuroidea | gastropod or bivalve | tetrapod, annelid, mollusc | teleost, tetrapod |
| Heronimoidea | gastropod | | tetrapod |

| Bucephaloidea | Gymnophalloidea | Paramphistomoidea | Pronocephaloidea | Haplosplanchnoidea | Echinostomatoidea | Opisthorchioidea | Apocreadioidea | Lepocreadioidea | Monorchioidea | Gorgoderoidea | Allocreadioidea | Plagiorchioidea | Microphalloidea |
|---|---|---|---|---|---|---|---|---|---|---|---|---|---|
| | | | | | | | | | or | or | or | | or |

150   *Reproduction and Development in Platyhelminthes*

- Most gorgoderids engage bivalves as FIH (see also Table 4.3).
- The widest range of molluscan hosts for a digenean species is an azygiid *Azygia bucci*, which employs 16 species in 12 genera and three families of gastropods including prosobranchs and pulmonates.
- Next to *A. bucci* is *Cyclocoelum mutabile*, which engages pulmonates belonging to 12 genera and three families.
- *Fasciola hepatica* is reported to engage *Lymnaea truncatula*, *L. cubensis* and *L. viatrix*, as FIH. In Australia, *Austropeplea tomentosa* was added, when *F. hepatica* was introduced with the arrival of Europeans.
- For many traits of their life cycle, sanguinicolids are unique as they (i) exploit a broad range of 45 families of fishes in 18 orders including holocephalans, elasmobranchs and teleosts as definitive hosts and (ii) engage an equally broad range of gastropods (including prosobranch and pulmonates), bivalves and polychaetes as FIH (as in bucephalids also).
- Allocreadiidae (20 orders of fishes), Clinostomidae, Fasciolidae (restricted to Lymnaeidae and Planorbidae) and Brachylaomidae (restricted to eupulmonate stylommaphorans alone) are families with a highly restricted range of FIH.

More information is available on trematode–mollusc association for the European freshwater host fishes. The findings of Faltynkova et al. (2016) on the association are: (1) The 'mollusc' dataset includes large number of pulmonates (29 species), 'prosobranchs' (15 species) and bivalves (11 species) serving as FIH for 171 trematode species in 89 genera and 35 families. Of these, 23 and 40 species employ freshwater fishes as DH and SIH, respectively. (2) The 'fish' database contains 8,202 records covering 122 trematode species in 49 genera and 19 families found in a total of 148 fish species in 21 families. Of these, 59 species infect fish species as DH. (3) The 'fish' dataset also includes 99 fish species in 63 genera and 19 families serving as SIH for 66 digenean species in 33 genera and 9 families. (4) Among the molluscs, the snails *Lymnaea stagnalis* (14 species), *Planorbis planorbis* (39 species), *Radix peregra* (31 speceis) and *R. ovata* (31 species) serve as FIH for the largest number of trematode species. Single trematode species is hosted by 14 host molluscan species.

Notably, the choice for FIH among gastropods is restricted to cerithioids alone for bivesiculoids and haplosplanchnoids, rissooids alone for apocreadioids, lymmaeoids alone for spirorchids and to eupulmonates alone for brachylaimoids (Table 4.3). Other parasite families have a choice from two or three taxa: paramphistomoids, echinostomatoids, gorgoderoids and plagiorchioids can select one or more of the four gastropod families. For SIH, the choice is restricted to vertebrates alone for bivesiculoids, bucephaloids and opisthorchioids. But the choice is broad among one or more hosts for gymnophalloids and lepocreadioids. A comprehensive picture of the choice for SIH taxa by digenean families is shown in

*Digenea* 151

Table 4.3. The choice for intermediate and DH is restricted to one family each in Transversotrematoidea and Haploplanchnoidea. Hence, the members of these two families may include oioxenics. Conversely, the hemiuroids, gymnophalloids and gorgoderoids with a broad spectrum for all three hosts may be euryxenics. Others like echinostomatoids and plagiorchioids have a fairly good choice to select one among three/four SIH taxa, one of the two DH taxa. Bivesiculoids and transversotrematoidids, bucephaloids, haplosplanchnoids, apocreadioids and lepocreadioids have a limited choice to select only one definitive host taxa.

To enhance the chances of infection, the digeneans exploit even a broader range of aquatic, amphibious and terrestrial fauna as SIH. The range includes gastropods, bivalves, aquatic (crustaceans) and terrestrial (e.g. ants) arthropods, polychaetes (see also Peoples, 2013), leeches (e.g. *Erpobdella octoculata, Helobdella stagnalis* by *Australapatemon* sp, Jocelyn, 2009), and echinoderms. The number of SIH employed by 15 digenean taxa decreases in the following descending order: vertebrates (221 species) > arthropods (142 species) > molluscs (102 species) > annelids (17 species) > echinoderms (7 species) > Cnidaria (5 species) > Chaetognatha (4 species) > platyhelminths (3 species) > Ctenophora (1 species). Cribb et al. (2003) estimated that in all, 556 species are engaged as SIH. However, their list is not complete. Even among invertebrate SIH like brachiopods (see Ching, 1995) are notably missing. According to Gibson and Bray (1994), the sequences of the second and third intermediate vertebrate host and to DH include 591 associations for fish + herptiles, 21 for fish + herptiles + birds, 36 for fish + herptiles + mammals, 257 for fish + birds + mammals, 512 for birds + mammals and 137 for herptiles + birds + mammals.

A compilation of values reported for species number in major digenean families by Littlewood et al. (2015) and the number of second intermediate host species in 22 families by Cribb et al. (2003) provides a rare opportunity for further analysis. Prior to the estimate made here, two points must be noted: (i) The SIH listed in Table 4.3 do not include cnidarians and chaetognaths; for Clinostomidae, no value is given. Hence, the number arrived here may still be a low estimate. (ii) However, the same host species may also serve as SIH for more than one fluke species. (iii) Besides, cercaria may directly be eaten by some bivesculid and azygiid species. Still, the estimate may open an avenue for further research in this area. The mean number of SIH awaiting or available for selection by digenean family members increases from zero in Schistosomatidae (with 453 species) to six in Lepocreadioidea; the choice is limited to one host species for Bivesiculoidea, Azygioidea, Brachylaimoidea, Bucephaloidea and Opisthorchioidea, two species for Apocreadioidea, 3 species for Diplostomoidea, Gorgoderoidea and Echinostomoidea, 4 host species for Hemiurioidea, Monorchioidea, Allocredioidea and Plagiorchioidea, 5 host species in Microphalloidea and 6 host species for Gymnophalloidea and Lepocreadioidea (Table 4.4). From this analysis, the following inferences can be made: (1) Information regarding SIH is known for 10,754 digenean

152    *Reproduction and Development in Platyhelminthes*

## TABLE 4.4

Estimates on distribution of fluke species/family over the number of second intermediate hosts (estimated from Cribb et al., 2003, Littlewood et al., 2015)

| Family | Species (no.) | Estimate on host species (no.) | Host species (no./family) |
|---|---|---|---|
| Families with no second intermediate host | | | |
| Schistosomatoidea | 453 | | |
| Transversotrematoidea | 30 | | |
| Heronimoidea | 1 | | |
| Paramphistomoidea | 431 | | |
| Pronocephaloidea | 293 | | |
| Haplosplanchnoidea | 50 | | |
| Total | 1258 | | |
| Families with 1 second intermediate host | | | |
| Bivesiculoidea | 28 | | |
| Azygioidea | 40 | | |
| Brachylaimoidea | 227 | | |
| Bucephaloidea | 416 | | |
| Opisthorchioidea | 839 | | |
| Total | 1550 | 1550 x 1 = 1550 | 310 |
| Families with 2 second intermediate host | | | |
| Apocreadioidea | 111 | 111 x 2 = 222 | 111 |
| Families with 3 second intermediate host | | | |
| Diplostomoidea | 797 | | |
| Echinostomatoidea | 1098 | | |
| Gorgoderoidea | 1084 | | |
| Total | 2979 | 2979 x 3 = 8937 | 993 |
| Families with 4 second intermediate host | | | |
| Hemiuroidea | 1334 | | |
| Monorchioidea | 336 | | |
| Allocreadioidea | 125 ? | | |
| Plagiorchioidea | 953 | | |
| Total | 2748 | 2748 x 4 = 10992 | 687 |
| Family with 5 second intermediate host | | | |
| Microphalloidea | 1335 | 1335 x 5 = 6675 | 1335 |
| Families with 6 second intermediate host | | | |
| Gymnophalloidea | 231 | | |
| Lepocreadioidea | 542 | | |
| Total | 773 | 773 x 6 = 4638 | 387 |
| **Grand total/Mean** | **10754** | **33014** | **3.5 host species/fluke species** |

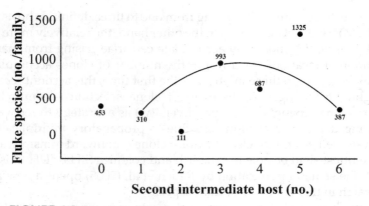

**FIGURE 4.3**

Number of fluke species/family distributed over the number of second intermediate host (drawn from data assembled in Table 4.4).

species in 22 major families; this value represents 89.5% of 12,012 digenean species. (2) Of 22 families, members of 6 families totaling to 1,258 species (or ~ 11.7% digeneans) have not incorporated SIH. (3) Of the remaining 16 families, ~ 88.3% of digeneans have incorporated a SIH; 5, 1, 3, 4, 1 and 2 fluke families engage one, two, three, four, five and six intermediate hosts, respectively (Table 4.4). (4) On the whole, 9,496 (88.3%) digenean species may engage 33,014 as SIH species; each fluke species has an average choice of 3.5 SIH either available or awaiting to be selected by a fluke species. In fact, the number of SIH per digenean family increases from 1-SIH species for 1,550 fluke species in five families to 3-SIH species for 2,979 fluke species in three families and then begins to decrease to 6-host species for 773 fluke species in two families (Fig. 4.3). Briefly, the incorporation of SIH, especially 3 to 5 SIH species/fluke species has enormously increased the scope for evolution to establish and enhance genetic (clonal) diversity by clonal mixing and lineage diversity in digeneans.

## 4.4 Clonal Selection

Unlike in free-living organisms, the flow of genetic diversity is restricted in endoparasitic digeneans. For, sexual reproduction is limited to gonochoric schistosomids (involving no second intermediate host) and among sexually active adults hermaphrodites localized in the same microhabitat within a host. This restricted opportunity for genetic exchange among the intra-population within a host is termed as 'nested hierarchy' (Esch et al., 2002). Localized in odd microhabitats like the air sacs of lungs, head cavities and eye, few digenean species may be selfers. Within the life cycle, the slow motile mollusc

154　*Reproduction and Development in Platyhelminthes*

may be infected by miracidia arising from one to three definitive hosts (Sire et al., 1999, Minchella et al., 1995). On the other hand, the relatively more mobile second intermediate host may accumulate cercariae arising from many FIH from different locations. As a result, the number of clones accumulated in FIH may be less than those in SIH. For the first time, this account recognizes and highlights the digenean discovery of 'clonal selection' (a comprehensive term for 'clone mixing' and 'clonal diversity') as a strategy to recover from inbreeding depression encountered due to propagatory multiplications in sporocysts/rediae and to select the fittest clone for onward transmission. For, only the fittest shall be able to establish and colonize SIH or DH. It is in this context, the seminal publication by Rauch et al. (2005) opens a new avenue for research in this vitally important aspect.

The eye fluke *Diplostomum pseudopathaceum* engages the freshwater snail *Lymnaea stagnalis* (life span: 2.5 years) and *Gasterosteus aculeatus* (life span: 4 years) as FIH and SIH; the snail emits between 7,000 and 37,000 cercariae/d for a period of two months (Rauch et al., 2005). The marine fluke *Maritrema novaezealandensis* uses the mud snail *Zeacumantus subcarinatus* and mud crab *Macrophthalus hirtipes* as FIH and SIH. Keeney et al. (2007) reported the occurrence of sexual reproduction in the fluke within the host intestine. The number of *D. pseudopathaceum* clones is 4.4/host in the FIH *Lymnaea stagnalis* and 21.9/host in the SIH *G. aculeatus*. Clearly, clonal diversity is increased more effectively in SIH (Rauch et al., 2005). Grazing on sediments, the mud snail ingests and accumulates eggs of *M. novaezealandensis*. With increasing size, the clone number of fluke is increased from ~ 2 in a small mud snail (8 mm) to 5 in a larger (13 mm) snail. The SIH, the mud crab, however, harbors as many as 24.2 clones/crab, irrespective of differences in size from 14 to 22 mm. Reporting the occurrence of clonal mixing from FIH to SIH, Rauch et al. (2005) found that of 44 different clones present in 218 SIH, 65% is shared by a single clone, 20% by another and the remaining by others. Clearly, the clones are not only mixed but also the best ones are selected to dominate the clonal population in SIH.

Notably, no information is available on clonal mixing in terrestrial SIH. In fact, Criscione and Blouin (2006) predicted that digeneans primarily employing aquatic transmission routes and multiple hosts may achieve more clonal selection prior to infection with DH than trematodes engaging fewer hosts in terrestrial transmission routes. Briefly, the incorporation of SIH in aquatic flukes has not only extended the life span of the infective larval stage but also clonal selection. This is an area urgently requiring more information.

## 4.5 Amphiparatenic Transmission

Paratenics are accidental hosts. With euryxenicity of digeneans, the incidence of paratenics is ubiquitous. Perhaps, Shoop (1988) was the first to describe

*Digenea* 155

amphiparatenics. In digeneans and perhaps among all the helminthic parasites, the strigeid *Alaria marcianae* stands as an amazing example for the magnitude of flexibility in incorporation of the largest number of paratenic hosts, retention of multiplication potency and to remain ever young (Shoop, 1988). The flowchart for the life cycle of *A. marcianae* proceeds from miracidium → sporocyst → cercariae → mesocercaria → metacercaria and involves the snail as FIH, tadpole as SIH and a felid or canid as a definitive host. However, amphibians, reptilians, avians and rodents may also serve as paratenic SIH. In a tadpole, the fluke continues to remain as mesocercaria, even when ingested by a snake and in turn, by an alligator. Following death, the infected alligator is scavenged by rodents and possibly avians, and it goes on *ad infinitum*. Despite all these transmissions through paratenic hosts, the mesocercaria remains as mesocercaria. In mammals, the somatic exploration of *A. marcianae* has landed it on the mammary glands. Utilizing milk as a medium, the mesocercaria has explored another generation and a novel route of transmammary transmission. Repeated experiments have revealed 100% infection in first litters born to infected rodents or carnivorous females (Shoop and Cockrum, 1983). Shoop (1988) named it as amphiparatenic transmission. Interestingly, the transmammary transmission is also reported for the trematode *Pharyngostomoids procyonis* (Harris et al., 1967), nematodes *Ancylostoma caninum*, *Neoascaris vitulorum*, *Strongyloides* spp, *Toxacara canis* and *Uncinania lucasi* (see Shoop, 1988).

## 4.6 Polyembryony and/or Parthenogenesis

The number of cercariae arising from a single miracidium varies widely from as few as 150–200 up to a million in strigeids (Hyman, 1951). Not surprisingly, the propagatory multiplication in digenean larval forms, more commonly but wrongly named as 'asexual reproduction', has remained the subject of ongoing debate since 1934 (Ishii, 1934), one group arguing in favor of polyembryony (e.g. Rees, 1940, Esch et al., 2002) and the other for parthenogenesis (Galaktionov and Dobrovolskij, 2003). Hence, it is a prerequisite to define the terms polyembryony and parthenogenesis. Polyembryony refers to the splitting of a sexually produced zygote or an embryo into many in the course of development. The definition and types of parthenogenesis have been described on p 64–65.

According to Rees (1940), a lineage of germinal or totipotent cells is established in the miracidium of the philopthalmid *Parorchis acanthus*, as a result of an asymmetric first cleavage of the zygote into one propagatory or germinal blastomere and the other as ectodermal or somatic blastomere. Subsequently, the germinal line is distinctly maintained from the somatic line through redia I and redia II. Typical of the totipotent cell, the germinal cell is characterized by a large nucleus and thin layer of cytoplasm.

156  *Reproduction and Development in Platyhelminthes*

Incidentally, these cells are histochemically identified by their great affinity for a hematoxylin stain (e.g. Ciordia-Davila, 1956). Each miracidium contains a dozen or more germinal cells and each of them generate > 12 redia I and in their turn, each redia I can generate ~ 20 rediae II. Incidentally, the life cycle of *P. acanthus* includes miracidium, redia and cercaria. Pearson (1972) considered that redia-like flatworm, which recapitulates more fully the digenean life cycle, may have been the ancestor. In *Echinostoma*, proliferating germinal cells are reported to be present as two morphologically distinguishable germinal cell subpopulations namely *nanos-2*+ and *nanos-2*- (Galaktionov and Dobrovolksij, 2003). With smaller cell size, the *nanos-2*+ is speculated to be the most undifferentiated cell type, whereas *nanos-2*- with restricted potency enters embryogenesis directly, indicating totipotency of *nanos-2*+ but the *nanos-2*- primed toward somatic fate (Rinaldi et al., 2012). If it is correct, it is not known how *nanos-2*- primed to become embryogenic revives *nanos-2*+ in daughter sporocyst and redia. Confirming the existence of two distinct germinal cell lineages in digenean *Schistosoma mansoni*, Wang et al. (2013) showed that the lineages differ in their proliferative kinetics and expression of a *nanos* ortholog. A *vasa/PL10* homolog is required for proliferation and maintenance of both the lineages. Serial transplantations of sporocyst into a naïve snail host, and continuation of sporocyst propagation and cercarial production is also cited as evidence in support of the totipotency of *nanos-2*+ germinal cells (Jourdane and Theron, 1980). However, it must be noted that all these evidences go to show the totipotency of the *nanos-2*+ germinal cells. But the asymmetrically dividing totipotent germinal cell producing a totipotent germinal cell and another somatic cell may not strictly be considered as an equivalent to splitting of a sexually produced zygote or an embryo.

From those in favor of parthenogenic multiplication, evidence is demanded to show the 'ovic' nature of the germ cell undergoing mitosis instead of meiosis (Clark, 1974). A computer search has revealed that as of now, convincing and conclusive evidence is not available. But there are some circumstantial evidences, which may be considered under three categories:

1. *Polar body*: the presence of smaller darkly stained cells in the germinal balls around developing embryos—interpreted as polar bodies—germinal cells undergoing 'meiosis' was first reported by Tennant (1906). The description of polar bodies in the redial and cercarial embryos of *Fasciola hepatica* was reported later (Bednarz, 1962). From their studies of germinal cell development in *Philophthalmus megalurus*, Khalil and Cable (1968) also arrived at a similar conclusion. Using elegant autoradiographic and DNA staining methods in the cell cycle of cercarial germinal balls (not from sporocyst, redia) of *Trichobilharzia ocellata*, Haight et al. (1977) reported that the claimed polar bodies or blebs possess a 2C DNA and may represent simple mitosis.

2. *Meiosis*: Describing an incomplete meiosis in each germinal sac generation of *P. megalurus* embryos, Khalil and Cable (1968) observed the pairing of highly contracted chromosomes to form the bivalents typical of

*Digenea* 157

the diakinesis stage in prophase but no reduction division and polar body formation. Understandably, meiosis does not proceed further and results in the formation of diploid eggs with 2C DNA, i.e. the occurrence of apomictic parthenogenesis. However, it must be noted that the observation of Khalil and Cable has not been confirmed in other digeneans.

3. *Adult parthenogenesis*: In almost all sexually reproducing digeneans, spermatogenesis generates 32 sperm and oogenesis results in the production of a single egg (e.g. *Halipegus eccentricus*, Guilford, 1961). Another form of parthenogenesis is generative (haploid) parthenogenesis, in which oogenetic reduction division occurs and an unfertilized haploid egg develops into a haploid male. This may not be immediately relevant to parthenogenic multiplication in larval forms but may hint at the parthenogenic ability of gonochoric digeneans. (3a) The rat schistosome *Schistosomatium douthitti* is a female heterogametic XY, rather ZW; in it, males can be either Z or ZZ. In unisexual infections, a mixture of haploid Z and diploid ZZ males as well as diploid ZW females is produced. Evidently, it produces generative parthenogenic diploid females as well as males, of which 95% are haploid males and 5% diploid females. In ZZ males, the haploid egg is diploidized by fusion of the nuclei from a haploid Z egg and Z from a polar body. On exposure of 1,000 naïve snails to 2,714 parthenogenic miracidia, 7 of 16 infections are by ZW females, 6 by ZZ males and 3 by Z males. Apparently, the haploid males are less infective (Short and Menzel, 1959). (3b) In Japan, diploid and triploid forms of *Fasciola hepatica* are considered as parthenogenic (Sakaguchi, 1980). The prevalence of triploid lung fluke *Paragonimus westermani* is widespread in Japan (Terasaki, 1980). Acquisition of polyploidy is often accompanied by better fitness. However, a comparative study of meiosis/mitosis in the diploid, triploid sporocyst is urgently required. (3c) Taylor et al. (1969) reported that a cross between *Schistosoma mattheei* female and *S. mansoni* male produced a large number of eggs, only a fraction of which are viable. Similar hybridization experiments led Basch and Basch (1984) to conclude that both diploid and haploid *S. mansoni* females are capable either of parthenogenic or bisexual reproduction, when appropriately paired. On pairing *S. japonicum* female with *S. mansoni* male, viable eggs were produced but no sperm; their cercariae developed into all females. The hybrids of the reverse cross produce sperm but only a few viable eggs; their cercariae develop into all males (Imbert-Establet et al., 1994). Evidently, females of *S. japonicum* are capable of parthenogenic reproduction. Conversely, small eggs produced by unisexually infected *S. mattheei* (Taylor et al., 1969), *Bilharziella polonica* (Brumpt, 1936) and *Heterobilharzia americana* (Armstrong, 1965) are not viable. Apparently, no diploid parthenogenic is produced in the unisexually infected schistosomes. (3d) Hybridization between the African *S. haematobium* and *S. mansoni* failed to produce parthenogenic eggs (Huyse et al. 2013).

Briefly, direct evidence for the existence of polyembryonic germ cell lineage from miracidium to rediae I and II is provided, *albeit* limited to histochemical and cytological evidence. 4. Circumstantial evidence for the 'ovic' nature of

158 *Reproduction and Development in Platyhelminthes*

the germ cells is provided by the presence of blebs interpreted as diploid polar body and incomplete meiosis within the sporocyst. However, these observations have not so far been confirmed in any other molluscs. It is difficult to comprehend incomplete meiosis as an evidence for ameiotic/apomictic parthenogenesis, in which diploid eggs are produced without production of the polar body. Therefore, 'more widespread and compelling evidence' (Whitfield and Evans, 1983) using modern molecular technique is not thus far available to confirm the occurrence of either apomictic parthenogenesis or polyembryony in the entire sequence of larval forms. Briefly, this debatable aspect continues to remain a virgin area for research. Incidentally, clonal multiplication has been convincingly demonstrated in cestode cysticerci (p 222) and by serial transplantation of sporocysts to naïve snail hosts (p 143). The transplantation of a single germinal cell from sporocyst/redial I to a naïve host may confirm clonal reproduction in digenean larvae. It is tempting to state that the propagatory multiplications in digenean larval forms are all clonal rather than parthenogenic or polyembryonic.

## 4.7 Sequence and Potency

Typically, the sequence of digeneans life cycle involves an alternation of generation from sexual oviparous reproduction in a definitive host to propagatory multiplicative or propagative viviparous reproduction in the first and second intermediate host, respectively. Among these larval forms, miracidium and cercaria are ubiquitous among digeneans. To compensate the risks encountered in transmission from one host to the next, propagatory multiplication occurs in sporocyst and/or redia and to further it, multiplication can also occur in daughter sporocyst and redia as well as metacercaria. The sporocyst undergoes a minimum of one (e.g. some heterophids, Hyman, 1951) to hundreds (e.g. *Maritrema novaezealandensis*, Keeney et al., 2007) of generations. Redia usually undergoes two generations (e.g. opisthorchids, Hyman, 1951) but it has the potency to undergo 27–40 generations (Donges, 1971). In gymnophallids like *Cercaria margaritensis* and *Parvatrema margaritense*, metacercaria undergoes two generations of multiplication; similar multiplication may also occur in mesocercaria of *Alaria marcianae* (Shoop, 1988). Hence, the potency for propagatory multiplication is retained in almost all larval forms from miracidium to metacercaria (Fig. 4.4). Remarkably, the gymnophallids engage bivalves and gastropods as the first and second intermediate host, respectively. It is not clear, whether an unknown substance present in the gastropod induces propagatory multiplication.

To reduce the risks in transmission, one or another larval form may be skipped; for instance, sporocyst is skipped by cyclocoelids (e.g. *Typlocoelum cybium*) or redia in many species (e.g. *Strigea subquetrus*). Interestingly,

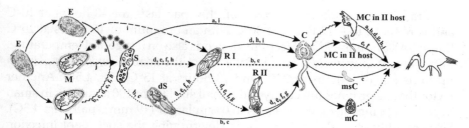

### FIGURE 4.4

The complicated life cycle of trematodes. E = egg, M = miracidium, S = sporocyst, dS = daughter sporocyst, R I = mother redia, R II = daughter redia, C = cercaria, mC = metacercaria, msC = mesocercaria. Direction of life cycle is shown by a straight midline with the arrow head. Wavy lines indicate free-swimming penetrative larval stages, dotted lines the occurrence of propagatory multiplication, wavy line mixed with straight line the trophic ingestion of free-swimming miracidium and discontinuous line the skipping of sporocyst stage. The line marked by • • with arrow head indicates the reverse direction of sporocyst developing into miracidium. Examples for the sequence, addition and reversion of larval forms are stated below:

1. M → S → R → C in a typical digenean life cycle, 2. M → S → R → M → mC *Gorgocephalus yaaji*, Huston et al. (2016), 3. M → S → dS → C → mC *Cercaria margaritensis*, Irwin et al. (2003), 4. M → S → C → msC → mC *Strigea subtriquetrus* (see Shoop, 1988), 5. M → S → R I → R II → C *Fasciola gigantica*, Phalee et al. (2015), 6. M → S → C → mC *Maritrema novalzealandensis*, Keeney et al. (2006), 7. M → S → dS → C *Rhipidocotyle papillosum*, Ciordia-Davila (1956), 8. M → R I → R II → C *Parorchis acanthus* (Rees, 1940), 9. M → S → M → C *Mesostephanus haliasturis*, Barkar and Cribb (1993). For more examples see James and Bowers (1967)

Examples at genus level: a = *Brachylaima*, b = *Diplostomum*, c = *Alaria*, d = *Echinostoma*, e = *Fasciolopsis*, f = *Paramphistomum*, g = *Nanophyetus*, h = *Heterophyes*, i = *Plaigiorchis* (from Toledo et al., 2006), j = *Mesostephanus haliasturis*, k = *Cercaria margaritensis*.

the sequence can also be reverted, from sporocyst to miracidium, as in *Mesostephanus haliasturis*. Briefly, the potency for reversion is also retained. As the digenean life cycle is highly complicated and an attempt has been made to depict it in Fig. 4.4.

## 4.8 The Transmission

In the digenean life cycle, the short-lived, free-swimming, non-feeding miracidium and cercaria are the most critical larval stages and involve the riskiest transmission from a definitive host to the first intermediate host and from the first intermediate host to the second intermediate/definitive host, respectively. In the miracidium, a limited quantum of incorporated glycogen serves as the main energy source to sustain survival and locomotion (Bryant and Williams, 1962). With progressive depletion of the energy source as age advances and rapid depletion due to elevated activity at higher temperatures, the life span of miracidium is restricted to a few hours from 24 to 36 hours (Esch et al., 2002). Limited information is available on the life span

160  *Reproduction and Development in Platyhelminthes*

of miracidium. In *Schistosoma mansoni*, the span lasts for 40 hours at 10°C and is reduced to ~ 12 hours at 30°C with an optimum of 16 hours at 15°C (Fig. 4.5A). The miracidial infectivity depends also on its age and temperature. Fifty percent of miracidia successfully infect the snail *Biomphalaria glabrata* up to its age of 9 hours at 20°C and 10 hours at 15°C (Fig. 4.5B). Another factor that determines infectivity is the snail size. From 100% success in small *B. glabrata* (1 mm size), it decreases to 0% in larger (19 mm) snail (Fig. 4.5C). Snail motility may also be a factor in determining the success of infection (Mouristen and Jensen, 1994, Lowenberger and Rau, 1994).

Miracidial host finding is an old topic of experimental interest in behavior parasitology. Despite availability of abundant literature, enough is not known to control or eliminate miracidial infection, especially the blood and liver flukes. The inverted conical body of miracidium is covered by ciliated epithelial cells, which provide a propulsive force for locomotion. Beneath the epithelial cells is a network of circular and longitudinal muscle fibers, which determine the body shape. Turning of miracidial body is effected by differential ciliary activity and bending of movement of its body. Hence, the miracidium can turn its body at an angle of its ability. However, the ability to turn progressively declines with decreasing temperature and it can only rotate its body slowly about the longitudinal axis instead of turning direction at temperatures below 12°C (Wilson and Denison, 1956). This may be a reason for the absence of *Schistosoma* in the European and North American waters.

In *Echinostoma ilocanum* miracidium, incubation period lasts for 18–30 days (Mehlhorn, 2016). Firstly, the miracidium is photopositive and geonegative. For example, it's swimming speed decreases from 2.4 mm/s at 100% illumination to 1.9 mm/s at 41% illumination. A radioisotope assay system using $^{75}$S-methionine labeled miracidia of *S. mansoni* has shown that the miracidium is geotactic and phototactic, and gravitates to the edges of shallow water bodies (see Christensen, 1980). Secondly, it is chemically sensitive to the snails and its response is chemoklinokinenic, i.e. it departs gradient dependently in a straight path but then begins to turn more and more frequently by taking a sharper angle of turning, as it approaches a host (Saladin, 1979). Hence, its first priority is to disperse from the site of origin. For example, the newly hatched miracidium of *S. mansoni* swims fast for ~ 1 hour and then slows down to a constant swimming level for up to the 5th hour and as its speed drops and angular speed increases (Mason and Fripp, 1976).

The swimming speed of *Fasciola hepatica* miracidium is determined by its age (Fig. 4.5D), temperature (Fig. 4.5E, see also Christensen and Nansen, 1976) and partial pressure of dissolved oxygen ($pO_2$) (Fig. 4.5F). It progressively decreases from 1.4 mm/s (37° angle of turning) in 0.4-hour old miracidium to 0.7 mm/s in 7.4-hour old one. But it progressively increases from 0.6 mm/s (32°) at 4.6°C to 1.4 mm/s (77°) at 20°C, and from 0.5 mm/s (108°) at 0.8 mm Hg to 1.10 mm/s (75°) at 159 mm Hg (Wilson and Denison, 1956). Notable is the increase in turning angles with advancing age and temperature elevation.

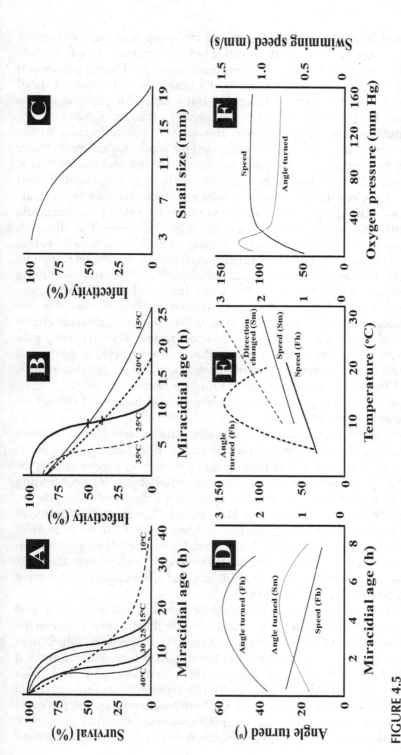

**FIGURE 4.5**

(A) Effect of temperature on survival of *Schistosoma mansoni* miracidium and (B) on infectivity of *Biomphalaria glabrata* miracidium at different temperatures, (C) Effect of *B. glabrata* size on miracidial infectivity (compiled and simplified from Anderson et al., 1982). Effect of (D) age, (E) temperature and (F) partial pressure of oxygen on miracidial swimming speed and angle of turning (direction change in E alone) in *Fasciola hepatica* (Fh) and *S. mansoni* (Sm) (compiled from Wilson and Denison, 1956, Mason and Fripp, 1976).

162 *Reproduction and Development in Platyhelminthes*

The trend and level of responses for swimming speed and angle of turning as function of age and temperature differ between miracidia of *F. hepatica* and *S. mansoni* are (Fig. 4.5D, E). The optimum for angle of turning is around 5 hours for both *F. hepatica* and *S. mansoni*. But the levels for swimming speed is faster for *S. mansoni* than for *F. hepatica* but temperature optima for the maximum angle of turning are higher for *F. hepatica* than for *S. mansoni*.

Several publications have indicated the attraction of miracidia in the presence of amino acids in water conditioned by snails; for example, water conditioned by *Lymnaea palustris* attracts *Schistosomatium douthitti* miracidia. Nineteen amino acids present in the Snail (*Biomphalaria glabrata*) Conditioned Water (SCW) are reported to attract *S. mansoni* miracidia (MacInnis et al., 1974). The swimming speed and direction changing rate of the miracidia increase with increasing concentration of SCW. However, the direction changing rate begins to decrease, when the age of SCW is increased beyond 30 days. Hass et al. (1995) demonstrated that (i) only *S. japonicum* miracidia display a characteristic orientation and are capable of directed swimming along chemical gradient toward the snail host and (ii) the miracidia of *S. haematobium* and *S. mansoni* are attracted by macromolecules like glycoconjugates with molecular mass of > 30 kDa, whose saccharide chains may be linked via serine and N-acetylgalactosoamine. The glycoconjugate is sensitive to lysozyme, which suggests that muramic acid, as gastropod specific component, is involved in the recognition process. Earlier reports have hinted that the specificity of miracidium to chemosensitivity is rather low, a reason that may explain why many lymnaeid or gastropod species are employable as the first intermediate host.

A large number of experimental studies elucidate the role played by other environmental factors influencing the miracidial host finding. In what is called the decoy effect, the host-finding and penetration processes may be interfered by the presence of others. The miracidial ability for host finding is reduced by a number of factors like increasing volume of water, depth, velocity and decreasing snail density. For example, the maximum depth, at which the miracidium can search for the host, is 9.1 and 5.1 m for *S. mansoni* and *S. haematobium*, respectively (Christensen, 1980). The pH range between 5.4 and 8.4 is favorable for miracidial host searching (Christensen et al., 1978). Using decoy and other mechanisms, molluscs and other organisms interfere with miracidial host finding (Table 4.5).

Relatively more information is available for cercarial emergence, survival and infectivity. Yet, only a few data are available for the duration between the entry of miracidium into the first host and cercarial emergence, which may be named as 'incubation period' in first intermediate host. For example, it lasts for ~ 48 days for *Azygia longa* (Sillman, 1962), 55 days at 20°C for *Fasciola hepatica* and *F. gigantica* (Abrous et al., 1999), 120 days for *Trichobilharzia ocellata* (Sluiters et al., 1980) and 245 days for *Diploproctodaeum arothroni*; the exact durations for the development of mother sporocyst, daughter sporocyst, redia I, redia II and cercaria are 7–14 days, 21–35 days, 42–56 days, 63–77 days

## TABLE 4.5

Molluscan species and others interfering with miracidial host finding (compiled from Christensen, 1980, Christensen et al., 1980)

| Trematode species | Interfering species |
|---|---|
| | **Decoy mechanism of interference** |
| *Fasciola hepatica* *Schistosoma mansoni* | *Lymnaea palustris, L. peregra, L. stagnalis* *Biomphalaria glabrata, Drepanotrema surinamensis, Physa marmorata, Pomacea glaucus, P. australis, T. granifera,* tadpoles of *Bufo marinus, Phyllomedus* sp |
| | **Filter feeding mechanism of interference** |
| *F. hepatica* *S. mansoni* | *Bithynia tentaculata, Sphaerium corneum, Daphnia pulex,* larvae of *Corethara* sp Larvae of *Culex pipens,* and *Aedes aegypti* |
| | **Predatory mechanism of interference** |
| *F. gigantica* *S. mansoni* | *Chaetogaster limnaei* *C. limnaei, Athya innocuous, Cyclops* sp*, Lebistes reticulates, Utricularia* sp |
| | **Toxic mechanism of interference** |
| *F. hepatica* *S. mansoni* | *Planaria* sp *Dugesia tigrina, Dendrocoelum lacteum, Planaria* sp 4 species |

and 77–93 days, respectively (Hassanine, 2006). In *Schistosomatium douthitti*, the minimum 'incubation period' lasts for 70 days; as mother and daughter sporocysts continue to produce daughter sporocysts and cercariae for > 150 days each, the longest period may last for 624 days (Loker, 1978). Strikingly, the minimum and maximum periods in *Posthodiplostomum cuticola* can be as short as 70 days or as long as 4.6 years (Donges, 1971). Clearly, the 'incubation period' is a highly flexible trait and depends upon the availability of an appropriate host in time and space. Hyman (1951) has also indicated that the period is variable depending on external factors, especially temperature, but generally lasts for some weeks and even months.

Cercarial emergence usually occurs during the night (e.g. *S. douthitti*, see Mehlhorn, 2016) and ~ 80% of the total cercariae emerge within 3 hours. In terms of number and pattern, it is temperature-dependent in *Plagiorchis elegans* (Fig. 4.6A). It is significantly higher at 20°C than at 15°C and 25°C. Surprisingly, the emergence window of *Himasthla quissetensis* from its FIH *Nassarius reticulatus* perfectly coincides with availability of SIH *Cerastoderma edule* (De Montaudouin et al., 2015). Cercarial survival decreases as a function of age. Notably, the decreasing trend is accelerated by decreasing relative humidity in terrestrial brachylaimid *Postharmostomum helicis* (Fig. 4.6D), temperature in *Echinostoma liei* (Fig. 4.6E) and increasing cadmium concentration in *Diplostomum spathaceum* (Fig. 4.6F). In fact, the reduction in life span with temperature elevation is a common phenomenon and has been demonstrated in many species, for example, *Schistosoma mansoni*

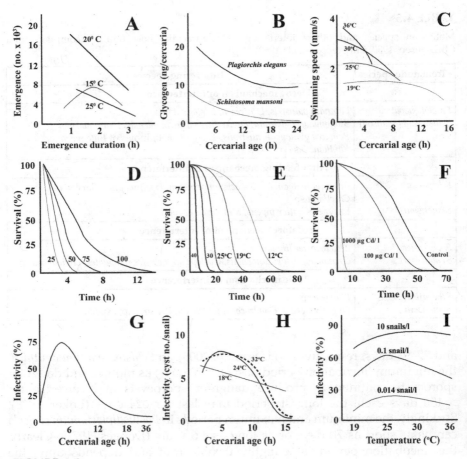

**FIGURE 4.6**

Emergence of *Plagiorchis elegans* cercariae in the first 3 hours following the reduction in light intensity (redrawn from Lowenberger and Rau, 1994). (B) Glycogen content of cercariae as function of age in *P. elegans* and *Schistosoma mansoni* (compiled from Lowenberger and Rau, 1994, Lawson and Wilson, 1980). (C) Swimming speed of *Echinostoma caproni* cercaria as functions of age and temperature (compiled from Meyrowitsch et al., 1991). Effect of (D) relative humidity (%) at 22°C, (E) temperature and (F) cadmium on survival of cercariae of *Postharmostomum helicis*, *Echinostoma liei* and *Diplostomum spathaceum*, respectively (simplified and redrawn from Barger, 2012, Evans, 1985, Morley et al., 2003). Effects of age on cercarial infectivity (G) in *Plagiorchis elegans* (redrawn from Lowenberger and Rau, 1994), (H) age and temperature in *Echinostoma trivolvis* (compiled from Pechenik and Fried, 1995) and (I) temperature and snail density in *Echinostoma caproni* (redrawn from Meyrowitsch et al., 1991).

(Lawson and Wilson, 1980), *Echinostoma caproni* (Meyrowitsch et al., 1991) and *E. trivolvis* (Pechenik and Fried, 1995). A major reason for the reduction in life span is a glycogen reserve, which serves as the main energy source, as in miracidium. Though the declining trend for the glycogen content of a cercaria with advancing age is the same for digenean species, the level of

Digenea 165

decline varies; for instance, it is from ~ 20 to 10 ng in *P. elegans* but from 8.5 to 0.5 ng in *S. mansoni* (Fig. 4.6B). Remarkably, advancing age and temperature elevation decrease the swimming speed of *E. caproni* cercaria. At faster swimming speed of ~ 3.8 mm/s at 36°C in *E. caproni*, glycogen is exhausted at the cercarial age of 6 days but it lasts for as long as 19 days at 19°C, at which the speed is reduced to less than half of that at 36°C (Fig. 4.6C). Notably, the speed of both miracidia and cercariae is almost equal at ~ 2 mm/s, irrespective of propelling cilia in the miracidium or the tail in cercaria. The duration for 50% cercarial infectivity lasts up to 10 days in *P. elegans* and infectivity remains highest from 24–32°C in *E. trivolvis* (Fig. 4.6G, H). A third factor that determines the successful infection is host density. At optimal 25°C, the infectivity of *E. caproni* cercaria decreases from ~ 80% with availability of 10 snails/l to ~ 25% at 0.014 snails/l (Fig. 4.6I). A fourth factor is the number of miracidial entry into a single intermediate host. For example, the *Tribilharzia ocellata* miracidial dose of one and four into first intermediate host *Lymnaea stagnalis* results in the production of 577 and 509 cercaria/day for a period of 120 days, respectively (Sluiters et al., 1980, cf Fig. 4.12F). Experimental exposure at different numbers (625, 1127) of *Dicrocoelum dentrictum* cercariae leads to the increased egg production from 1,043/day at lower dose to 2,256/day at higher dose in sheep, and from 1,043/day to 1,987/day in cattle (Beck et al. 2015). The fifth factor is the susceptibility of the host. *Echinostoma* infection occurs in the second intermediate anuran host during the larval development from pronephric (tadpole stage 25–27) to mesonephric (stage 32–33) and metanephric (stage 37–39) stages. Only a fewer cercariae successfully encyst in the later tadpole stage than in early tadpole stages. At the cercarial exposure dose of 50, susceptibility decreases from ~ 29% at pronephric stage to 0% at the mesonephric stage indicating immunity development. With increasing cercarial dose from 20 to 100, infectivity, however, increases from 14 to 42% and 6 to 32% at the pronephric and mesonephric stage, respectively (Schotthoefer et al., 2003). The sixth factor can be dual infection. Temperature optima for miracidial infection differ from one parasite species to other. Interestingly, the dual infection of *Lymnaea truncatula* with *Paramphistomum daubneyi* and *Fasciola hepatica* miracidia results in 71% of the snail producing *F. hepatica* cercariae at 20°C but 62% producing *P. daubneyi* cercariae at 6–8°C. However, 26–27% of the snails produced cercariae of both parasite species (Abrous et al., 1999). As in miracidium, some environmental factors interfere with cercarial host-finding. Among the interfering organisms, some are predators (e.g. cyclops, daphnids); others secrete cercarial toxins (e.g. *Bufo bufo*) (Christensen, 1979, Christensen et al., 1980).

From bits and pieces information reported by Hyman (1951) and others, the penetrative and/or trophic modes of transmission by miracidium and cercaria in 26 digenean families are broadly classified into four major groups (Table 4.6). Group 1: Penetrative miracidium and cercaria occur in three families namely Sanguinicolidae, Spirorchidae and Schistosomatidae.

166  *Reproduction and Development in Platyhelminthes*

## TABLE 4.6

Penetrative or trophic mode of transmission by digenean miracidium and cercaria (compiled from Hyman, 1951, † Gonchar and Galaktionov, 2017, * Blackwelder and Shepherd, 1981, ** Madhavi, 1980, †† Ubelaker and Olsen, 1972).

| |
|---|
| **1. a. Penetrative miracidium and cercaria** |
| Sanguinicolidae, Spirorchidae, Schistosomatidae (1 intermediary host only) |
| **1. b. Penetrative miracidium and cercaria + trophic meso-metacercaria** |
| Strigeidae (2–3 hosts) |
| **2. Penetrative miracidium and trophic cercaria** |
| Paramphistomidae, Clinostomidae, Fasciolidae (1 host)<br>Bucephalidae, Zoogonidae, Cyclocoelidae, Echinostomatidae (2 hosts)<br>Allocreadiidae**, Troglotrematidae, Notocotylidae† |
| **3. Trophic miracidium and penetrative cercaria** |
| Opisthorchidae |
| **4. a. Trophic egg/cercaria*, trophic miracidium and metacercaria in cercarial cyst** |
| Heronimidae*, Dicrocoeliidae (1 host) |
| **4. b. Trophic miracidium and encysted cercaria* or metacercaria**** |
| Didymozoonidae*, Microphallidae, Brachylaimidae (1 host only)*<br>Azygiidae, Hemiuridae, Gorgoderidae ††, Heterophyidae, Plagiorchidae** (2 hosts) |

Group 2: In 10 + 1 (Strigeidae) families, penetrative miracidium is followed by trophic cercaria/metacercaria. Group 3: Opisthorchiidae seems to be the only family, in which trophic miracidium and penetrative cercaria are present. Group 4: However, as many as 10 digenean families include trophic miracidium and trophic cercaria/metacercaria. Hence, a majority (15 of 26) of digenean families opt for at least one penetrative either miracidium or cercaria. At generic and species levels, there are many variations. For example, the hatched miracidium of *Schistosomatium douthitti*, following ingestion of its egg, penetrates internally in *Lymnaea catascopium* prior to its arrival in the development site. In many digenean taxa like Brachylaimoidea (e.g. *Leucochloridium*, Hyman, 1951), Hemiuroidea, Opisthorchioidea and Ponocephaloidea (Cribb et al., 2003), the molluscan host eats the eggs and the hatched miracidium then penetrates internally (Fig. 4.4). Hence, this grouping may be arbitrary and may not hold but it paves the way to identify routes and modes of infecting the definitive host.

Cribb et al. (2003) recognized seven routes and modes through which cercaria (C) is transmitted; however, this account recognizes nine routes and modes (Fig. 4.7): (1) C remains in FIH and awaits trophic ingestion by DH (e.g. Heronimidae, Cyclocoelidae, Table 4.6), (2) C directly attaches to the surface of DH (e.g. Transversotrematidae), (3) C is directly eaten by DH (e.g. in some bivesiculids, azygiids), (4) Following emergence, the encysted C in open as metacercaria (MC) awaits trophic ingestion by DH (e.g. Paramphistomatidae,

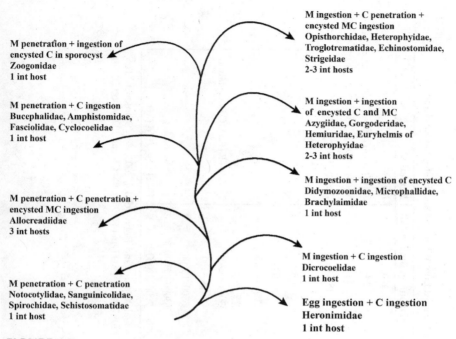

**FIGURE 4.7**

Transmission modes in some digenean families. Transmission modes through ingestion are shown on the right side and those involving penetration + ingestion, or penetration + penetration are shown on the left side (information from Hyman, 1951, Anderson et al., 1965, Madhavi, 1978) M = miracidium, C = cercaria, MC = metacercaria, int = intermediate.

Haplosplanchnoidea, some echinostomatoids). (5) C is ingested by SIH but penetrates into the host gut and as MC awaits trophic transmission to DH (e.g. Hemiuroidea, some gorgoderoids, azygiids, bivesiculids), (6) A most common transmission process among the digeneans involves the cercarial penetration into SIH and MC awaits trophic transmission to DH (e.g. Diplostomidae, Bucephaloidea, Ophistorchoidea, Echinostomatoidea, Leprocreadioidea) and (7) C directly penetrates the DH (e.g. Schistosomatoidea, Spirorchiidae, Sanguinicolidae). Known combinations of trophic, trophic cum penetrative, penetrative cum trophic and penetrative transmission modes for both miracidium and cercaria are depicted in Fig. 4.7 (see also Table 4.6).

Some representative cercarial output rates are listed in Table 4.7. The number ranges from 1–3 cercariae/host/day in *Azygia longa* to 17,000 cercariae/host/day in *Cotylurus flabelliformis*. A number of compensatory measures are adduced to explain the wide variations in cercarial production: (i) The cercarial production rate is as low as 14/f (fluke)/day in *Fasciola hepatica* but as high as 7,000–37,000/f/d in *Diplostomum pseudopathaceum*; however, the fecundity is 25,000 eggs/f/day and 10–21 eggs/f/d in the former and latter, respectively. (ii) The low fecundity and 23% non-viable

**TABLE 4.7**

Cercarial outputs of selected digeneans and their respective hosts (condensed from Whitefield and Evans, 1983 and added)

| Parasite species | Host species | Cercarial output (no./host/d) | Remarks |
|---|---|---|---|
| *Azygia longa* | *Amnicola limosa* | 1–3 | R + C |
| *Fasciola hepatica* | *Lymnaea truncatula* | 14 | 25000 eggs/f/d |
| *Asymphylodora dollfusi* | *Bithynia tentaculata* | 15 | M+S+R+DR+C+MC |
| *Trasversotrema patialense* | *Melanoides tuberculata* | 18–24 | 2 eggs/f/d |
| *Notocotylus attenuatus*\*\* | *Lymnaea peregra* | 30 | M+MS+DS+C |
| *Echinoparyphium recurvatum*\* | *L. peregra* | 390–560 | |
| *Trichobilharzia ocellata* | *L. stagnalis* | > 500 | |
| *Himasthla quissetensis* | *Nassarius obsoletus* | 1000–2700 | |
| *Biomphalaria alexandria*β | *Schistosoma mansoni* | 958 | |
| *Cryptocotyle lingua*☼ | *Littorina littorea* | 3300 or 830†† | M+MS+R+C+MC Encystation |
| *Plagiorchis micracanthos* | *Stagnicola exilis* | 3800–5400 | |
| *Diplostomum flexicaudum*† | *S. emerginata* | 5500–10000 | |
| *Cotylurus flabelliformis*\*ω | *L. stagnalis* | 13000–17000 | |
| *Diplostomum pseudopathaceum*† | *L. stagnalis* | 7000–37000 | 10–21 eggs/f/d (M) |

† = *D. spathaceum, D. parviventosum* and *D. volvens*, \* = natural, \*\* = Galaktionov and Dobrovolskij, 2003, ☼ = Stunkard, 2005, ω = Kocur, 2011, †† = @ 830 cercariae/f/d amounting to 1 million/y for 7 years, see Hyman (1951), β = Chu and Dawood (1970).

egg production in *Transversotrema patialense* are compensated with broad spectrum of eight DH fish species with infection success up to 73% for the characinid *Cheirodon* sp and short life span of 6–12 weeks (Whitfield et al., 1986). In *Trichobilharzia ocellata*, the low cercarial production (> 500/f/d) is compensated by a longer period of 120 days emittence. Cercarial output is altered by (i) the number of miracidial infection in *Lymnaea stagnalis* infected with *Trichobilharzia ocellata* (Sluiters et al., 1980), (ii) quantity and quality of food offered to the snail *Helisoma anceps* infected with *Halipegus occidualis* (Keas and Esch, 1997) and (iii) with increasing body size of *Biomphalaria glabrata* infected with *Schistosoma mansoni* (Sturrock, 1966).

Transmission efficiency refers to the fraction of successful infection by cercaria/metacercaria considering the number of eggs produced as hundred. In a review, Marcogliese (1995) summarized the then available values on transmission efficiency of digeneans facilitated by zooplanktonic second intermediate hosts (Table 4.8). These values range from 0.0005% for *Paronatrema* sp with euphausiids as SIH to 0.3% for *Hemiurus* sp with calanoid copepods serving as SIH. For a freshwater digenean, Dronen (1978) reported the efficiency value of 0.305% for *Haematoloechus coloradensis* with *Physa virgata* serving as FIH in the Colorado River. It is not clear whether the efficiency through oceanic zooplankton is lower than that involving benthic larger dragonfly nymph, as in *H. coloradensis*. More information is required prior to making generalizations. Incidentally, the efficiency from egg to successful miracidial infection is 3.25%; the cercarial infection on SIH, the dragonfly nymph is 1.55%; finally, from an egg to DH, the efficiency is 0.03% for *H. coloradensis* (Dronen, 1978, see also p 34). Still, these low efficiency values suggest the need for truncating the life cycle into a shorter one.

*Attractive strategies*: To evoke a successful infection, cercaria and metacercaria adopt different (1) structural, (2) locational and (3) behavioral adaptive strategies. Some of these strategies are described hereunder: (1a) Most cercariae mimic the food items consumed by the potential hosts like insect larvae, aquatic crustaceans and so on. (1b) Metacercariae of many species eliminate the protective coloration of SIH in such a way as to make it vulnerable to predation by DH (see Esch et al., 2002). (1c) The chaetognaths infected with digeneans are more conspicuous and more susceptible to predation than non-infected individuals (Pearre, 1979). (1d) *Podocotyle stenotometra* metacercaria engages the coral polyps as SIH. It slows down the colonial coral growth. In defence, the inflammated coral becomes bigger and pink; the visibly larger and pink polyp is more often eaten by DH (Lefevre et al., 2008). (1e) The cercarial infection of *Ribeiroia ondatrae* causes severe limb malformation ranging from missing limbs to multiple supernumerary limbs. These malformations may increase susceptibility of infected amphibians to predation by DH (Redmond et al., 2011). (2a) Cercaria of *Paratimonia gobii* develops in the bivalve *Abra* within its autotomic siphon; the siphon floats in water resembling the movements of benthic animals and attracts feeding of

170 *Reproduction and Development in Platyhelminthes*

**TABLE 4.8**

Role of zooplankton as second intermediate host in transmission of digeneans (rearranged from Marcogliese, 1995, data from Dronen, 1978* are added)

| Parasite species | Intermediate host | Transmission (%) | Location |
|---|---|---|---|
| *Paronatrema* sp | Euphausiids | 0.0005 | NW Pacific |
| *Syncoelium* sp | Euphausiids | 0.003 | NW Pacific |
| *Pseudopecoelus japonicus* | *Euphausia similis* | 0.004 | NW Pacific |
| *Derogenes varicus* | *Sagitta setosa* | 0–0.001 | English Channel |
| *Bunodera luciopercae* | *Mesocyclops oithonoides* | 0.005 | Lake Druzno |
| *D. varicus* | *S. elegans* | 0.01–0.06 | White Sea |
| *Brachyphallus crenatus* | *S. elegans* | 0.01–0.16 | White Sea |
| *Hemiurus* sp | Calanoid copepods | 0.04–0.3 | White Sea |
| *Haematoloechus coloradensis** | *Physa virgata* | 0.305 | Colorado River |

the infected siphon harboring metacercaria (Combes, 1995). (2b) *Diplostomum* metacercariae are located in the eye lens of infected fish. They debilitate the vision and cause blindness. Unable to orient, the blind fish is more readily predated by DH. For example, *D. baeri* causes blindness in *Perca flavescens* (Ubels et al., 2018). (3a) The clinostomatid cercariae, known as 'yellow grubs' encyst on fish; the obviously visible large grubs attract the definitive wading avian hosts (see Hyman, 1951). (3b) Another classic case of behavioral strategy is associated with *Dendrocoelum dentricticum*. The parasite irritates the snails causing it to produce and release small sticky balls of slime from its salivary glands. In the slimy balls, the cercariae are bathed in a mini aquatic habitat. These balls are delicacies for ants. In infected ants, some cercariae migrate to the subesophageal ganglia and lock the jaw at declining temperatures. Such ants amidst grasses are readily ingested by DH. (3c) The brachylaimid *Leucochloridium* metacercariae are developed in the germinal sacs within the sporocyst of the snail *Succinea*. Some of these highly branched larger sacs store the encysted metacercaria in the terminal end of enlarged tentacle. These brightly colored pulsating sacs in the tentacle attract the definitive bird host (Niewiadomska and Projmanska, 2011).

## 4.9 'Short is Sweet'

Among platyhelminths, monogeneans complete the life cycle via a single host and are designated as simple life cycle parasites (SLPs). But complex life cycles (CLPs) are hallmarks of digeneans and cestodes (see Parker et al., 2015a, b). The evolutionary transmission from SLP to CLP, i.e. the Upward Incorporation (UI) involves multiple host species often in an ontogenetic sequence to complete the life cycle. Alternatively, some digeneans have adopted the Downward Incorporation (DI) (Auld and Tinsley, 2015). The UI life cycle evolves, when intermediate hosts are more abundant in terms of number of species and their population density as well as greater susceptibility than those for the definitive host, as in many digeneans. As a consequence, the UI in digeneans facilitates longer longevity, larger body size (see Fig. 1.9) and increased fecundity (see Fig. 1.10A, B) and thereby increases the chances of transmission, perhaps at reduced cost. However, each transmission event to a subsequent host poses an obstacle for completion of life cycle (Choisy et al., 2003). An alternative is DI, in which one or more larval forms are truncated to shorten the cycle by progenetic larva in SIH and skipping DH. Typically, the complex digenean cycle has to encounter obstacles at three stages for the cycle to be completed: (i) the eggs/miracidia released from DH must find a suitable host, (ii) the short-lived cercariae emerging from FIH must locate and attach to or penetrate into a suitable SIH, and (iii) the metacercariae must be ingested along with SIH by an appropriate DH. The digeneans have, however, developed adaptive strategies to counteract the risks involved in these transmission events. They are (a) adult high fecundity, (b) propagatory multiplication in all first and rarely SIH, (c) efficient host-finding mechanisms and (d) incorporation of many alternate host species during transmission from FIH to SIH and to DH.

Poulin and Cribb (2002) accomplished an onereous task of collating the incidences of truncation in the digenean life cycle of 47 genera and 31 families. The truncation is achieved through (i) abbreviation of the life cycle by elimination of DH, (ii) inclusion of facultative or obligate progenesis resulting in FIH acting as SIH or SIH serving also as DH. Incidentally, progenesis refers to the precocious development of the functional reproductive system leading to early sexual maturity and egg production in the juvenile stage. The truncation of life cycle may occur with (A) use of FIH as SIH (e.g. *Microphallus*, Deblock, 1980) or (B) progenesis in SIH (e.g. *Orthotrotrema monostomum*, Madhavi and Swarnakumari, 1995) or (C) SIH as DH (e.g. *Opisthoglyphe* Grabda–Kazubska, 1976). Some examples are provided hereunder:

(1) In a 3-host life cycle, the macroderoidid *Alloglossidium* with catfish as DH passes through a mollusc, crustacea as FIH and SIH, respectively. Of 15 macroderoidid species analyzed, 5 species undergo 3-host life cycle. Following a switch from crustaceans to leeches, SIH serves as a definitive host in the truncated 2-host life cycle of *Alloglossidium hamrumi*, *A. hirudicola*, *A. macrobdellensis*, *A. schmidti*, *A. turnbulli* and *Hirudicolotrema richardsoni*. The crustaceans themselves have begun to serve as DH in 2-host life cycle of *A. corti*, *A. greeri*, *A. renale*, *Alloglossoides caridicola* and *Ao. dolandi* (Smythe and Font, 2001).

(2) The truncation occurs also in some micophallids, lissrorchiids, fellodistomids and gymnophallids. In *Gymnophallus choledochus*, a season-

172   *Reproduction and Development in Platyhelminthes*

based abbreviation occurs. The fluke follows the normal 3-host cycle in summer but switches to 2-host cycle during winter.

(3) The plagiorchiid *Opisthioglyphe ranae* cercaria encyst in the superficial epithelial skin of tadpole. When the frog's skin sloughs, the frog swallows the sloughed skin together with the encysted metacercaria and thereby becomes the DH (Grabda-Kazubska, 1976). Similarly, the macroderoidid *Haplometra cylindracea* cercariae penetrate into the buccal mucosa and encyst as metacercariae. After a few days, the cyst bursts and the fluke migrates into the lung (Grabda-Kazubska, 1976).

(4) For the opecoelid *Coitocaecum parvum*, the hydrobid mud snail *Potamopyrgus antipodarum*, amphipod *Paracalliope fluviatilis* and the bully *Gobiomorphus cotidianus* serve as FIH, SIH and DH, respectively. In the absence of a chemical cue from the bully, some progenetic cercariae develop into metacercariae in the amphipod itself. Normal and progenetic metacercariae of *C. parvum* can co-exist in the amphipod. The progenetic metacercariae sexually mature and produce self-fertilized eggs (Poulin, 2001, Lagrue and Poulin, 2009) and the SIH amphipod begins to serve as DH in the 2-host facultative life cycle of *C. parvum*.

(5) The life cycle of another opecoelid *Plagioporus sinitsini* represents a combination of truncation by abbreviation and progenesis. *P. sinitsini* passes through miracidium to mother sporocyst, daughter sporocyst, cercaria and finally to metacercaria, and involves the prosobranch *Elimia symmetrica* and the benthic feeding rosyside dace *Clinostomum funduloides* as FIH and SIH, respectively and reaches the DH, when the dace is ingested. The fluke has three alternate stages, in which it can be released from the snail. The first one is the usual 3-host life cycle. In the second, cystophorous cercaria, instead of being shed, encysts inside the daughter sporocyst and develops into metacercaria within *E. symmetrica* itself. The daughter sporocyst containing the encysted metacercaria is voided with feces of the snail. On ingestion, the dace becomes the DH, from which the fluke's eggs are voided and free-swimming miracidia are hatched. Hence, the 3-host life cycle is abbreviated to the 2-host cycle. In the third one, the most intriguing of the three alternatives, the cercaria within the snail itself progenetically sexually matures and produces hatchable eggs as well as releases miracidia, like an adult. Briefly, the 3-host life cycle is abbreviated to a 1-host life cycle (see Esch et al., 2002).

Progenesis eliminates the need for a definitive host and thereby increases the chances of completion of the life cycle. But the release of self-fertilized eggs to the exterior becomes a problem, especially in the facultative truncated life cycle. Encystation within the gonad enhances the chances of their eggs being released along the host's gametes, when the host spawns. Conversely, the progenetic flukes located elsewhere have to await the death of SIH. Progenetic flukes adopt different strategies to release the eggs like (i) fluke migration to the gut prior to the release of progenetic eggs, (ii) virulence-mediated death of the host to disperse the eggs (e.g. *Aphalloides codomicola*), (iii) reduced virulence to maximize fecundity, and (iv) gonadal location to gain exit of the eggs along with the host's gametes. (i) *Alloglossidium macrobdellensis* migrates into the gut of its second intermediate hirudinean host *Macrobdella ditetra* and subsequently begins progenetic reproduction (Corkum and Beckerdite, 1975). (ii) High virulence of *Aphalloides codomicola* and its association with a tissue-liquefying myxozoan accelerate the death of SIH, the gobiid *Pomatoschistus microps*, and then disperse progenetic eggs (Pampoulie et al., 1999, 2000). (iii) With low virulence of the fluke in the amphipod *Paracalliope fluviatilis*, progenesis facilitates *C. parvum* to grow and produce maximum number of progenetic eggs. Incidentally, the conditioned water by the bully reduces egg production by the fluke in the amphipod with single or double infection (Poulin, 2001, Lagrue and Poulin, 2009). (iv) The fluke *Stegodexamiene anguillae* also passes through the snail *P. antipodarum* to the bully *Gobiomorphus/Galaxias* sp and to *Anguilla dieffenbachia*. In the bully, the fluke intensity ranges from 7.7 to 52.2 flukes/host. Of them, 10.2% metacercariae undergo progenetic development. The majority of these progenetic flukes are located in the gonads of the bully and more are located in the ovary than in the testis. Besides supplying ample resources, the gonad also provides an exit for the fluke's egg along with the host's gametes. In the bully, temperature and photoperiod regulate the spawning events. Synchronizing with the host spawning events, the fluke also releases its eggs by increasing the proportion of a progenetic worm (Herrmann and Poulin, 2011a).

## 4.10 Prevalence and Intensity

For definitive vertebrate hosts, the available information is mostly related to epidemiology. Information on intensity of infection is limited; the intensity ranges from 3/host in *Aponurus* sp (Lo et al., 1998) to 52/host in *Stegodexamene anguillae* (Monteiro and Brasil-Sato, 2010). Hence, this account is limited to the biological aspects of intermediate hosts. Experimental studies have shown that the development of resistance in digenean infection causes reduction in the host's fecundity (e.g. Webster and Woolhouse, 1999). Despite the molluscan susceptibility to many digeneans, the flukes seem to limit their choice to one or two species as first intermediate host(s) but broaden to as many as SIH. The limitation of the choice may be related to host ability to supply the nutrient requirement to sustain propagatory multiplication in the FIH. The SIH affords mostly protection to the encysted cercaria/metacercaria. For example, the snail *Helisoma anceps* is susceptible to as many as eight (Crews and Esch, 1986) and nine (Esch et al., 2002) digenean species (Table 4.9). Of 806 *H. anceps* specimens collected from the Charlie's Pond, USA, the infection was limited to 518 snails. Of them, 85% (or 55% of

174 *Reproduction and Development in Platyhelminthes*

**TABLE 4.9**

Prevalence and cercarial shedding from *Helisoma anceps* in Charlie's Pond from November 1984 to March 1985 (from Crews and Esch, 1986, *calculated)

| Parasite species | Prevalence | | Cercarial shedding | |
|---|---|---|---|---|
| | as % of total | as % of infected* | as % of total | as % of infected* |
| *Halipegus occidualis* | 54.6 | 84.9 | 31.4 | 48.8 |
| *Diplostomulum scheuringi* | 3.7 | 5.8 | 3.1 | 4.8 |
| *Zygocotyle lunata* | 3.6 | 5.6 | 2.6 | 4.1 |
| *Glypthelmins quieta* | 0.5 | 0.8 | 0.4 | 0.8 |
| *Spirorchis* sp | 0.3 | 0.3 | 0.2 | 0.4 |
| *Clinostomum* sp | 0.3 | 0.3 | 0.2 | 0.4 |
| Unknoan (2 species) | 1.5 | 1.5 | 2.1 | 0.2 |
| Total infected host (no.) | 518 | | 308 | |
| Total infected host (%) | 64.3 | | 38.2 | |

the total number) specimens were infected with *Halipegus occidualis*. Each of ~ 6% *H. anceps* was also infected with *Diplostomulum scheuringi* and *Zygocotyle lunata*. In a subsequent publication, Esch et al. (2002) reported that of 4,899 *H. anceps* investigated, only 30% were infected and infection with two digenean species was limited to 7 (or 0.5% of the infected) snails only. The fact that the snails serve as FIH mostly for a single digenean species also holds true for the bivalves. About 46% of *Pisidium amnicum* is infected with *Bunodera lucipercae*, *Phyllodistomum elongatum* and *Palaeorchis crassus*. Considering this 46% prevalence as 100, 76, 13 and 11% prevalence are shared by *B. lucipercae*, *Ph. elongatum* and *Pa. crassus*, respectively (Rantanen et al., 1998). These observations are also consistent with the prevalence in other molluscan hosts (see Crews and Esch, 1986).

For prevalence of digenean infection in FIH and SIH, some representative values are summarized in Table 4.10. These values range from 5% for snail *Zeacumantus subcarinatus* to 100% in *H. anceps* for FIH and from 0.8% in the polychaete *Hydroides dianthus* to 100% in *Diopatra neapolitana* for the SIH. Host body size is perhaps the most important factor that determines prevalence and intensity. Many molluscs ingest and accumulate eggs/miracidia, as age advances or body size increases (e.g. *Z. subcarinatus*, Fredensborg et al., 2005). Rediae of *B. lucipercae* consume sporocysts of *Ph. elongatum* and *Pa. crassus*; this observation may explain the relatively higher prevalence of *B. lucipercae* in larger clams (Rantanden et al., 1998). In Russian waters, prevalence of *B. lucipercae* increases from 1% in a 1-year old clam to 2.4% in a 3-year old clam (Zhokhov, 1991). This holds true also for the presumably penetrative (Madhavi and Jhansilakshmibai, 1994) miracidia of *Transversotrema patialense* infecting *Cerithidea californica* (Fig. 4.8A). Body size is also an important factor in deciding the prevalence and intensity of intermediate hosts like *Stegastes*

# TABLE 4.10

Prevalence of trematode infection on molluscan and second intermediate hosts. * data from field investigations, † = definitive host.

| Parasite species | Intermediate host | Prevalence (%) | Remarks | Reference |
|---|---|---|---|---|
| **First intermediate molluscan host** | | | | |
| *Schistosoma mansoni* | *Biomphalaria glabrata* | 10–51 | In different populations | Webster and Woodhouse (1999) |
| Parasites | *Littorina saxatilis* | 45 | 1000–2000 parasites/$m^2$ | Granovitch et al. (2009)* |
| *Cryptocotyle lingua* | *L. littorea* | 19 | Females alone infected | Pechenik et al. (2001)* |
| *Austrobilharzia vargilandis* | *Ilyanassa obsoleta* | 17 | | |
| *Maritrema novaezealndensis* | *Zeacumantus subcarinatus* | 5–75 | Increasing shell height 8–15 mm increases prevalence | Fredensborg et al. (2005)* |
| *Ribeiroia ondatrae* | *Helisoma trivolvis* | 90–100 | At 50 or 400 eggs exposure | Redmond et al. (2011) |
| | *B. glabrata* | 100 | 65 or 650 eggs exposure | |
| *Halipegus occidualis* | *Helisoma anceps* | 80–100 | Peeks in May and October | Crews and Esch (1986) |
| *H. occidualis* | *H. anceps* | 15–45 | Peeks in May and June | Negovetich and Esch (2007)* |
| *Saccocoelioides nanii* | *Prochilodus argenteus* | 49, 74 | Rainy, dry seasons | Monteiro and Brasil-Sato (2010)*† |
| **Second intermediate non-molluscan host** | | | | |
| *Diplostomum gasterostei* | *Gasterosteus aculeatus* | 56 | Fish | Pennycuick (1971) |
| *Cercaria loossi* | *Hydroides dianthus* | 0.8 | Polycheate | see Peoples (2013) |
| *Gymnophalus choledochus* | *Diopatra neapolitana* | 100 | Intensity: 200/host | |
| Opecoelid E | *Abarenicola affinis* | 48 | Metacercaria | |
| Opecoelid E | *Heteromastus filiformis* | 62 | Metacercaria | |
| *Plagiorchis elegans* | *Aedes egypti* | 44–92 | Metacercaria in mosquito larva | Lowenberger and Rau (1994) |
| *Coitocaecum parvum* | *Paracalliope fluviatilis* | 16–28 | Increases with size 3–3.5 mm, encysted metacercaria in amphipod | Poulin (2001) |

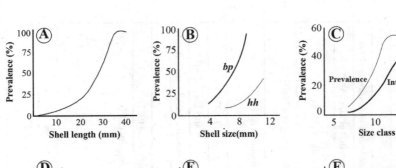
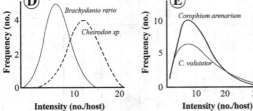
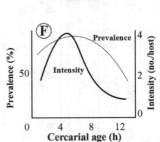

**FIGURE 4.8**

(A) Prevalence of the ectoparasite *Transversotrema patialense* cercariae as a function of shell length in *Cerithidea californica*, (B) *Halipegus occidualis* in the snail *Helisoma anceps* (hh) and *Bunodera luciopercae* in the bivalve *Pisidium amnicum* (bp), (C) Prevalence (P) and intensity of *Aponurus* sp in different size classes of *Stegastes nigricans*. (D) Cercarial intensity as a function of frequency in *Brachydanio rerio* and *Cheirodon* sp infected with *Transversotrema patialense* cercariae and (E) *Corophium arenarium* and *C. volutator* infected with *Microphallus claviformis* cercariae. (F) Prevalence and intensity of *Plagiorchis elegans* metacercariae in fourth instar of *Aedes aegypti* larvae (modified and redrawn from Sousa, 1983, Negovetich and Esch, 2007, Lo et al., 1998, compiled from Whitefield et al., 1986, Jensen et al., 1998, Rantanen et al., 1998, Lowenberger and Rau, 1994).

*nigricans* (Fig. 4.8C). In their review, Sorensen and Minchella (2001) noted a positive relation between snail size and digenean prevalence in 21 of 30 species. The relation is also apparent in 68 and 73% of infected snails from freshwater and marine habitats, respectively.

Apart from size, host susceptibility may also play a role; for example, the bivalve *Pisidium amnicum* is more susceptible to *Bunodera luciopercae* than the snail *Helisoma anceps* to *Halipegus occidualis* (Fig. 4.8B). With increasing intensity, the incidence frequency of *T. patialense* begins to decrease in fishes serving as SIH beyond 8 cercariae/*Brachydenio rerio* host and 13 cercariae/*Cheirodon* sp host (Fig. 4.8D). But the amphipod *Corophium* spp serving as SIH supports ~ 8–9 *Microphallus claviformis* cercariae/host up to the incidence frequency of ~ 6 in *Corophium volutator* and 10 in *C. arenarium* (Fig. 4.8E). With advancing age of *Plagiorchis elegans* cercaria in *Aedes aegypti*, both prevalence and intensity decline, albeit at different ages and levels, Fig. 4.8F).

Other factors like geographical location (e.g. *Biomphalaria glabrata*, *Littorina sexatilis*), sex (e.g. *L. littorea*, *Ilyanassa obsoleta*), shell size (e.g. *Z. subcarinatus*), calendar month (e.g. *Helisoma anceps*), and rainy and dry seasons (e.g.

*Prochilodus argenteus*) may also modify the prevalence level (Table 4.10). In New Zealand, the prevalence of *Maritrema novaezealandensis* in the snail *Zeacumantus subcarinatus* range from 4.4% in the Dowling Bay to 75% in the Oyster Bay. Interestingly, it is increased with increasing host biomass, indicating parasitic castration as well as decreasing prevalence beyond 12 mm host size in the Oyster Bay (Fredensborg et al., 2005). The freshwater ovoviviparous *Potamopyrgus antipodarum* serves as FIH to at least 14 digenean species. It is composed of both obligatory sexual diploid and parthenogenic triploid females. Prevalence of *Microphallus* infection is positively correlated with male size but is negatively correlated with the mean size of brooding females. Understandably, parthenogenesis occurs in snails found in deeper waters, where the DH water fowl spends very little time (Jokela and Lively, 1995). Dividing the rough winkle *Littorina saxatilis* into moderately (prevalence 10–15%) and heavily (50–80%) infected populations of the White Sea, Sokolova (1995) found (i) higher prevalence amidst winkles attached on macrophytes than on a stone surface, (ii) no sex-specific difference in prevalence, and (iii) increased fecundity but fewer juveniles, perhaps due to reduced recruitment in a heavily infected population.

## 4.11 Causes and Losses

The successful establishment of a digenean is associated with remarkable changes in activity, survival, growth and reproduction in the molluscan host, as well as alteration of sex ratio in crustacean, condition factor in fish and changes in limb and its component in anurans.

### 4.11.1 Activity and Survival

Negative effects of digenean infection commences with reduced feeding and locomotion. Besides castration, the flukes infecting the digestive gland affect the digestive capacity. For example, the periwinkle *Littorina littorea* infected with *Cryptocotyle lingua* feeds on less number of algae than the naïve winkle (Wood et al., 2007). Infected with *Renicola reoscovita* or *Himasthala elongata*, *L. littorea* reduces consumption of macrophytic green alga to 69% of that of naïve winkle (Clausen et al., 2008). Crawling velocity of *Hydrobia ulvae* is reduced from 0.6 cm/minute in control to 0.5 or 0.4 cm/minute in snails infected with microphallids or *Himasthala continua* (Mouritsen and Jensen, 1994). At 20°C, mobility of *Stagnicola elodes* infected with *Plagiorchis elegans* is reduced from ~ 80 cm in a naïve snail to < 10 cm in an infected snail from the time of infection up to 180 minutes and possibly longer. This finding holds true for the snails at 15° and 25°C also (Lowenberger and Rau, 1994).

Intra-molluscan digenean parasitism is associated with host survival, growth and reproduction. Host survival depends on its (i) age (Fig. 4.9),

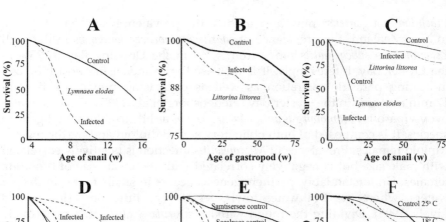

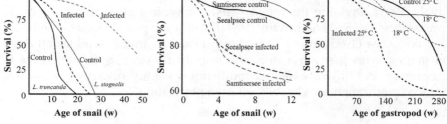

**FIGURE 4.9**

Survival of control and infected first intermediate gastropod host. (A) Survival of *Lymnaea elodes* infected with *Echinostoma revolutum* and (B) *Littorina littorea* infected with *Cryptocotyle lingua*. (C) A + B combined. (D) Survival of *Lymnaea truncatula* infected with *F. hepatica* and *L. stagnalis* infected with *Trichobilharzia ocellata*. (E) Survival of *L. peregra* infected with *Diplostomum phoxini* in Seealpsee and Samtiersee snails. (F) Effect of temperature on survival of *Zeacumantus subcarinatus* infected with *Maritrema novaezealandensis* (modified and redrawn from Sorensen and Minchella, 1998, Huxham et al., 1993, Wilson and Denison, 1980, Hodasi, 1972, McClelland and Bourns, 1969, Ballabeni, 1995, Fredensborg et al., 2005).

(ii) life span and reproductive pattern (Fig. 4.9A, B), (iii) castration and gigantism (Fig. 4.9D, F) as well as (iv) specific host-parasite combination. Virulence-mediated host death occurs to disperse progenetic eggs in *Aphalloides codomicola*. However, tissue destruction is the main cause for increased mortality of an infected host. Progressive mortality occurs, as age advances either rapidly in freshwater pulmonate like *Lymnaea elodes* (Fig. 4.9A) or slowly in marine prosobranch like *Littorina littorea* (Fig. 4.9B). Most freshwater inhabiting lymnaeid and planorbid pulmonates have a short life span lasting from one to three years and some of them are semelpares (e.g. *Lymnaea peregra*). Conversely, the marine snails have a long life span of > 7 years (e.g. *Littorina littorea*). In the smaller, short-lived lymnaeids and planorbids, the rapidity, at which tissues are destroyed for cercarial production, is faster than in the large and long-lived littorinids. The number of *Halipegus occidualis* cercariae emerging from *Helisoma anceps* is ~ 400/snail/day and it may be at the cost of rapid tissue destruction. Contrastingly,

*Digenea* 179

the affordable slow rate of tissue destruction in a relatively larger, long-living *L. littorea* infected with *Cryptocotyle lingua* allows cercarial emission at the rate of 830/snail/day lasting for a period of 7 years (see Hyman, 1951). This is also true of some bivalve hosts. With a life span of 14 years, *Anodonta piscinalis*, on infection with *Rhipidocotyle fennica*, continues to live and emit cercariae for years until its natural death (see Rantanen et al., 1998). Its survival is equal in both naïve and infected clams. The digenean *Bunodera luciopercae* infects *Pisidium amnicum* at its age of 14 months, when the clam has already spawned once or twice and it dies naturally at the age of 38 months, whether it is infected or not but after infection at the age of the 14th month, the castrated clam does not spawn. *L. stagnalis* infected with *Trichobilharzia ocellata* lives longer than the naïve snail (Fig. 4.9D). With early infection of *T. ocellata* and the consequent complete castration, the snail continues to survive up to the age of 45 weeks. Hence, the infection and consequent gigantism is another important factor that determines survival. The semelparous snail *L. peregra* is an annual. Ballabeni (1995) limited his observation to 12 week periods on survival of naïve and infected *L. peregra* with *Diplostomum phoxini* at two habitats. With incomplete castration, the snail exhibits a negative relation between survival and age (Fig. 4.9E). Had Ballabeni reared the snail up to 1-year, the infected snail could have also outlived the naïve ones. Temperature profoundly alters the negative survival-age relationship in *Zeacumantus subcarinatus* infected with *Maritrema novaezealandensis* (Fig. 4.9F). Interestingly, survival of *Biomphalaria glabrata* is decreased with increasing number of (*Schistosoma mansoni*) miracidial infection. On the 126th day after infection, it decreases from 90% to 60, 25 and 10% in the snails infected with 1, 2, 3 and 4 miracidia, respectively (Makanga, 1981).

## 4.11.2 Growth

In intestinal parasitic worms, intermediate metabolism has received considerable attention (e.g. Tielens et al., 1992). Under hypoxic/anoxic conditions prevailing in the host gut, glycolysis yields three molecules of ATP/gluclose molecule (see Pandian, 1975). In aerobic animals, the combined glycolytic and tricarboxylic acid pathways yield 36 molecules of ATP per glucose molecule. There is convincing evidence that helminth parasites localized in hypoxic or anoxic conditions have evolved anaerobic pathways capable of increasing the ATP yield, while maintaining a favorable redox potential (see Calow, 1987). A common mechanism that retrieves energy from the reduced NAD (Nicotinamide-Adenine-Dinucleotide) generated in the initial glycolytic pathway is to use an electron transport phosphorylation with fumerate or succinate as the final electron acceptor. This is associated with $CO_2$ fixation and yields extra four to six ATP molecules (see Pandian, 1975, Calow, 1987). Glycogen serves as the main energy source for the fluke and its miracidium and cercaria (e.g. Jokela et al., 1993, see Fig. 4.6B).

180   *Reproduction and Development in Platyhelminthes*

Strikingly and surprisingly, most digenean experts have not recognized the dynamics of the growth process in infected molluscs. For the correct estimate of growth in infected molluscs, the following must be considered, *albeit* problems related to the presently available methods: (i) altered input of food resource due to partial destruction of the digestive gland, (ii) metabolic losses via feces, urine, aerobic/anaerobic respiration, (iii) its own gametes, and (iv) cercarial output by the fluke(s) from the molluscan body. Cercarial shedding, for example, is commenced from the 5th and 6th week following infection in juvenile and adult *Biomphalaria glabrata*, respectively. Likewise, eggs are laid from the 5th week after infection in juveniles but only up to 6th week in adults (Gerard and Theron, 1997). However, no author has ever considered growth losses due to host gametes and cercarial shedding by the fluke through the host.

In most snail-digenean combinations, larval development and propagatory multiplication occur in the gonad-digestive gland complex of the host snail. The complex extends into the coiled body delimited by a solid shell. Two patterns of spatial integration of the digeneans are recognized: (i) an integration by '*substitution*', where the volume of parasite is accommodated within the space created by the destruction of the host tissue or (ii) an integration by '*addition*', where the parasite is accommodated by a faster shell growth. Spatial integration by substitution within the host, there are three possibilities: (a) no change in the digestive gland but the gonad is completely consumed by the fluke, e.g. *Segmentina trochoides, Fasciolopsis buski*, (b) partial destruction of the digestive gland but partial/total inhibition of gonad growth and egg production by altering the host's hormonal balance, e.g. *Biomphalaria glabrata, Schistosoma mansoni* and (c) partial destruction of the digestive gland and gonad followed by reduced fecundity, e.g. *Littorina littorea, L. saxatilis* (Probst and Kube, 1999). Regarding hormonal imbalance, the molluscan genome is reported not to contain genes for enzymes that are involved in biosynthesis of vertebrate type steroids. The implicated role of steroids absorbed from the surrounding water is a matter of debate (see Pandian, 2017). Evidently, neuroendocrines alone regulate gametogenesis and spawning in molluscs. For example, schistosomin, a neuropeptide antagonizes female gonadotropic calfluxin in schistosomids alone (De Jong-Brink et al., 1991, Schallig et al., 1991).

The classical evidence elucidating the antagonistic interaction involves the digenean-mediated changes in host fecundity and somatic growth. Terms commonly used to describe these changes are castration, gigantism and stunting of the host. Castration involves partial or complete reduction in fecundity, as a consequence of partial or complete gonad destruction. Stunting and gigantism refer to hosts that are atypically small (e.g. *Lymnaea elodes* infected with *Echinostoma revolutum*, Krist and Lively, 1998, *Potamopyrgus antipodarum* infected with *Microphallus*) or large (e.g. *L. stagnicola* infected with *Trichobilharzia ocellata* (see also Gerard and Theron, 1997). Digenean castration also occurs in many bivalves, e.g. *Ostrea lutaria, Crassostrea virginica, C. gigas, C. madrasensis* and *Pinctata radiata* (see Hassanine, 2006). With the antagonistic interaction, the host has access only to the resources remaining after the digenean has taken its lion's share. Complete castration may leave a little more resource that can profitably be channeled into somatic growth resulting in gigantism.

With reference to substitution, the following observations are relevant: the propagative multiplication in sporocysts and rediae progressively invade the host body like the cockle *Cerastoderma edule* to such an extent that the fluke weighs ~ 34% of the host weight (Dubois et al., 2009). In *Hydrobia ventrosa*, the gonad is tightly packed with 300 sporocysts of *Microphallus* (each measuring 350 x 85 μm), which occupy a volume of 6.4 mm³ space (Probst and Kube, 1999). Firstly, considerable efforts have been made to estimate growth by measuring different host traits like shell length (e.g. *Lymnaea truncatula*, Wilson and Denison, 1980), shell diameter (e.g. *B. glabrata*, Gerard and Theron, 1997), shell height (e.g. *Hydrobia ulvae*, Mouritsen and Jensen, 1994), live weight/ dry weight (e.g. *H. ulvae*, Mouristen and Jensen, 1994) and shell, dried body and gonad weights (e.g. Wilson and Denison, 1980). Even in a carefully designed experiment in *Anodonta piscinalis* infected with *Rhipidocotyle fennica*, Jokela et al. (1993) separated the foot, mantle, gill blades, gonad and "the rest of the visceral mass (= 'body') of each clam" for measurements. Incidentally, 'the branching sporocyst tubules of *R. fennica* invade mainly the gonad of the clam. Still, the quantum of branching sporocyst tubules remaining within the 'body' has not been estimated. Hence the parasite tissue integrated by replacement or substitution has not so far been taken into consideration in estimates of host growth. Regarding integration by addition or positive growth resulting in gigantism too, there are contradictory observations. Finding no evidence for gigantism from their mark-recapture experiments in the fields, Fernandez and Esch (1991) described it to a laboratory artifact. In *B. glabrata*, the difference in shell diameter was wide between naïve and infected juveniles but the difference became neutralized in adults of infected *L. stagnalis* (Gerard and Theron, 1997). More interestingly, Mouritsen and Jensen (1994) found that growth in shell height was higher in infected than in naïve *H. ulvae*. But dry weight was less in infected snails than in the naïve ones.

Secondly, the infected hosts express different levels of gigantism corresponding with destruction level of the gonad-digestive gland complex. With no reduction in incoming food resource in an infected snail suffering no destruction of the digestive gland complex, the snail may respond differently from those, in which the input of food resources is reduced due to partial destruction of digestive gland complex, as in *Littorina littorea* (see also Probst and Kube, 1999). Thirdly, both gigantism and stunting are dynamic processes. Interestingly, Probst and Kube also found almost no difference between the ratio of soft tissue biomass; but the shell mass is consistently greater in infected than in naive *H. ventrosa*. This observation clearly indicates that in substitution, the volume of space created in the host is replaced by the

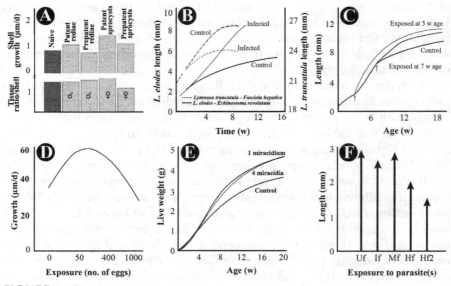

**FIGURE 4.10**

(A upper) Shell growth in *Hydrobia ventrosa* infected with sporocysts or rediae during prepatent and patent periods, (A lower) ratio between soft tissue mass and cell mass of naïve and infected *H. ventrosa* male and female (compiled from Probst and Kube, 1999), (B) Contrasting effects of digenean infection on growth of *Lymnaea truncatula* and *L. elodes* (compiled from Wilson and Denison, 1980, Sorensen and Minchella, 1998). (C) Effect of age of infection of *Schistosoma mansoni* on *Biomphalaria pfeifferi* (arrows indicate age of infection, redrawn from Sturrock, 1966). Effect of exposure to different number of (D) eggs of *Ribeiroia ondatrae* to *Helisoma trivolvis* (from Redmond et al., 2011), (E) *Trichobilharzia ocellata* miracidia to *Lymnaea stagnalis* (from Slutiers et al., 1980) and (F) light (If), moderate (Mf) or heavy (Hf) infection of *Rhipidicotyle fennica* and *R. fennica* + *R. campanula* (Hf2) on *Anodonta piscinalis* (modified and redrawn from Slutiers et al., 1980, Redmond et al., 2011, Taskinen, 1998).

sporocysts and/or rediae but in addition, the shell grows faster in infected snails to provide additional space (Fig. 4.10A).

In their review limited to publications in the English language between 1970 and 2000, Sorensen and Minchella (2001) collected 41 publications reporting 113 fields and laboratory experiments involving 30 host snail species and 39 digenean species. This account, however, has also not considered those left by them and those published after 2000. The following may be generalized from their review: (1) Of 29 infected freshwater snails from field studies examining the life history traits, (a) *H. ulvae* infected with *Microphallus claviformis* or *M. pirum* or various species and (b) *H. ventrosa* infected with *Bunocotyle progenetica* (at low snail density but not at high density) or *Cryptocotyle* sp are the only two snails, in which the growth is positive. But growth is negative in *Lymnaea elodes* infected with various species and *Helisoma anceps* infected with *Halipegus occidualis* alone. In six others, growth is equal between naïve and infected snails. Briefly, *Hydrobia* spp alone are subjected to castration and gigantism. (2) Of 27 infected marine snails from field studies examining the life history traits, *Microphallus pseudopygmaeus* infecting *Onoba aculeus* alone induces positive growth. In *Littorina littorea*, *Catatropis* or *Echinoparyphium* infected with *Cerithidea californica* is the only combination, in which the snail's growth is negative. Concurrent infection by two or more digeneans consistently induces negative growth. (3) Of 33 infected snails from laboratory studies examining the life history traits of infected prosobranch and planorbid snails, growth is positive in *Biomphalaria glabrata – Schistosoma mansoni* combination, irrespective of age/patency and presence or absence of redia (Thornhill et al., 1986). However, there are other reports (e.g. Gerard et al., 1993) indicating negative growth for the same combination. Negative growth results in at least four combinations: (i) *Microphallus – Potamopyrgus antipodarum*, (ii) *S. mansoni – B. glabrata*, (iii) *S. margrebowei – Bulinus natalensis* and (iv) *Halipegus occidualis – Helisoma anceps*.

An array of factors like parasite prevalence and intensity as well as life span, cercaria output rate, age, density of the host and their interactions result in varied levels of gigantism or stunting. (i.a) With occurrence of infection during meiosis of host germ cells, the germinal epithelium in the gonad is almost completely destroyed and the space is occupied by sporocysts and rediae during the patent period (e.g. *H. ventrosa*, Probst and Kube, 1999). (i.b) Infection of *S. mansoni* on immature (5-week aged) *B. glabrata* delays growth of the ovotestis and inhibits the development of albumin gland and the female organ so that eggs are not produced but the snail grows to the largest size. However, oviposition occurs at the cost of somatic growth of *B. pfeifferi* infected with *S. mansoni*, when maturing (7-week aged) snail is infected (Fig. 4.10C, see also Meier and Meier-Brook, 1981, Theron and Gerard, 1994). (i.c) If infection occurs during the patent period, some host germ cells may be saved resulting in reduced fecundity, as in *B. glabrata* infected at the juvenile stage. (ii) Regarding growth, molluscan species with daughter sporocyst differs from those with a daughter redia; the infection site within the snail's digestive gland-gonad complex also alters growth (Probst and Kube, 1999). In snails involving rediae, the frequency of gigantism is 8, 60 and 58% for infection at pre-patent, patent and post-patent period, respectively. The corresponding values for sporocysts are 57, 58 and 42%. Hence, redia introduces a higher frequency of gigantism during the patent and post-patent periods of infection. (iii) The levels of positive or negative growth accompanying the digenean infection vary widely. For example, the positive growth and consequent gigantism ranges from ~ 15% of live weight of *Lymnaea elodes* infected with *Echinostoma revolutum* (Sorensen and Minchella, 1998) to 68% of weight of *Biomphalaria glabrata* infected with *Schistosoma mansoni* (Gerard and Theron, 1997). Similarly, the negative growth and consequent stunting may be as small as 11% in *Potamopyrgus antipodarum* infected with *Microphallus* (Krist and Lively, 1998) and as high as 25% in *L. truncatula* infected with *Fasciola hepatica* (Wilson and Denison, 1980) and *L. stagnalis* infected with *T. ocellata* (McClelland and Bourns, 1969). (iv) The

184   *Reproduction and Development in Platyhelminthes*

life cycle of both schistosomes and fasciolids involves only one molluscan intermediate host. With transmission involving penetrative miracidium and cercaria, the snails infected with schistosomes are more fecund than fasciolids, in which the transmission depends on penetrative miracidium but trophic cercaria. (v) Regarding the contrasting responses of *Lymnaea* (Fig. 4.10B), the cercarial transmission is direct from *L. truncatula* to a definitive host but *Echinostomum* requires an additional second intermediate host (Table 4.6). Consequently, the cercaria output rate is higher in the latter than in the former. As a result, *L. elodes* incurs complete gonad destruction by the 6th week after infection and begins to display gigantism then onwards (Fig. 4.10B). (vi) Differences in the age of infection induces different levels of positive growth and consequent gigantism in *Biomphalaria pfeifferi* (Fig. 4.10C) and *L. truncatula* infected with *F. hepatica* (Hodasi, 1972) (vii) Experimental studies have consistently shown that (a) exposure to increase in number of parasite eggs (Fig. 4.10D), miracidia (Fig. 4.10E) and intensity (Fig. 4.10F) results in reduced growth or almost stunting in the host snails and bivalve. (viii) Parallel dual or multiple infections occur but less often (see Sorensen and Minchella, 1998). Concurrently infected with *Rhipidocotyle fennica* and *R. campanula, Anodonta piscinalis* suffer more severe negative growth than when heavily infected with *R. fennica* alone (Fig. 4.10F). Notably, *Ilyanassa obsoleta, Littorina littorea, Lymnaea elodes* and *Helisoma anceps* suffer negative growth on infection with various digenean species (Sorensen and Minchella, 1998). However, there can also be competitive exclusion of one or more digenean species by other species, especially by *Halipegus occidualis*, in which the rediae may tear and swallow not only host tissues but also the rediae of other parasite species. For example, Fernandez and Esch (1991) reported that *Helisoma anceps* was found infected with *Halipegus occidualis, Echinostoma trivolvis* and *Diplostomum scheuringi* during August in Charlie's Pond but *E. trivolvis* and *D. scheuringi* were eliminated during October and subsequent May, respectively (Fig. 4.11A).

With competitive exclusion of *E. trivolvis* and *D. scheuringi,* the specific growth rate was stabilized at a lower level in presumably mature snails from a higher level (Fig. 4.11A). Contrastingly, Sousa (1983) reported striking differences in specific growth rates between immature and mature *Cerithidea californica* infected with digenean parasites (Fig. 4.11B). The differences observed from all these field investigations can be explained from an important contribution by Gorbushin (1997). He demonstrated the growth rate of *H. ulvae* is specific to parasite species and host maturity status (Fig. 4.11C). (i) The snail growth rate decreased with increasing host size (shell diameter). (ii) Infected with *Cryptocotyle* sp, microphallids and *Notocotyle* sp, it decreased with increasing parasite size in both immature and mature snails; the values for the three digenean species are widely scattered in immature snails but species specific trends become apparent in mature snails. (iii) The slope was at the highest level for *Cryptocotyle* sp and at the lowest for microphallids. The growth rate of *H. ventrosa* infected

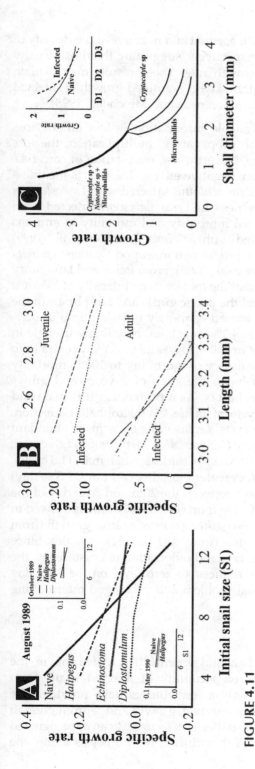

**FIGURE 4.11**

(A) Effect of competitive elimination and stable specific growth rate of *Helisoma anceps* in August, October and May (compiled from Fernandez and Esch, 1991). (B) Effect of maturity status on specific growth rate of infected *Cerithidea californica* (compiled from Sousa, 1983). (C) Parasite species specific and host maturity status specific effects on growth rate as a function of size in *Hydrobia ulvae* infected with 2 (microphallids), 3 (*Cryptocotyle* sp) and 4 (*Notocotylus* sp); window shows the effect of *H. ventrosa* density (D1, D2 and D3 represent 2,600, 7,400 and 19,000 snails/m², respectively) on growth rate (compiled and redrawn from Gorbushin, 1997).

186  *Reproduction and Development in Platyhelminthes*

with *Bunocotyle progenetica* was positive, equal and negative at the density of 2,600, 7,400 and 19,000 snails/m$^2$, respectively, suggesting that host density considerably influences the host growth rate. These results may explain different conclusions arrived by authors, who examined growth of infected snails from the fields (see Tables 1, 2 of Sorensen and Minchella, 1998).

*Second intermediate hosts* (SIH): Available information is less on the loss incurred by SIH. In the absence of 'propagatory' multiplication, the loss incurred by them can be less severe. Nevertheless, encystation of cercaria/ metacercaria in them can also induce negative effects. The effects include a reduced condition factor in *Gasterosteus aculeatus* infected with *Diplostomum gasterostei* (Pennycuick, 1971), blindness in *Perca flavescens* infected with *D. baeri* (Ubels et al., 2018), increased morbidity and mortality in anurans like the treefrog *Hyla regilla* infected with *Ribeiroia* (Johnson et al., 1999) and reduced survival, growth and reproduction in isopod *Cyathura carinata* infected with microphallids (Ferreira et al., 2005). From fields and laboratory studies, Johnson et al. (1999) found that the metacercaria intensity of *Ribeiroia* was the highest in the tissues around the pelvic girdle and hind limbs of the treefrog. In the field, the infection caused ectromely (missing limb in 6.5% frogs), ectrodactyly (missing digit in 4.2% frogs), taumely (bony triangle in 3.4% frogs), polydactyly (extra digit in 4.8% frogs) and polymely (extra limb in 16.7% frogs). With increasing frequency of morbidity to 100%, mortality also increased to 60%. *C. carinata* has a life span of 2 years and grows continuously to generate a single cohort/y. As a protandric, it is male and female during the first and second year of its life. On microphallid infection, it hosts 4.2 cysts/isopod. Of these cysts, *Levinseniella* sp is more abundant (2.4 cysts) than *Maritrema subdolum/Microphallus claviformis* (1.8 cysts). Infected female hosts more cysts (2.9) of *Levinseniella* sp than male (1.8 cysts). Consequently, infected females suffer greater mortality (> 1 but < 1.5%/day) than the infected males (< 0.5%/day) between the 20th and 80th day of the experiment. With more females suffering mortality, the sex ratio is skewed in favor of the male. Infected females also suffer greater negative growth (from 0.71 to 0.65 cephalic lengths) than that (from 0.81 to 0.71) of males. Since molting is a prerequisite for fertilization and oviposition in crustaceans, the number of the ovigerous female is reduced to zero level on the 45th day of the experiment due to an ecdysial problem and prolonged intermolting frequency.

### 4.11.3 Reproduction

The advantages of a short life span and amenability to rearing in the laboratory have facilitated studies on fecundity of infected snails from freshwater habitats. Nevertheless, only a few authors have provided the required information on all aspects of survival, growth and reproduction in infected snail species (Table 4.11). Two patterns of fecundity can be recognized among infected freshwater snails: With inclusion of sporocysts alone, the

## TABLE 4.11

Comprehensive summary of the effect of digenean prevalence on survival, growth and reproduction in some molluscan hosts. C = control/uninfected, I = infected, † = Loy and Haas (2001), †† = Schweizer et al. (2007)

| Host-Parasite, Reference | *Lymnaea stagnalis-Trichobilharzia ocellata* (McClelland and Bourns, 1969) | *L. elodes-Echinostoma revolutum* (Sorensen and Minchella, 1998) | *L. truncatula-Fasciola hepatica* (Wilson and Denison, 1980) | *Helisoma anceps-Halipegus occidualis* (Crews and Esch, 1986, Keas and Esch, 1997) |
|---|---|---|---|---|
| Prevalence | 4%† | 14% in immature ~ 25% in adult | 7%†† | 55% |
| Survival | C = 0% on 23rd week I = 0% on 44th week | C = 40% on 15th week I = 0% on 11th week | Equal in C & I up to 16th week | ~ 70% for I |
| Growth | C = 45 mm, I = 36 mm from 26th week | 15% higher on 10th week | C = 9 mm, I = 6.5 mm up to 16th week | C = 27 mm/week I = 0.17 mm/week |
| Fecundity | C = 57 eggs/snail I = 0.9 egg/snail | Egg laying ceased from 7th week | C = 175 eggs/snail I = 25 eggs/snail | C = 38 eggs/week I = 4.4 eggs/week |

schistosomids seem to let the infected host survive long and oviposit almost as long as the naïve snail. But the fasciolids do not allow the host to survive long or too long and produce eggs. Briefly, the schisotosomids are consumers, which decrease their host reproduction to different levels, whereas fasciolids are castrators, which terminate reproduction (Hall et al., 2007) or increase fecundity at the cost of survival and growth of the host. In the schistosomid pattern, both survival and oviposition in *Biomphalaria glabrata* infected with *Schistosoma mansoni* continued up to the age of the 20th week, as in control but fecundity was reduced to 60% of the control (Minchella and Loverde, 1981). In *Lymnaea stagnalis* infected with *Trichobilharzia ocellata*, the infected snail survived longer (44th week) than the control (23rd week) but grew 20% slower and produced 0.9 egg/week, in comparison with 57 eggs/week in the control (Fig. 4.12A). However, the age of exposure causes different levels of fecundity loss. In *B. pfeifferi* exposed to *S. mansoni* at the snail's age of 1st, 3rd, 5th, 7th and 16th week, fecundity decreased, irrespective of the differences in

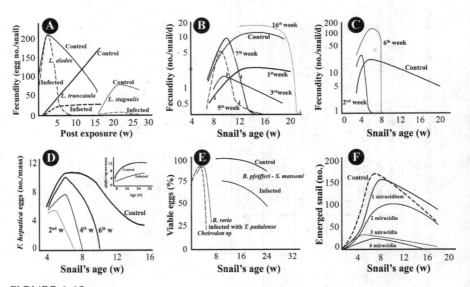

### FIGURE 4.12

(A) Fecundity as function of post exposure age in *Lymnaea elodes*, *L. stagnalis* and *L. truncatula* (compiled from Sorensen and Minchella, 1998, McClelland and Bourns, 1969, Wilson and Denison, 1980). (B) Effect of age of infection on fecundity of *Biomphalaria pfeifferi* exposed to *Schistosoma mansoni* at 1st, 3rd, 5th, 7th or 16th week (recalculated data compiled from Sturrock, 1966) and (C) *Lymnaea truncatula* infected with *Fasciola hepatica* at the age of the 2nd or 6th week (drawn from data reported by Hodasi, 1972). (D) Effect of age of infection on fecundity/sac in *L. truncatula* infected with *F. hepatica* (drawn from Hodasi, 1972), D-Window: Effect of *S. mansoni* infection on fecundity in egg number/sac in 3 week old *B. pfeifferi* (drawn from Sturrock, 1966). (E) Viable fecundity of *B. pfeifferi* infected with *S. mansoni* (drawn from data reported by Sturrock, 1966) and the ectoparasite *Transversotrema patialense* from *Brachydanio rerio* and *Cheirodon* sp (redrawn from Whitefield et al., 1986). (F) Survival of young *B. pfeifferi* infected with *S. mansoni* as function of host age and miracidial density (compiled from Makanga, 1981).

age of exposure; oviposition ceased at the age of 20th, 18th, 12–16th week in those exposed at the age of 16th, 5th and 7th week, respectively (Fig. 4.12B). Thornhill et al. (1986) reported that *B. glabrata* infected with *S. mansoni* also continued oviposition from 14th to 35th day after infection but at reduced fecundity (83% of control) and growth (95% of control). Hence, the number of sporocysts may not be adequate to destroy the entire gonad up to the age of 12 weeks. For, exposure on the 16th week, when sporocysts seem to have increased in adequate numbers, oviposition ceased within 2 weeks, indicating the destruction of the entire gonad in *B. glabrata* infected with *S. mansoni* (Sturrock, 1966). Parasitic strains may also reduce or prolong the duration of oviposition. In *B. glabrata*, infected with *Spirorchis scripta* involving only two generations of sporocyst (Hyman, 1951), the oviposition lasted for 6 and > 10 weeks in different Georgia and Minnesota strain, respectively (Hosier and Goodchild, 1970).

In the fasciolid pattern, the infected snails have one of the following options: (1) Reductions in survival, growth and reproduction. In *L. elodes* infected with *Echinostoma revolutum*, survival was reduced from 15 weeks to 11 weeks (Fig. 4.9A), growth (Fig. 4.10B) and reproduction from 16th to 8th week (Fig. 4.12A). (2) Maintenance of equal or slightly less survival at the cost of slow growth but with greater fecundity. Survival was almost equal (Fig. 4.11) but growth (28%) and fecundity (86%) were reduced in *L. truncatula* infected with *L. truncatula*: in fact, the infected *L. truncatula* ceased oviposition from the 6th and 8th week after infection but the control continued to reproduce until the 20th week(Fig. 4.12A). In *Helisoma anceps* infected with *Halipegus occidualis*, the reductions were ~ 30, 99 and 88% for survival, growth and fecundity, respectively (Table 4.11, Keas and Esch, 1977).

In these two patterns, egg number decreased within an egg mass in *B. pfeifferi* infected with *S. mansoni* (Fig. 4.12D-Window) and egg strip in *L. acuminata* infected with *F. hepatica* (shodhanda.inflibnet.ac). In *L. truncatula* infected with *F. hepatica*, the egg number/sac not only decreased with advancing host age as well as age of infection (Fig. 4.12D).

With advancing time, viability of digenean eggs decreases. A rare publication reports that the ectoparasite *Transversotrema patialense* resides in spaces beneath the scales of *Brachydanio rerio*, *Cheirodon* sp and others. Its life span is 6 weeks in relatively a smaller fish *B. rerio* and at its intensity of 6.9/host lays 200 eggs. But it is 12 weeks in a larger fish *Cheirodon* sp and at its intensity of 12.5/host spawns 333 eggs. Viability of *T. patialense* eggs decreased from 80–85% at the fluke's age of 2 weeks to ~ 20 and ~ 25% in 4 and 5 weeks-old flukes, respectively (Fig. 4.12E). The life span, density and age of the fluke seem to exert profound effects on egg viability. In naïve snails, egg viability fluctuated between 99 and 69% in *B. pfeifferi*. But on infection with *S. mansoni*, it progressively decreases from 75% at the snail's age of 8 weeks to 48% in 24 week old snails (Fig. 4.12E). The exposure of different *S. mansoni* miracidial density has a profound effect on the emergence and survival of the young *B. pfeifferi*. With increasing density of miracidial exposure, the

190    *Reproduction and Development in Platyhelminthes*

number of emerging young snails progressively decreases as functions of the miracidial density and snail age (Fig. 4.12F). Survival of these young snails also decreases to 95, 90, 60, 25 and 10% with increasing density from 0 to 1, 2, 3 and 4 miracidia, respectively.

Prevalence (P) of digenean infection may considerably alter recruitment, and population structure and size. Field investigations on population dynamics of the infected snails usually involve the examination of more than one population from different locations within a geographical area (e.g. Fredensborg and Poulin, 2006, Chapuis, 2009) and years of analysis (e.g. Granovitch et al., 2009). Not surprisingly, only a few publications are available on this topic. Investigations by Chapuis (2009) covered 15 freshwater snail populations of *Galba truncatula* infected with *Fasciola hepatica* in and around Switzerland. The hermaphroditic *G. truncatula* reproduces exclusively (90%) by selfing but is widely dispersed by its definitive hosts, the aquatic birds. In this species, propagatory multiplications include sporocysts and rediae. Fredensborg and Poulin (2006) undertook investigations on P of the microphallids (60% P) and four other trematodes in the marine prosobranch *Zeacumantus subcarinatus* (annual) populations at three locations, namely, Turnbull (TB, 13% P), Company Bay (CB, 30% P) and Lower Portbellow Bay (LPB, 86% P) in New Zealand. In *Z. subcarinatus*, propagatory multiplication of the microphallids includes sporocysts and rediae (Fredensborg et al., 2005). Undertaking long-term investigations on two populations of long-living gonochoric, prosobranch *Littorina saxatilis* infected with 11 microphallid species, Granovitch and Sergievisky (1990) and Granovitch et al. (2009) investigated the effect of snail density on P, and intensity as well as snail age on fecundity. *L. saxatilis* is ovoviviparous and propagatory multiplication of the microphallids includes sporocysts. Earlier, Sokolova (1995) also undertook a detailed investigation on lightly and heavily infected *L. saxatilis* on surfaces of macrophytes, gravel and stone in the White Sea.

With regard to parasitic effect, *G. truncatula* infected with *F. hepatica* may be expected to follow the earlier described fasciolid pattern. However, the results reported by Fredensborg and Poulin (2006) and others clearly indicated that there were wide differences in host-parasite interactions between experimental observations on host physiology- (individual-) based **proximate** response and ecological life history based **ultimate** response at population level (e.g. Granovitch et al., 2009). Hence, the altered life history traits in the infected snail at population level is summarized in Table 4.12. Results reported in these publications are briefly be summarized hereunder:

1. *G. truncatula – F. hepatica* (Table 4.12, Fig. 4.13B)

(A) Prevalence does not exceed 50%, (B) With increasing prevalence, (i) maturity size is increased, i.e. maturity is postponed and (ii) fecundity is progressively decreased.

## TABLE 4.12

Effect of digenean prevalence on survival, growth and reproduction in some molluscan hosts. C = uninfected, I = infected, * = Vignoles et al. (2003), ** = Huxham et al. (1993)

| Host-Parasite, Reference | *Galba truncatula-F. hepatica* (Chapuis, 2009) | *Zeacumantus subcarinatus* – microphallids (Fredensborg and Poulin, 2006) | *Littorina saxatilis*-castrating digeneans (Granovitch et al., 2009) |
|---|---|---|---|
| Life history traits | | | |
| Life span Sexuality | 1–1.5 years Hermaphrodite | 1 year Gonochoric | 5 years Gonochoric, ovoviviparous |
| Larval forms | Sporocysts, rediae | Sporocysts, rediae | Sporocysts |
| Prevalence (P) | 1% to 51% | Up to 75%, with increasing shell height | 35–75% in different populations |
| Survival (S) | C = 98%*, I = 75% | C = 78%, I = 50% at 18°C | I = 86%** |
| Growth (G) | C = 3.55 mm at 0% P I = 3.9 mm at 50% P | Infected continues to grow up to 300 days | C = 8.5 mm at 5 years I = 6.5 mm at 5 years |
| Fecundity (F) | 0% P = 9 eggs/30 days 50% P = 4 eggs/30 days | F increases with increasing age and size of infected | 0% P, 200 eggs 70% P, 125 eggs |

192  *Reproduction and Development in Platyhelminthes*

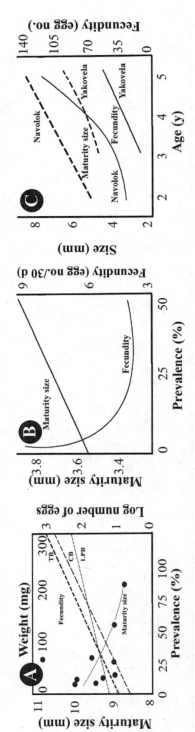

### FIGURE 4.13

(A) *Zeacumantus subcarinatus* infected with microphallids. Maturity size decreases with increasing body size. Fecundity is increased as functions of body weight and prevalence in lightly, moderately and heavily infected populations of *Z. subcarinatus*. Values for prevalence from Fredensborg et al. (2005) are extrapolated. (B) *Galba truncatula* infected with *Fasciola hepatica*. With increasing prevalence, maturity size is increased and fecundity is decreased. (C) *Littorina saxatilis* infected with 11 microphallid species: growth (in size) and fecundity as a function of age in different populations (compiled and redrawn from Fredensborg and Poulin, 2006, Chapuis, 2006, Granovitch et al., 2009).

*Digenea* 193

2. *Z. subcarinatus* – microphallids (Table 4.12, Fig. 4.13A)

(A) Contrastingly, size at sexual maturity is decreased, i.e. sexual maturity is advanced with increasing prevalence. (B) Prevalence is increased up to 100%, in comparison to 50% by *F. hepatica* in *G. truncatula*. (C) Infected populations continue to grow until the age of the observed period of 300 d. (D) Fecundity is progressively increased with increasing size and advancing age. However, the level of the slope for fecundity vs prevalence relation is higher for the lightly infected TB population than that for the heavily infected LPB population, suggesting the cost of parasitic intensity on fecundity. (E) Interestingly, there is a tradeoff between TB and LPB populations. The former produces more number of small eggs and small sized (182 μm) embryo and the latter smaller number of large sized eggs and embryos.

3. *Littorina saxatilis* – Castrating microphallids (Table 4.12, Fig. 4.13C)

(A) In the investigated Navolok and Yokovela populations, age at sexual maturity is postponed, in contrast to the observation in *L. saxatilis* infected with *M. novaezealandensis* (Fig. 4.13A). (B) However, fecundity is increased with advancing age (size) of the snail, confirming the observation in *Z. subcarinatus* and *M. novaezealandensis*. (C) No marked change is indicated in fecundity with increasing prevalence.

4. *L. saxatilis* lighthly or heavily infected (Sokolova, 1995)

(A) Prevalence is limited to < 75% and ranges from 5% in lightly infected snails to 75% in heavily infected ones. (B) Life span is extended from 7 y in the former but to 10–12 y in the latter, i.e. increasing prevalence extends the life span of the snail. (C) With increasing prevalence, mean fecundity is increased from 24 offspring in the lightly infected snail to 144 offspring in the heavily infected snail, at the comparative age of 6 y. Clearly, increasing in prevalence extends life span as well as produces more offspring, in contrast to those reported for freshwater snail infected with *F. hepatica*.

Comparison of the ultimate responses of the snails suggests the following: 1. The infected snails continue to grow with increasing size and advancing age, suggesting the occurrence of gigantism. However, the response of *L. saxatilis* to the castrating microphallids represents stunting. 2. With increasing prevalence, fecundity is reduced in the freshwater almost annual (1.5 y) oviparous hermaphroditic *G. truncatula* but increases in the marine long-living ovoviviparous gonochoric *L. saxatilis*. Hence, the ultimate response depends on sexuality, life span and mode of reproduction. 3. Size or age at sexual maturity also differs with reference to these factors, besides parasitic nature of the digenean. For example the microphallids lets the hosts to reproduce but *F. hepatica* castrates the host.

# 5

## Cestoda

### Introduction

The cestodes are endoparasitic tapeworms devoid of a epidermis, mouth or digestive tract. Their body consists of an anterior attachment organ, the *scolex* and an elongated ribbon-like body divided by transverse constrictions into segments known as *proglottids*. However, the monozoic cestodes have a simple, undivided body resembling flukes in appearance. The ribbon-like body is opaque white or yellowish (or rarely pinkish in *Diphyllobothrium latum* due to selective absorption of vitamin B12, Marchiondo et al., 1989) in color, ranges from 1 mm to 10–12 m in length and consists of 3 (e.g. *Echinococcus*) to 4,000 (e.g. *Diphyllobothrium latum*) proglottids. Typically, it consists of a small knob-like or clevate head, the scolex bearing organs of attachment like hooks and suckers in *Taenia solium* or bothridia or phyllidia in tetraphyllids (e.g. *Spinometra mansonoides*), followed by a relatively short undivided region, the *neck*, from which the proglottids proliferate and succeed by a long chain of proglottids called the *strobila*. All the non-segmented tapeworms were earlier considered as Monozoa and the segmented ones as Polyzoa. But this classification is no longer valid. For example, the 'monozoic' caryophyllaeids are now classified within the order Pseudophyllidea. Incidentally, the segmentation is limited to the anterior strobila in *Ligula* (Hyman, 1951). Consequent to the absence of a mouth and gut, they are unable to exploit diverse habitats such as the lung, kidney or circulatory system and are relegated to the host gut as the sole habitat. Nevertheless, the (i) osmotrophic absorption of the already digested superabundant nutrients from the gut, (ii) easier access for the passage of their eggs to the exterior through host excreta and (iii) reduced host immune-responsiveness for parasites in the gut have greatly enhanced their success and evolution (Mackiewicz, 1988). Processing of the superabundant nutrients for the gainful production of enormous number of eggs by the unique repetitive proglottids each with dual reproductive systems has neutralized the obstacles encountered during trophic transmission by one or more of their larval forms to the next host. In cestodes, the first intermediate host is more often an arthropod and less often a mammal, in some of which clonal multiplication occurs (Table 5.7).

## 5.1 Taxonomy and Diversity

On the basis of larval characteristics and morphological features like the scolex (in cestodes hosted by elasmobranchs, Caira et al., 2014), the class Cestoda is divided into two subclasses Cestodaria and Eucestoda. In turn, the former is divided into two orders: (i) Amphilinidea and (ii) Gyrocotylidea with 0.2 and 0.3% of overall cestoda species, respectively. The latter is divided into nine orders by Hyman (1951) but the number is increased. According to the molecular phylogeny of Littlewood et al. (2015), the class Cestoda comprise 19 orders, 72 families, 867 genera and 4,671 species (Table 5.1). With highly motile tetrapods in promoting diversity, cestodes are hosted by 29% of families, 55% of genera and 49% of species. In aquatic habitats, fish serve as a definitive host for 46% of cestodes. The marine elasmobranchs host 53% of families, 24% of genera and 28% of species. The mainly freshwater teleosts host cestodes in 17% of families, 21% of genera and 18% of species (Littlewood et al., 2015). More than 150 elasmobranch species are known to host 400 species of adult cestodes representing five orders (Campbell, 1983). As in digeneans (pp 21–23), host motility has been the driving force to promote diversity and speciation in cestodes.

DNA techniques have resolved a few issues on classification at order and species levels. For example, an apparent conflict between scolex morphology and proglottid anatomy impeded the assignment of many genera to families in the order Lacanicephalidea. Employing two nuclear markers (D1–D3 of *lsrDNA* and complete *ssrDNA*) and two mitochondrial markers (partial *rrnl* and partial *cox1*) maximum likelihood and Bayesian analyses were made for 61 lacanicephalidean species representing 23 of 25 valid genera. The analysis (i) confirmed monophyly of the order, (ii) synonymized and placed sesquipedalian among the species of *Anteropora*, (iii) maintained the names of existing four families but proposed new names for four others and (iv) showed that three of four orders of Batoidea exclusively host the lacanicephalideans but no skate (order Rijiformes) hosts any of them (Jensen et al., 2016). *Cyclops abyssorum praelpinus* and Arctic charr *Salvelinus umbla* serve as first and second intermediate hosts for the tapeworms *Triaenophorus crassus* and *T. nodosus*. The use of Ban I restriction sites within 207 bp – 18s rRNA-amplified fragment yielded two and three species specific products to distinguish between *T. crassus* and *T. nodosus* (Boufana et al., 2011). For tapeworm identification, Schmidt (1986) may be consulted.

## TABLE 5.1

Phylogeny of major Cestoda groups indicating numerical diversity of families, genera and species. Molecular phylogenies for cestodes are derived mostly from ssrDNA (18S) and lsrDNA (28S) ribosomal RNA genes (condensed from Littlewood et al., 2015)

| Phylogeny | Taxonomic group | Family (no.) | Genera (no.) | Species (no.) |
|---|---|---|---|---|
| | Gyrocotylidea | 1 | 1 | 10 |
| | Amphillinidea | 2 | 6 | 15 |
| | Caryophyllidea | 4 | 51 | 167 |
| | Spathebothriidea | 2 | 5 | 6 |
| | Haplobothriidea | 1 | 1 | 2 |
| | Diphyllobothriidea | 3 | 15 | 129 |
| | Diphyllidea | 1 | 5 | 58 |
| | Trypanorhyncha | 21 | 79 | 389 |
| | Bothriocephalidea | 4 | 57 | 263 |
| | Litobothriidea | 1 | 2 | 9 |
| | Lecanicephalidea | 4 | 32 | 161 |
| | Rhinebothriidea | 1 | 15 | 114 |
| | Cathetocephalidea | 2 | 3 | 10 |
| | Tetraphyllidea | ? | 4 | 20 |
| | Phyllobothriidea | 2 | 34 | 164 |
| | Tetraphyllidea | ? | 7 | 51 |
| | Onchoproteocephalidea | 5 | 78 | 658 |
| | Tetraphyllidea | ? | 9 | 70 |
| | Nippotaeniidea | 1 | 2 | 6 |
| | Mesocestoididae | 1 | 2 | 32 |
| | Tetrabothriidea | 1 | 8 | 73 |
| | Taenioidea | 15 | 151 | 2264 |
| | **Total** | **72** | **867** | **4671** |

## 5.2 Reproductive Systems

According to the description of Hyman (1951), the rudiments of the reproductive system appear in young proglottids shortly behind the neck. As most cestodes are protandrous, the proglottids with (i) mature male organs, (ii) fully developed systems of both sexes and (iii) ripe or gravids are arranged in the antero-posterior sequence. However, all the proglottids mature simultaneously in many pseudophyllids. In taenioids like *Cotugni* and *Moniezia* (Fig. 5.1), trypanorhynchids like *Dibothriorhynchus* and pseudophyllids like *Diplogonoporus* (*Diphyllobothrium*), each proglottid contains bilaterally arranged two complete reproductive systems. In the taenioid, *Paronia* (Fig. 5.1) and *Cittotaenia* have double sex organs (Hyman, 1951).

In each proglottid, the testes are usually small, numerous, rounded bodies up to several hundreds or even thousands in number, as in tetraphyllids (Fig. 5.1). But they are reduced to fewer (1–3) and larger, as in *Hymenolepis*.

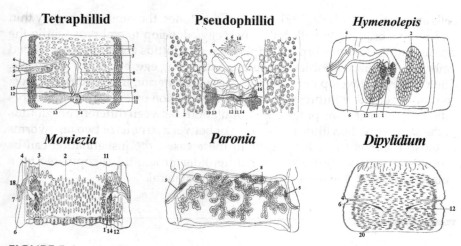

**FIGURE 5.1**

Reproductive system of selected cestodes. 1–yolk glands, 2–testes, 3–sperm duct, 4–cirrus sac, 5–cirrus, 6–vagina, 7–female gonopore, 8–uterus, 9–yolk ducts, 10–common yolk duct, 11–ovary, 12–seminal receptacles, 13–ovicapt, 14–shell gland(s), 15–uterus duct, 16–prostatic gland, 17– ootype, 18–antrum, 19–oviduct. Note the presence of dual reproductive systems in *Moniezia*, double male systems and horse-shoe shaped uterus in *Paronia* and uterus capsules following breaking up of the uterus in *Dipylidium* (free hand drawings from Hyman, 1951).

They are strewn dorsally in the central mesenchyme in tetraphyllids. Fine sperm ductules join to form the sperm duct, which terminates in the cirrus sac holding the cirrus. Prior to entry into the cirrus sac, the sperm duct may enlarge to form seminal vesicle. Prostatic glands are usually present.

The single ovary is bilobed and is located ventrally and posteriorly in the central mesenchyme (Fig. 5.1). Each lobe is subdivided into subsidiary lobules. The yolk glands occur as numerous follicles arranged in lateral bands. In the clonally reproducing taenioids alone, the yolk gland is reduced to a small body. The ductules of the follicular yolk glands unite to form the yolk duct, which eventually enters the oviduct. The oviduct originates from the interconnecting ovarian bridge of the lobes and enters the vagina, which terminates in an enlarged shell gland. Following entry of the glands, the oviduct is called an ovovitelline duct, which is then enlarged to form the uterus. The most characteristic feature of the cestode reproductive system is the uterus, which develops into a median large tube in tetraphyllids (Fig. 5.1) and trypanorhynchids or slender median tube with lateral branches in taenioids or large spirally coiled tube in pseudophyllids or horseshoe-shaped tube in *Paronia* (Fig. 5.1). As the taeniod proglottid ripens, the uterus fills the entire proglottid, while all other organs degenerate.

The eggs are sucked into the oviduct by the ovicapt (Fig. 5.1). Following fertilization, they are enclosed with yolk cells in a shell or capsule. For example, each capsule of caryophyllidean *Khawia sinensis* contains six yolk

198   *Reproduction and Development in Platyhelminthes*

cells and one egg (Scholz, 1991b). In some cestodes, the capsules are very thin and receive only a few yolk cells. Their eggs develop to embryos within the uterus and these tapeworms shed the ripe proglottids containing developed embryos. Others with thick capsules contain one egg but many yolk cells. Their embryos develop only after reaching the exterior. In cestodes, the most common method of impregnation is self fertilization by eversion of the cirrus into the vagina of same proglottid. Copulations between different proglottids of the same strobila within a tapeworm or between strobila of two tapeworms can also occur but less frequently. In these cases, the insemination can be mutual or bilateral. But when a mature older proglottid is copulated by a young proglottid, the insemination can be unilateral.

## 5.3  Life Cycles

In cestodes, the indirect life cycle includes two to three distinct larval stages and involves one to two intermediate hosts and a vertebrate as a definitive host. Unlike digeneans, which are tied to a mollusc as First Intermediate Host (FIH), the cestodan choice is wide spread through the entire spectrum of arthropods, annelids and tetrapods, as well. With exception of a very few taenioids, no 'clonal' multiplication occurs in any intermediate host of cestodes. In fact, the digenean life cycle is more complicated with a minimum of four larval forms and one to three intermediate hosts including the ubiquitous 'propagatory' multiplication in a molluscan host. Nevertheless, information on the complete life cycle is described for a relatively far less number of cestodes than that for digeneans (Fig. 1.3B). In *Mesocestoides*, the inability of the hexacanth to infect a definitive dog host has been proved. But the missing intermediate host that links the tetrathyridium and the dog has not been identified since the 1930s (see Hyman, 1951). In *Bothriocephalus claviceps* too, the linking second intermediate host between first intermediate copepod host and definitive eel (*Anguilla* spp) host remains to be discovered (Scholz, 1997). For the first time, an attempt has been made to trace the ontogenetic pathways of cestodan life cycles. It has brought to light that two intermediate hosts are required for completion of the life cycle commencing with coracidium/lycophore but one or none with oncosphere/hexacanth.

### 5.3.1  Larval Forms

A comparison of eggs, larval forms and ontogeneses suggests that the cestodan life cycle is divisible into aquatic and terrestrial patterns. The former includes (a) the oncosphere and (b) coracidium types, i.e. the cycle commences with oncosphere in the former but coracidium in the latter, respectively. The terrestrial pattern covering mostly the taenoids includes three types; on the basis of larval sequence combination, the pattern is divisible into (a) hexacanth-cysticercoid, (b) hexacanth-tetrathyridium and (c) hexacanth-cysticercus or the bladder worm types.

*Eggs*: In the aquatic oncosphere type, the egg capsule is ovoid in shape and relatively smaller in size, for example, 43 x 30 µm in *Khawia sinensis* (Scholz, 1991b). In this type, egg size ranges from 25 µm in *Proteocephalus osculatus* to 49 µm in *P. torulosus*. The corresponding values for the oncosphere range from 16 to 23 µm (Table 5.2). These values suggest that yolk in the yolk cells is converted into the oncosphere at the efficiency of ~ 61%. Incidentally, the yolk is converted into larva at 60–80% efficiency in many aquatic free-living crustaceans (Pandian, 1967, 1970 a,b). In the coracidium type, the egg is also ovoid in shape and is relatively larger in size, e.g. 60 x 38 µm in *Bothriocephalus claviceps* (Scholz, 1997). Osmotic uptake of water inflates the egg volume to facilitate its floating in *P. percae* (Fig. 5.2G). Others like *Diorchis* sp and *D. myrocae* mimic diatoms (Mackiewicz, 1988). In the aquatic pattern, the first larva is followed by procercoid and plerocercoid stages in both the types. In the oncosphere type, either a copepod/gammarid or an aquatic oligocheate serves as an intermediate host, in which the development of both procercoid and plerocercoid is completed. But two different hosts are required to complete the cycle commencing with coracidium.

*Oncosphere* is a rounded or oval mass of cells without an outer epithelium but is provided with three pairs of hooks (Fig. 5.2I). The *lycophore* of cestodarian has five pairs of hooks (Fig. 5.2B). Only when the capsule is ingested by an appropriate host, the oncosphere emerges and forms into various organs.

*Coracidium*: On enclosing the oncosphere in a one layered ciliated (e.g. *Bothriocephalus claviceps*, Fig. 5.2D) or unciliated (pseudophyllid *Eubothrium*) membrane, it becomes a spherical coracidium. Not much information is yet available on its life span and swimming capability. The ciliated coracidium swims to maintain its pelagic level but the unciliated one falls to the bottom (Hyman, 1951). It is hatched from the egg of trypanorhynchan *Lacistorhynchus tenuis* after ~ 8 days at 11–15°C but 5 days at 19°C in 75% sea water and survives for 60 days at 15°C in 75% sea water (Sakanari and Moser, 1985).

## TABLE 5.2

Size of egg and oncosphere of *Proteocephalus* spp (simplified from Scholz, 1999)

| Species | Egg (µm) | Oncosphere (µm) | Oncosphere as % of egg (%) |
|---|---|---|---|
| *P. osculatus* | 23–26 | 14–18 | 65 |
| *P. macrocephalus* | 25–31 | 16–20 | 64 |
| *P. cornuae* | 27–36 | 19–22 | 68 |
| *P. longicolis* | 41–52 | 23–32 | 59 |
| *P. torulosus* | 45–53 | 21–25 | 47 |

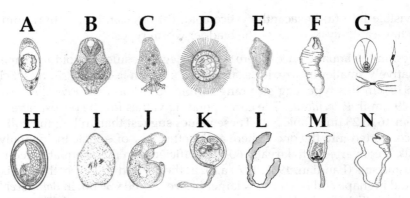

**FIGURE 5.2**

Cestode eggs and larvae: (A) Capsule, (B) lycophore and (C) procercoid larvae of the cestodarian amphilinidean *Amphilina foliacea*. (D) coracidium, (E) procercoid and (F) plerocercoid larvae of aquatic *Dipyllobothrium latum*. (G) Large floating eggs of *Proteocephalus* and *Diorchis* (drawn from Jarecka, 1961). (H) hexacanth, (I) oncosphere (J) developing and (K) completely developed cysticercoid of *Hymenolepis exigua*. (L) *Spirometra* larva, (M) *Caryophyllaeus* larva from a tubificid and (N) Tetrathyridium larva of *Mesocestoides* from a lizard (free hand drawings mostly from Hyman, 1951).

*Procercoid* has a solid but slender body and bears hooks on the posteriorly located cercomer region, used for clinging (Fig. 5.2E).

*Plerocercoid* has an elongate solid body, which bears an adult scolex (Fig. 5.2F).

*Hexacanth* is the first larva in taeniods. Unlike the oncosphere, it is covered with one to three acellular membranes. As in oncosphere, it bears the characteristic three pairs of hooks (Fig. 5.2H).

*Cysticercoid* succeeds the hexacanth. It has a single non-invaginated scolex withdrawn into a small anterior vesicle. It has a posterior tail-like region with persisting larval hooks (Fig. 5.2J, K).

*Cysticercus* is characterized by a single scolex invaginated into its fluid-filled large bladder. It is equipped with suckers and hooks.

*Tetrathyridium* has an elongate, solid body with a deeply invaginated acetabular scolex and occurs in *Mesocestoides* and *Cylindrotaenia* (Fig. 5.2N).

### 5.3.2 Ontogeneses and Life Cycles

In cestodes, the egg is enveloped with a few yolk cells in a capsule with or without a lid. Rarely, many embryos are packed in a single capsule (e.g. linstowiinid *Inemicasifer*). In some cestodes, the egg with > 6 yolk cells (see Scholz, 1991b) is enveloped in a thick scelerotic capsule. Apparently, these eggs can be alive and generate embryos that can remain infective for a long duration. Conversely, the taenioid egg is enclosed with a single yolk cell in a thin evascent capsule. While still in the worm's uterus, it develops into a typical (oncosphere) hexacanth. The ripe gravid proglottid filled with hexacanths is shed. In still others like some pseudophyllids, the egg is enclosed with < 6 yolk cells (see Scholz, 1991b) in a thin capsule. This egg type displays different adaptive strategies. For example, *Proteocephalus* eggs are not enclosed in tanned capsules. In lotic waters, the readily inflatable (due to osmotic uptake of water) *P. osculatus* egg quickly swells, floats and facilitates its capture by planktonic copepods. Living in riverine waters with strong currents, *P. torulosus* egg is less inflatable and sinks to be fed by demersal copepod *Cyclops strenuus* (Scholz, 1999). Incidentally, the shedding process of reproductive products may have implications to iteroparous or semelparous life history and fecundity, as well. The cestode eggs are shed as eggs or proglottid (iteroparity) or by the voluntary exit of the worm, as in cestodarian *Gyrocotyle rugosa*, a rarity among cestodes or by rupturing the worm body, as in *Archigetes* (see Mackiewicz, 1981). The bothriocephalid *Triaenophorus* is a semelparous annual; following egg release, the tapeworm dies (Hyman, 1951).

For cestodarians, the complete life cycle is known for *Amphilina foliacea* with the definitive host *Acipenser*. The thin shelled capsule (Fig. 5.2A) is released through the abdominal pore to the exterior. During its passage along the long uterus, the egg within the capsule develops into cestodarian larva *lycophore* with ciliated epidermis and five pairs of hooks (Fig. 5.2B). However, the lycophore is hatched only after the capsule is ingested by first intermediate host *Gammarus* or *Dikerogammarus* (Fig. 5.3A). It develops into procercoid and then to plerocercoid. In *G. rugosa* with definitive elephant fish host (*Gnathonemus petersii*), the complete cycle remains to be described.

In the oncosphere type, it becomes a *procercoid*, when covered with a ciliated or non-ciliated one layer. One generation of life cycle is completed, irrespective whether a crustacean (*Cyclops*) (Fig. 5.3C) or an aquatic oligochaete (Fig. 5.2B) serves as an intermediate host. Strikingly, despite their relatively smaller size than that of *Tubifex/Limnodrilus*, the copepods are able to support the complete development of procercoid and plerocercoid. In aquatic pattern, the oncosphere limits the number of intermediate host to one only.

Conversely, the coracidium type involves two intermediate hosts. The procercoid of *Triaenophorus crassus*, for example, grows to lengths of 176–295 μm in ~ 26 days in *Cyclops bicuspidatus thomasi* (Rosen and Dick, 1983). When the infected copepod is ingested by a fish, the second intermediate host, develops into a plerocercoid. The cycle of marine otobothriid *Poecilancistrium caryophyllum* (Fig. 5.4A), diphyllobothriid *Diphyllobothrium latum*, trypanorhynchid *Haplobothrium globuliforme*, tetraphyllidean *Acanthobothrium coronatum* (zoology.uok.edu.in), diphyllobothriid *Ligula intestinalis* (Loot et al., 2001) and bothriocephalid *T. crassus* (Rosen and Dick, 1983) commence with free-living coracidium; they engage a copepod and a small fish to complete procercoid and plerocercoid stage, respectively.

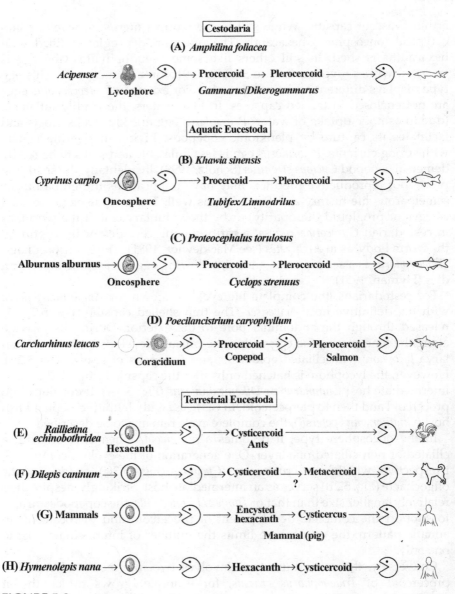

**FIGURE 5.3**

Ontogenetic pathways of some cestodes.

For example, *T. crassus* engages *Cyclops bicuspidatus thomasi* and *Coregonus clupeformis* as intermediate hosts and *Esox lucius* as a Definitive Host (DH). Similarly, *A. coronatum* engages small crustaceans and sardines as well as sharks as intermediate hosts and DH, respectively. The ontogenetic sequence of *L. intestinalis* is coracidium → procercoid in copepod *Eudiaptomus gracilis*→

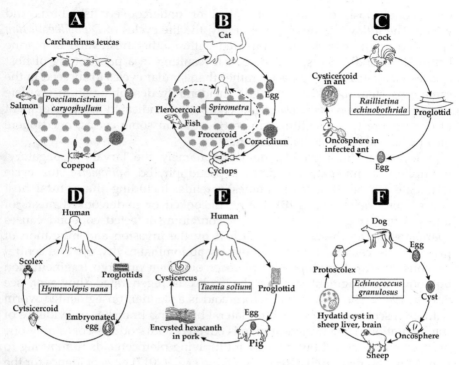

## FIGURE 5.4

Representative life cycles of cestode. (A) *Poecilancistrium caryophyllum*, (B) *Spirometra*, (C) *Raillietina echinobothrida*, (D) *Hymenolepis nana*, (E) *Taenia solium* and (F) *Echinococcus granulosus*. Note the cycles in A & B is more of aquatic but those of C to F are terrestrial. In D, the cycle involves no intermediate host and in F, it involves 'asexual' cloning in hydratid cyst, a larval form.

plerocercoid in roach *Rutilus rutilus* → adult in grebe *Podiceps cristatus*. Within the aquatic cestode pattern, the circumboreal *D. latum* eggs pass from coracidia through procercoids in *Cyclops* or *Diaptomus* and plerocercoids in perch (*Perca fluviatilis*) or pike (*E. lucius*) to the piscivorous birds as DH. Hence, switching to terrestrial DH occurs in diphyllobothriids.

Some diphyllobothriids provide examples for switching to terrestrial DH. Incidences are known for the palaearctic *D. dendrictum* adults from fishes in north Europe, but from birds in USA and arctic fox *Alopex lagopus* in Alaska. Reports are also available for the incidence of *D. ursi* from polar bear, *D. stemmacephalum* from dolphin and *Diplogonoporus balaenopterae* from whales (see Kuchta et al., 2013) as well as *D. grandis* from whales and seals in Japan (Hyman, 1951). *D. dendrictum* has been introduced in Argentina and Chile, perhaps by migrating birds like *Sterna paradisea* and *Larus pipixum* and with the introduction of rainbow trout *Oncorhynchus mykiss* at the beginning of the 20th century. Humans are also susceptible and acquire the disease

204　*Reproduction and Development in Platyhelminthes*

diphyllobothriosis by eating uncooked or undercooked fish, birds and whale. The following are notable: Firstly, the life cycles of *Diphyllobothrium* spp need not necessarily be completed within either freshwater or marine habitats. Secondly, the spatial dispersal resulting in a partitioning of host species facilitates several species rather than similar cycles to succeed in the same ecosystem (Mackiewicz, 1981). Incidentally, describing an incomplete life cycle of *Bothriocephalus claviceps*, Scholz (1997) indicated the development of oncosphere to coracidium, suggesting that in some cycles, oncosphere may be hatched prior to its ingestion.

In the coracidium type, a deviation namely the larval 'propagatory' multiplication must be noted. In a pseudophyllid *Spirometra*, the cycle terminates in adult wild or domestic felids, including the natural host bobcat *Lynx rufus* (Fig. 5.4B). Eating uncooked or undercooked meats of infected fish/frog or drinking water containing infected copepod causes sparganosis. The disease is established by the invasion and migration of pro- or plero-cercoids into subcutaneous, abdominal cavity, eye and central nervous system. Pseudophyllid plerocercoid often undergo fragmentation but only the fragment bearing the scolex is regenerated (however, see Smyth, 1949). The *Spirometra* plerocercoid is a slender, unsegmented worm without a scolex and often exhibit lateral buds and branches or evidence of fragmentation. The infected human is covered with nodules containing one or more encapsulated but actively multiplying plerocercoids numbering to thousands (Hyman, 1951). Recently, Okino et al. (2017) gave evidence for the occurrence of 'apomictic reproduction' in triploid *S. erinaceieuropaei*. In fact, apomictic parthenogenesis has earlier been demonstrated in pseudophyllids *Glaridacris luruei*, *G. catostomi*, *Isoglaridacris bulbocirrus* and *Atractylytocestus huronensis* (Mackiewicz, 1981). But it is not clear whether the plerocercoid undergoes clonal fragmentation or sexual apomictic reproduction.

The ontogenetic pathway of terrestrial cestodes, especially the taenioids, the cycle commences with (oncosphere) hexacanth and includes three sequences of larval combinations described earlier (Fig. 5.3). The families Devaineidae (e.g. *Raillietina echinobothrida*, Figs. 5.3F 5.4C), Dilepididae (e.g. *Dilepis caninum*) and Hymenolepidae (e.g. *Hymenolepis exigua*) are characterized by hexacanth-cysticercoid combination. For *R. echinobothrida*, *D. caninum* and *H. exigua*, ants (*Pheidole vinelandica*/*Tetramorium caespitum*), flea (*Ctenocephalus canis*) and gammarid (*Orchestia*) serve as intermediate host, respectively. Their DH is chicken, dog and Hawaiian chicken, respectively. In families Mesocestoididae (e.g. *Mesocestoides*) and Nematotaeniidae (e.g. *Nematotaenia*), the hexacanth is followed by a larva called tetrathyridium. In the former with a dog as a definitive host, an unknown intermediate host serves and in the latter, for example, *Nematotaenia* with frogs as definitive hosts, an intermediate host is totally eliminated from the cycle; this holds also true for *Hymenolepis nana* (Fig. 5.4D). From a decade long survey of 706 amphibians represented by eight species in Iowa State of USA, Ulmer and James (1976) recorded the incidence of 13.6% cestodes infected with nematotaeniids. Notably,

*Cestoda* 205

the incidences are 14 for *Mesocestoides* tetrathyridium and *Cylindrotaenia americana*. The family Taeniidae includes a combination of hexacanth-cysticercus. In them, different species of mammals serve as an intermediate host and a definitive host. The hexacanth usually remains encysted in the musculature of intermediate host (Figs. 5.3G, 5.4E). Clonal multiplications occur in the hydratid cyst of a few taeniids (e.g. *Echinococcus granulosus*, Fig. 5.4F) and is elaborated later.

The penetration process of oncosphere was described by Scholz (1997) from the gut to hemocoel of *Cyclops strenuus*. It required 60 minutes for completion of the process. In all *C. strenuus* examined (but not all the exposed), the oncospheres were found in hemocoel between 75 and 120 minutes after the commencement. However, information was not provided on the fraction of oncospheres that were unsuccessful and dead. Invertebrates are not known to secrete pepsin (Mukhin et al., 2007). They may maintain a neutral or slightly alkaline gut. Further, the half digested food in them is subjected to intracellular digestion (Pandian, 1975, Section 1.5). Hence, the brief passage of oncosphere in the invertebrate gut may not pose a great challenge. The procercoids are clothed with a thick cuticular 'skin' to enable them to osmotrophically absorb nutrients from their hosts. Vertebrates including fish secrete acid and pepsin in the stomach to facilitate chemical digestion of food at acidic pH (Pandian, 1975). *Ligula intestinalis* procercoid is reported to secrete acid to drop pH in the culture medium from 8 to 5 (Smyth, 1949). It is not known how the procercoid/plerocercoid encounters the acidic gut of fish. The taeniid hexacanth also encounters a similar 'acidic climate' during the brief passage from the mammalian gut into the blood stream or other organs.

Representative examples for the durations of embryonic (oncosphere) development and larval forms are listed in Table 5.3. *Spirometra* sp requires 130 days to complete ontogenetic development from an egg to plerocecoid (Kavana et al., 2014). The duration is species specific, larva-specific and within each of which temperature plays a profound role. Available values indicate that infectivity of larval forms is also species and larva specific and is a temperature-dependent trait in aquatic cestodes but temperature + moisture-specific trait in terrestrial cestodes.

The time required for completion of embryonic development is reported to depend on temperature, copepod sex and host density. In the oncosphere type, it ranges from 16 days at 23°C to 57 days at 11°C for oncosphere development in *Khawia sinensis* (Table 5.3, Fig. 5.5A). Similarly, the duration required for completion of both procecoid and plerocercoid stages of *Proteocephalus neglectus* in *Cyclops strenuus* is 30 days at 6°C to 18 days at 21°C (Fig. 5.5B). In the coracidium type, the procercoid of *Triaenophorus crassus* requires ~ 26 days in *C. biscupidatus thomasi* but its body growth depends on copepod sex. With intense competition, growth in the coracidium type *T. crassus* procercoid is slower in *C. biscupidatus thomasi* males than in the less intensely competed copepod females (Fig. 5.5C). To complete larval development, the

## 206 *Reproduction and Development in Platyhelminthes*

## TABLE 5.3

Representative examples for oncosphere development and larval forms as well as their infectivity duration

| Larval form | Remarks |
|---|---|
| | **Duration required for development** |
| Oncosphere | Embryonic development: 2 days in *Diphyllobothrium sebago* (Meyer and Vik, 1963), 9 days in *Spirometra* sp (Kavana et al., 2014), *Khawia sinensis* 16 days at 23°C but 57 days at 11°C (Scholz, 1991b), *Bothriocephalus claviceps* 20 days at 10°C but 5 days at 21°C (Scholz, 1997) |
| Coracidium | 2 days in *B. claviceps* (Scholz, 1997) |
| Procercoid | 12 days in *Spirometra* sp (Kavana et al., 2014), 14 days in *D. sebago* (Meyer and Vik, 1963), 28 days in *Triaenophorus crassus* (Rosen and Dick, 1983) |
| Plerocercid | 30 days in *Spirometra* sp (Kavana et al., 2014), 113 days in *D. sebago* (Meyer and Vik, 1963) |
| Procercoid + plerocercoid | *Proteocephalus torulosus* 9–12 days at 21°C, 28 days at 10°C (Scholz, 1993) |
| Cysticercoid | *Oochoristica osheroffi* 15 days at 25°C, 10 days at 30°C (Widmer and Olsen, 1967) |
| Cysticercus | 56 days in *Monoecocestus sigmodontis* (Melvin, 1952) |
| | **Infectivity duration** |
| Egg | *Echinococcus multilocularis* infective for 240 days (Veit et al., 1995) |
| Oncosphere | *P. neglectus* 25 days at 10°C, 20 days at 5°C, 10 days at 21°C, *P. torulosus* 35 days at 6°C, 12 days at 11°C, 8 days at 21°C (see Scholz, 1999) |
| Hydatid | *Echinococcus* 10 years (Mackiewicz, 1988) |

*Diphyllobothrium sebago* procercoid requires 14 days in *C. biscuspidatus thomasi* and plerocercoid 113 days in *Salmo salar* (Meyer and Vik, 1963).

Infectivity of larva awaiting transmission from one to the next host is a critically important event. The oncosphere and hexacanth types have to overcome two such critical events and the coracidium three such events. In the oncosphere type, oncosphere development time depends on egg age, contact duration between the egg and host, and temperature, as well. *P. neglectus* eggs remain infective to *C. strenuus* up to the age of 25 and 20 days at 5 and 10°C, respectively but only up to the age of 5 days at 21°C (Fig. 5.5E). The infectivity increases with increasing contact duration up to 250 minutes in *P. neglectus* eggs to *C. strenuus*. But it is limited to 150 minutes for *P. torulosus* eggs to infect *C. strenuus* (Fig. 5.5D-window). Notably, it remains far higher (64%) and longer (240 minutes) in the lotic *P. neglectus* but far lower (27%) and shorter (150 minutes) in the riverine *P. torulosus*. It is likely that the strong current in the riverine waters brings the egg-host contact earlier than in standing waters.

In the hexacanth type, the infection and establishment of *Hymenolepis diminuta* cysticercoid in flour beetle *Tenebrio molitor* increase linearly and proportionately on exposure to increasing number of eggs (Fig. 5.5F), indicating the inability

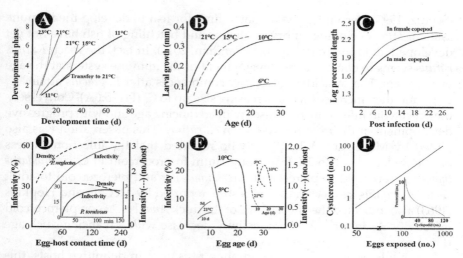

## FIGURE 5.5

(A) Effect of temperature on the development phase of *Khawia sinensis* oncosphere in tubificid intermediate host: 2 = embryo occupies 1/3 of egg length, 4 = embryonal hooks anlagen, 6 = developed penetration glands. Note the change on transfer from 11°C to 21°C at phase 3 (redrawn from Scholz, 1991b). (B) Effect of temperature on larval (procercoid + plerocercoid) growth of *Proteocephalus neglectus* in *Cyclops strenuus* (modified and redrawn from Scholz, 1991a). (C) Growth of *Triaenophorus crassus* procercoid in male and female *Cyclops biscupidatus thomasi* as function of post infection day (modified and redrawn from Rosen and Dick, 1983). (D) Effect of *P. neglectus* egg and *C. strenuus* contact duration on infectivity and intensity of *P. neglectus* oncosphere. Window: the same in *P. torulosus* (compiled and drawn from data of Scholz, 1991a, 1993). (E) Effect of *P. neglectus* egg age on its oncosphere infectivity and density (window) in *C. strenuus* (drawn from data of Scholz, 1991a). (F) Effects of *Hymenolepis diminuta* egg density on cysticercoid density in flour beetle (redrawn from Dhakal et al., 2018) and window: Successful *T. crassus* plerocercoid number as a function of *C. bicuspidatus thomasi* copepodid number (modified and redrawn from Rosen and Dick, 1983).

of the beetle to resist infection intensity. In the coracidium type *T. crassus*, the number of successful plerocercoid, however, decreases with increasing *C. biscupidatus thomasi* cyclopodid number (Fig. 5.5F–window), indicating the reduced infection and colonization of plerocercoids. Rosen and Dick (1983) have not provided any reason for this negative relation.

In the coracidium type, transmission events are more critical, as they involve coracidium to procercoid in immunologically less resistant invertebrate copepod hosts and procercoid to plerocercoid in immunologically more resistant second intermediate fish host. A preamble is required prior to understanding the larval infectivity in the coracidium type. Firstly, development of procercoid into infective plerocercoid, the copepod may require a minimum size like the $C_4$ stage in *Eudiaptomus gracilis*. In copepods, sex differentiation commences at $C_5$ stage and is completed in the adult (see Pandian, 2015). In fact, the infectivity of *P. torulosus* egg on *C. strenuus* changes from 10.7% in copepodid stage to 12.2% in males and 5.9% in females

(Scholz, 1993). Secondly, it takes some time for fish to develop the immune system. The developing embryos, alevins and hatchling of fish have not yet developed an immune system (see Pandian, 2011). In fact, the roach *Rutilus rutilus* seems to complete the development of the immune system at the age of > 2 years. The coracidial infectivity commences after *E. gracilis* attains $C_4$ stage and then increases with increasing size up to $C_5$ (Fig. 5.6A). Conversely, the procercoid infectivity decreases with advancing age and with progressive development of the immune system in *R. rutilus*. This observation was also hinted by Meyer and Vik (1963), who reported that the infection success of ~ 46, ~ 8 and ~ 1 to 2% from coracidium to procercoid in *C. biscupidatus thomasi*, procercoid to plerocercoid in *Salmo salar* and plerocercoid to adult *Larus argentatus*, respectively. The results reported by McCaig and Hopkins (1963) also indicated the effect of age of the duck *Anas boschas* on plerocercoid infectivity. As in a roach, the infectivity decrease with advancing duck age (Fig. 5.6B).

To provide an idea about the immune resistance in definitive hosts, this account provides only a couple of examples. A single infection by *Hymenolepis nana* confers immunity to succeeding infection for a long time. The immunity is passed on to a young one by the infected mother *in utero* and with the milk. The intensity may reach 7,000 for *H. nana* with no intermediate host in omnivorous human, 782 for *H. erinacei* with no intermediate host in insectivorous hedgehog but only 12 for *Dipylidium caninum* with flea as the intermediate host in a carnivorous bulldog (see Hyman, 1951).

According to Hyman (1951), the complete life cycle is not known for any species in the orders Tetraphyllidea, Lecanicephalloidea, Diphyllidea and

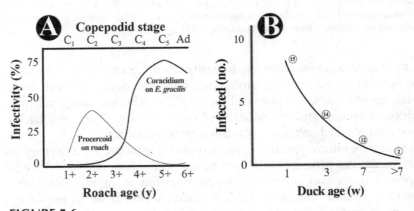

**FIGURE 5.6**

(A) Relationship between *Ligula intestinalis* coracidium infectivity and copepodid stages of *Eudiptomus gracilis* as well as *L. intestinalis* procercoid infectivity and age classes of roach *Rutilus rutilus* (drawn from data of Loot et al., 2001). (B) Infection success of *Schistocephalus* plerocercoid on domestic duck *Anas boschas*. Values in circle indicate the number of plerocercoid to which the duck was exposed (drawn from data reported by McCaig and Hopkins, 1963).

*Cestoda* 209

Nippotaeniidea. A computer search has also indicated no new description of complete life cycles for the species belonging to these orders. The only cycle known for the order Trypanorhyncha is for *Haplobothrium globuliforme* and it is of the coracidial type. Through an experimental study, Sakanari and Moser (1989) described the complete life history of another trypanorhynchan *Lacistorhynchus dollfusi* in the leopard shark *Triakis semifasciata*. The sequence in its experimental life cycle includes coracidium → procercoid in copepod *Tigriopus californicus* → plerocercoid in *Gambusia affinis* (Sakanari and Moser, 1989). Hence, the trypanorhynchans is also of the coracidium type. Thanks to Dr. T. Scholz, the cycles are known for a number of species belonging to the genus *Proteocephalus* in the order Proteocephalidae and they are all of oncosphere type.

Using relevant information scattered in Hyman (1951) and others gathered from a computer search, an attempt has been made to estimate the number of cestode taxa with coracidium, oncosphere or hexacanth type. Notably, the cestode life cycle comprising coracidium requires two intermediate hosts but only one in others for completion of a generation. Hence, the estimates represent the first step for further analysis. However, the available data are indeed very limited for each major taxon, especially like Tetraphyllidea. It is not known whether the ontogenetic combination of larval sequence is a common feature of cestodan orders. Yet, almost the available data, for example, for different species in *Proteocephalus* is of oncosphere type (Scholz, 1991a, b, 1993, 1999). It may also be true for many other taxa. For 4,761 cestode species, the assembled data in Table 5.4 represent 92% of the known cestode species. The share for coracidium, oncosphere and hexacanth types is 29.5, 17.0 and 53.5%, respectively. In other words, the coracidium and oncosphere in aquatic habitat make up 46.5% and the hexacanth 53.5%. This 47% value for aquatic cestode is little less than those (46%, p 195, Littlewood et al., 2015) arrived from another analysis (see Table 7.4, p 243). Hence, the values assembled in Table 5.4 may vary only a little but not too much.

### 5.3.3 Intermediate Hosts

Despite critical risks involved in transmission from one host to the next in the life cycle, cestodes and digeneans have incorporated one to three intermediate hosts. From oionoxenic monogeneans, host specificity is progressively diluted with increasing levels of euryxenicity in digeneans and cestodes. With 'propagatory' multiplication restricted to only a few teanioids, the cestodes are expected as euryxenics. Information scattered in Hyman (1951) indicates that cestodes have exploited one or other taxa from medusa to mammal as intermediate host (Table 5.5). Many of these 'intermediate hosts' are likely to be accidentals or paratenics. Nevertheless, they seem to suggest that mobility and consequent dispersal of the host has been the selection criteria; for example, the mobile scyphozoan medusae are exploited but the sessile hydrozoans and anthozoans are avoided.

# Reproduction and Development in Platyhelminthes

## TABLE 5.4

Taxonomic distribution of larval forms in cestodes (compiled from Hyman, 1951, Meinkoth, 1947, Sakanari and Moser, 1985, Tyler II, 2006, Planetary Biodiversity Inventory).

| Major taxa | Species (no.) | Coracidium | Oncosphere | Hexacanth |
|---|---|---|---|---|
| Caryophyllidea | 165 | 165 | – | – |
| Spathebothriidea | 13 | 13 | – | – |
| Haplobothriidea | 2 | 2 | – | – |
| Diphyllobothriidea | 129 | 129 | – | – |
| Trypanorhyncha | 389 | 389 | – | – |
| Bothriocephalidea | 263 | 263 | – | – |
| Phyllobothriidea | 164 | 164 ? | – | – |
| Tetraphyllidea | 141 | 141 ? | – | – |
| Gyrocotylidea | 10 | – | 10 | – |
| Amphilinidea | 15 | – | 15 | – |
| Diphyllidea | 58 | – | 58 | – |
| Oncoproteocephalidea | 658 | – | 658 | – |
| Nippotaeniidea | 8 | – | – | 6 ? |
| Mesocestoididae | 32 | – | – | 32 |
| Taeniioidea | 2,264 | – | – | 2,264 |
| Tetrabothriidea | 73 | – | – | – |
| Litocephalidea | 9 | – | – | – |
| Lecanicephalidea | 161 | – | – | – |
| Rhinebothriidea | 114 | – | – | – |
| Cathetocephalidea | 10 | – | – | – |
| Total (no.) | 4,678 | 1,266 | 731 | 2,302 |
| As % of 4,311 species | | 29.5% | 17.0% | 53.5%% |

The choice of 23 bird species by *Myxolepis collaris* and 40 bird species by *Ligula intestinalis* are good examples for euryxenicity of cestodes to DH. Similarly, 51 cyprinid species in 40 genera as the choice for pleroceroids in the coracidium type confirms the euryxenicity to the highly mobile SIH. However, the level of euryxenicity is limited to the less mobile invertebrate intermediate hosts. For example, it is limited to 7 naidid species for 5 cestode species and 15 tubificid species to 53 cestode species. Considering, the genus *Proteocepalus*, for which the more information is available, the choice is wider for *P. longicolis* but is limited to *Eudipatomus gracilis* alone for *L. intestinalis*. Table 5.5 indicates the need for discovery of natural hosts for *P. cernae*, *P. macrocephalus* and *P. osculatus*, and *P. thymalli*. Notably, *P. torulosus* can complete procercoid and plerocercoid development only in *Cyclops strenuus* but not in *Mesocyclops leuckarti*, although infectivity of the latter can be 50%

*Cestoda* 211

## TABLE 5.5

Euryxenic host specificity of cestodes for intermediate and definitive hosts. * From Scholz (1999)

| Taxa/species | Intermediate hosts, remarks | | |
|---|---|---|---|
| *Myxolepis collaris* | Eggs: 12 species of copepods, ostrocods, amphipods<br>Cysticercus: 2 lymnaeid snails<br>Adults: 26 bird species, of which 23 are ducks (Mackiewicz, 1988) | | |
| *Ligula intestinalis* | Procercoid: *Eudiaptomus gracilis, Mesocyclops leuckarti*<br>Plerocercoid: *Rutilus rutilus, Alburnus alburnus*. 13 host fish species<br>Adults: 40 bird species, mainly *Podiceps cristatus* (Loot et al., 2001) | | |
| Caryophyllaeids | Procercoid: 7 naidid species host 5 cestode species; 15 tubificid species host 53 cestode species<br>Plerocercoid: 51 cyprinid species in 40 genera, 25 catostomid species in 9 genera, 5 clariid species and 4 cobitid species (Mackiewicz, 1982) | | |
| *Diphyllobothrium* Scholz et al. (2009) | FIH: 40 species in *Acanthodiaptomus, Arctodiaptomus, Diaptomus, Eudiaptomus, Eurytemora, Boeckella, Cyclops*<br>SIH: Freshwater: *Perca fluviatilis, Esox lucius, Lota lota, Sander canadiensis, S. vitreus*<br>Salmonids: *Oncorhynchus gorbuscha, O. keta, O. masou, O. nerka*<br>Brackish water: *Liza haematocheila, Centropomus undecimalis*<br>Marine: *Salmo salar, Engraulis japonica, Sardinops melanosticta* | | |
| Lecanicephaloidea | Pearl oyster, bivalves, copepods, fishes, birds (Hyman, 1951) | | |
| Diphyllidea | Marine molluscs, crustaceans (Hyman, 1951) | | |
| Trypanorhyncha | Encysted plerocercoids in medusa, holothurians, snails, bivalves, cephalopods, crustaceans, teleosts (Hyman, 1951) | | |
| Taenioidea | Ants, fleas, louses, butterflies, dragonflies, earthworms, centipedes, millipedes, mites, mammals (Hyman, 1951) | | |
| *Proteocephalus** | **Natural hosts** | **Natural + Exp hosts** | **Experimental hosts** |
| *P. ambiguus* | *Eudiaptomus gracilis* | – | – |
| *P. cernuae* | – | – | *Cyclops strenuus* |
| *P. filicollis* | *E. gracilis* | *Mesocyclops oithonoides* | *C. strenuus, Eucyclops serrulatus* |
| *P. longicollis* | *E. graciloides, E. zachariasi, C. lacustris, Macrocyclops albidus, Mesocyclops oithonoides* | *C. kolensis, C. strenuus, C. vicinus, C. scutifer* | *C. furcifer, Eucyclops serrulatus* |
| *P. macrocephalus* | – | – | *Acanthocyclops vernalis, C. strenuus C. abyssorum* |
| *P. osculatus* | – | – | *C. strenuus* |
| *P. percae* | *E. graciloides, C. kolensis, C. vicinus, Megacyclops gigas* | – | *C. agilis, M. viridis, Mesocyclops leuckarti* |
| *P. thymalli* | – | *Epischura baicalensis* | *C. kolensis, C. vicinus* |
| *P. torulosus* | *E. gracilis, Heterocope appendiculata* | *C. strenuus, Cyclops* sp, *Eucyclops serrulatus* | *Diaptomus castor* |

212 *Reproduction and Development in Platyhelminthes*

but only 12% in the former (Scholz, 1993). Hyman (1951) also hinted that pseudophyllids require a specific copepod species. Briefly, the cestodes are more of stenoxenics to FIH but euryxenics to SIH and DH, as has been noted for digeneans (see p 173–174). The choice for the euryxenic cysticercoids of *Raillietina echinobothrida* ranges widely; *Pheidole* sp, *P. dentate*, *P. fervida*, *P. vinelandica*, *Onthophagus ater*, *O. viduus*, *Tetramorium caespitum* serve as intermediate host in USA (Bartel, 1965). To *Hymenolepis furcata* cysticercoid too, the choice is widespread in beetles as an intermediate host and changes from country to country; 6 beetle species in 5 genera in Czechoslovakia, 2 species in 2 genera in Poland and in France, but the beetle is replaced by a millipede. An extreme example is the dog, which serves for 7 species of *Taenia*, 4 of *Multiceps* and one each for *Echinococcus* and *Alveococcus*, in addition to serving in life cycles of *Hydatigera* (Mackiewicz, 1988). Being on the top of food chains, predatory carnivores become a key parasitic sink.

## 5.4 Neoteny—Progenesis

As in some digeneans, the neotenic or progenetic strategy is also adopted by some that are not adequately recognized coracidial type. Strikingly, it occurs only in freshwater cestodes and is restricted to two pseudophyllid families alone namely Bothriocephalidae (*Bothriocephalus*, *Cyathocephalus*) and Caryophyllaeidae (*Archigetes*, *Biacetabulum*, *Brachyurus*, *Caryophyllaeus*). In a cyathocephalid, *Botrhionomus*, Sandeman and Burt (1972) found that its larvae often attained sexual maturity and produced eggs, while still in the gammarid intermediate host. However, the eggs neither from the neotenic nor from normal definitive fish host could infect the gammarid. Apparently, the gammarid was a paratenic host. In the ligulid *Schistocephalus* and *Ligula*, the plerocercoid grows to an unusual length of 40 cm in freshwater fish serving as SIH. In DH of aquatic birds, it attains sexual maturity but lives for a short time only. For example, *S. solidus* lives as an adult only for 11 days in the natural duck host (Fig. 5.7A) or 6, 10, 14 and 18 days in experimental host rat, chicken, pigeon and hamster, respectively (Morag et al., 1963).

The true neotenic/progenetic abbreviation of the life cycle occurs only in the coracidium type. For example, the *Bothriocephalus* procercoid develops into an advanced plerocercoid in the first intermediate host *Cyclops* itself and the neotenic advanced larvae directly infects DH and thereby eliminates SIH from its life cycle (Fig. 5.7B). Similarly, *Cyathocephalus* procercoid in *Gammarus* develops into plerocercoid that can directly infect a definitive host (Fig. 5.7C). In the first intermediate tubificid host itself, the procercoid of caryophyllids *Biacetabulum*, *Brachyurus* and *Caryophyllaeus* develop reproductive system and become infective (Fig. 5.7D) (Hyman, 1951). The increasing progenetic tendency culminates in *Archigetes* (see Mackiewicz, 1981). As a consequence, a fraction of the *Archigetes* population undergoes progenetic life cycle with one

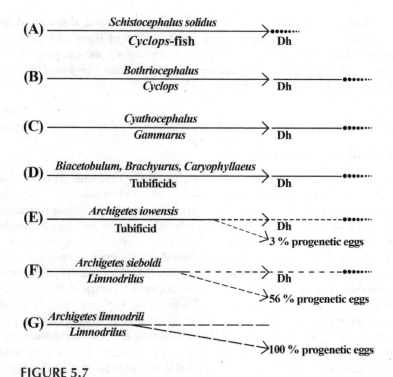

**FIGURE 5.7**
Ontogenetic pathways of some freshwater pseudophyllid cestodes. Dh = definitive host (source: Hyman, 1951, Mackiewicz, 1981).

intermediate host, while the other the normal life cycle with two intermediate hosts. The fraction of one host cycle ranges from 3% in *A. iowensis* with *Tubifex* as intermediate host (Fig. 5.7E) to 56 and 100% in *A. sieboldi* (Fig. 5.7F) and *A. limnodrili* (Fig. 5.7G) with *Limnodrilus* as an intermediate host, respectively. Hence, most worms of *A. iowensis* shall pass through a 2-host cycle but *A. limnodrili* with a 1-host cycle (Calentine, 1963). Drying and freezing of some lotic waters have enforced selection pressure to reduce two intermediate host cycle in the coracidium type to explore different strategies (i) to reduce the adult life span, (ii) to eliminate SIH and (iii) complete the life cycle with a plerocercoid stage at (a) column-inhabiting *Cyclops* and *Gammarus* and (b) sediment-inhabiting aquatic oligochaetes.

In the terrestrial pattern too, the tendency for abbreviation of life cycle is reported. However, progenesis does not occur but the abbreviation is restricted to the aquatic medium alone. In the nematotaenioniid *Cylindrotaenia*, the cycle involves ingestion of the discharged ripe proglottids directly by a definitive frog host. In the intestinal mucosa of the frog, the hexacanth cyst develops into a tetrathyridum, which escapes into the intestine, where it becomes an adult. A more known classical example for abbreviation with no intermediate

214    *Reproduction and Development in Platyhelminthes*

host is *Hymenolepis nana* (Fig. 5.4D). The egg ingested through drinking contaminated water develops into a hexacanth and then to a cysticercoid in the intestinal villi (thereby escapes from the digestive propulsion) in ~ 4 days. Subsequently, the cysticercoid passes into the intestinal lumen, where it grows to maturity (Hyman, 1951).

## 5.5 Prevalence and Intensity

In comparison to digeneans, available information on this aspect is limited for cestodes; however, there are volumes of epidemiological and clinical reports on diseases caused to poultry and mammals including humans (e.g. Eckert and Deplazes, 2004). Still, the cestodes cover ~ 30% of helminthes of North American freshwater fish, 31% of ducks and 33% of cats and dogs. For individual host species, they constitute ~ 18% from humans, 61% from spiny dogfish and 100% from round stingray (Mackiewicz, 1988). In general, prevalence decreases with advancing developmental stage in the oncosphere, coracidium and cysticercercus types (Table 5.6). For example, it decreases with advancing copepodid stage from 35 to 6% in oncosphere type (e.g. *Proteocephalus neglectus*, Scholz, 1991a), including *Khawia sinensis* oncosphere infecting aquatic oligochaetes (Scholz, 1991b), from 57 to 26% in coracidium type (e.g. *Ligula intestinalis*, Loot et al., 2001) and also in cysticercus type (e.g. *Pheidole* sp ant, Harkema, 1943). In fact, the adults of *Pheidole* spp and other ants are not susceptible at all (Horsfall, 1938). Prevalence of plerocercoid of the coracidium type in tetraphylleadean *Scolex pleuronectis* ranges from 4.8% in the Atlantic horse mackerel *Trachurus trachurus* to 55% in ground goby *Neogobius melanostomus* (Guneydag et al., 2017). Notably, the values are in the range of those found for *L. intestinalis* in *Alburnus alburnus* (Table 5.6). Secondly, prevalence is the highest for coracidium type (57–26%) but lowest for cysticercoids type (9.3–2.8%). Clearly, the motile coracidium is more successful than the less motile egg/oncosphere. In terrestrial habitat, successful infection of egg/hexacanth on invertebrate intermediate host seems to be less. In fact, the infectivity of cyticercus for *Ecchinococcus granulosus* is the least with 1.3% success in deer populations (Cerda et al., 2018).

With regard to infection intensity, two observations may be noted: (i) it is limited to 1.6–2.9/host in copepodid of cyclops but 5–15/host in relatively larger aquatic oligochaetes *Limnodrilus* and *Tubifex* (Table 5.6). For *Hymenolepis diminuta* cysticercoids, it is 18/host and 2/host in starved and satiated coleopteran beetle *Tribolium confusum*, respectively. Apparently, healthy beetles resist infection (cf p 103). Incidentally, it differs from the hymenopteran ants with regard to the infective stage. In the context of infection intensity, the responses of an established tapeworm against the entry of potentially competitive new worms can be considered under the

## TABLE 5.6

Prevalence and intensity of cestode infection

| Host species | First intermediate host | Second intermediate host |
|---|---|---|
| *Proteocephalus neglectus*<br>Scholz (1991a) | *Cyclops strenuus*: 35%, 2.7/host | |
| *P. torulosus*<br>Scholz (1993) | Copepodid: 11%. 2.5/host<br>♂: 13%, 2.9/host<br>♀: 6%, 2.4 /host<br>*Mesocyclops leuckarti*: 50%, 1.6 host | |
| *Khawia sinensis*<br>Scholz (1991b) | *Limnodrilus*: 13%, 5/host<br>*Tubifex*: 14%, 15/host | |
| *Traenophorus crassus*<br>Rosen and Dick (1983) | *C. b. thomasi*: 77.5%, 4.9/host | Fish: 30%, 2.6/host |
| *Ligula intestinalis*<br>Loot et al. (2001) | *E. gracilis*: C3 0%, C4 57%, C5 41%, Adult 26% | *Alburnus alburnus*: 1+year 10%, 2+years 41%, 3+years 26%, 4+years 8% |
| *Lacistorhynchus tenuis*<br>Sakanari and Moser (1985) | Procercoid: 4.3%, 1/host at 19°C in 75% sea water | |
| *Raillientnia echinobothrida*<br>*R. loeweni*, Bartel (1965) | *Pheidole* sp ant 2.8% *P. s. campestris* 9.3% *P. bicarinata* 0.8% | |
| *Hymenolepis diminuta*<br>Keymer and Anderson (1979) | Starved *Tribolium confusum*: 18/ starved host, 2/satiated host | |
| *Echinococcus granulosus*<br>Cerda et al. (2018) | Deer 1.3% | |

following three types: (1) Resistance and possible elimination of the new entrants. In *Gyrocotyle rugosa, G. urna* and *G. fimbriata,* Lynch (1945) found the occupancy of only large mature tapeworm or only small juveniles in rat-fish *Hydrolagus collici.* This finding suggests that through the maximum occupiable space or an unknown pheromonal mechanism, infection at different intervals is eliminated. (2) In the so called 'crowding effect' of *Schistocephalus solidus* plerocercoids in *Gasterosteus aculeatus,* multiple entries are allowed but only one may grow the largest among competitors. As a result, the cumulative mass of the plerocercoid as well as the largest plerocercoid mass decrease with increasing intensity (Fig. 5.8A). (3) Allows free entry to new worms resulting in increased intensity. For example, Guneydag et al. (2017) summarized the prevalence and intensity of *Scolex pleuronectis* plerocercoid in the Atlantic Ocean, Mediterranean Sea, Black Sea, Red Sea and so on. Some of the highest values listed by Gunydag et al. are 87% and 17/host in pylorus for *Coryphaenoides mediterraneus,* 83% and 77/host for *Lepidopus caudatus* in its pyloric caeca, intestine and gall bladder. When the values reported by Guneydag et al. (2017) were plotted for the relationship between prevalence and intensity, a positive trend became apparent; up to 30% prevalence, intensity increased linearly to 10 plerocercoids/fish host and subsequently levels off (Fig. 5.8B). It may be recalled that the same

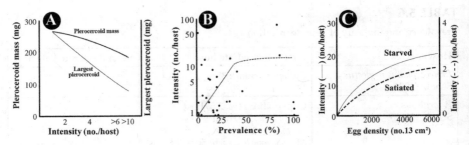

**FIGURE 5.8**

(A) Effect of *Gastrosteus aculeatus* intensity on *Schistocephalus solidus* pleroceroid mass and largest plerocercoid mass (compiled from Heins et al., 2002). (B) Effect of prevalence on intensity of *Scolex pleuronectis* plerocercoid in 40 fish species (drawn using data from Guneydag et al. 2017). (C) Effect of infective cysticercoid *Hymenolepis diminuta* eggs on infection intensity in starved and satiated *Triborium confusum* (compiled from Keymer and Anderson, 1979).

trend between prevalence and intensity has also became apparent for monogeneans (see Fig. 3.2E). In fact, the relationship between intensity and density of infective cysticercoids containing eggs of *H. diminuta* also displays a similar asymptotic trend, *albeit* the level is higher for a starved beetle than for the satiated ones (Fig. 5.8C). In large *Taenia* spp with many proglottids, the worm usually occurs singly, i.e. intensity is 1 worm/host. On feeding, 7,000 *H. diminuta* hexacanths, a maximum of 383 worms developed into adults but beyond 15 worms/rat, reproductive output was grossly reduced to a fewer larvae (Hager, 1941). Hence, investigations are required to study the effect of *S. pleuronectis* intensity on reproductive output of *C. mediterraneus* and *L. caudatus* populations.

As in digeneans, cestodes also manipulate the behavior of the intermediate host to attract definitive hosts. *L. intestinalis* plerocercoid modifies the normal swimming behavior of *R. rutilus* into a jerky movement to attract the bird's attention (Loot et al., 2001). In terrestrial cycles, the heavily infected (by *Multiceps coenuri*) alters the fur color of snow shoe rabbit and thereby renders the rabbit as an easy prey for dogs. On infection of the brain or spinal cord, the *M. coenuri* can cause an array of symptoms from staggering to paralysis and thereby render the host vulnerable for easy predation (see Mackiewicz, 1988).

## 5.6 *Schistocephalus solidus*

The cestodes are protandrics. Among selfing cestodes, insemination can be achieved between different proglottids within a strobila or within the same mature proglottids. However, authors like Williams and McVicar (1968), who listed selfing in a dozen cestode species, did not distinguish these two

*Cestoda* 217

possibilities. Smyth and Smyth (1969) found exclusive self-insemination in *Echinococcus granulosus*. In *S. solidus*, selfing occurs *in vitro* only under compression in cellulose tube. Using ³H thyrimidine labeling, Nollen (1975) provided direct evidence for selfing in singly infected *Hymenolepis diminuta*. However, 92% of cross inseminations occurred in multiply infected worms.

In its coracidium type life cycle, the polyzoic pseudophyllid *S. solidus* passes through the procercoid stage in the hemocoel of copepod *Macrocyclops albidus* and plerocercoid stage in the peritoneum of stickleback *Gasterosteus aculeatus*; the plerocercoid becomes infective only after a period lasting for 1–3 months. Its definitive hosts are piscivorous birds like heron *Ardea conera*, cormorant *Phalacrocorax carbo* or kingfisher *Aledo atthis* (Scharer and Wedekind, 2002, Milinski, 2006). Its adult life span lasts for hardly 10 days, of which oviposition is limited to the initial 6.6 days. On transmission, 60–100% of worms are successfully established as adults (Scharer and Wedekind, 2001) and grow to 50–1,029 mg body weight (Luscher and Wedekind, 2002). Plerocercoid growth depends on single or multiple infection(s) on *G. aculeatus*. For example, it grows to a body weight of ~ 160 mg on the 9th and 14th day, when infected singly and triply, respectively (Barber and Svensson, 2003). All the larval stages of the worm are amenable to experimental rearing. Tempted by the rare, shortest adult life span, Smyth (1946, 1954) developed *in vitro* bird culture system, which was then modified by Wedekind (1997) and improved by Scharer and Wedekind (1999). For details on *in vitro* bird culture system, Luscher and Wedekind (2002) may be consulted. Within the *in vitro* bird tube, a single worm can be reared in isolation or in pairs; the isolation enforces selfing; paired rearing provides an opportunity for the worms to select either outcrossing or selfing, or selfing cum outcrossing. The *in vitro* bird culture system therefore provides a unique opportunity to investigate sperm storage and competition, sex allocation and reproductive output as well as mating system in this hermaphroditic parasitic worm.

With the development of successful *in vitro* system for *S. solidus*, a large number of publications are available. The results reported in these publications are comparatively analyzed and a comprehensive summary is provided hereunder:

(1) Worm size is progressively decreased with increasing parasitic density either as cumulative mass, as in nature (Fig. 5.8A) or *in vitro* culture system (Fig.5.9A). Allocation for sexual development increases with worm age (Scharer et al., 2001). Expected of a protandric, the allocation is (disproportionate to increasing body size) progressively increased for the female function, especially for yolk glands. However, the corresponding increase for the male function is far less. This expandingly decreasing allocation can be further complicated with the social situation like the parasite intensity (Fig. 5.9A). For example, the seminal vesicular sperm volume is progressively decreased from ~ 1.0 mm³ in singles to 0.4 mm³ in triplets (Scharer and Wedekind, 2001). When reared in singles, individual worm stores more seminal vesicular

218 *Reproduction and Development in Platyhelminthes*

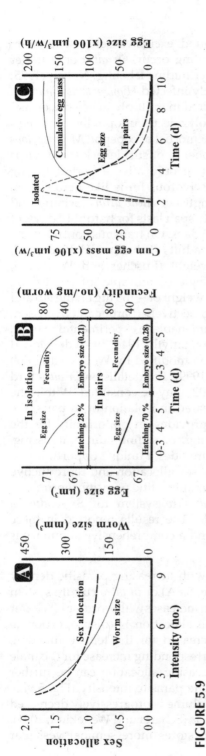

**FIGURE 5.9**

*Schistocephalus solidus* reared in *in vitro* bird. (A) Effect of parasite intensity on worm size and sex allocation. (B) Decreasing trends for egg size and fecundity with advancing age of the worms reared in singles and pairs. In them, hatching success and embryo size are also indicated. (C) Egg mass and cumulative (cum) egg mass produced by worms in isolation and in pairs as function of time with advancing age (compiled and redrawn from Scharer and Wedekind, 1999, 2001, Wedekind et al., 1998).

Cestoda   219

sperm (to be used for insemination). However, the paired and triplet worms store more sperm in seminal receptacles (sperm received via insemination). The values for sperm storage in the receptacles are 0.18 and 0.48 mm$^3$ in pairs and triplets, respectively (Scharer and Wedekind, 2001). As a consequence, the worms reared in pairs and triplets do not exclusively self fertilize.

(2) The egg size of worms reared in singles and in pairs did not differ significantly (Christen and Milinski, 2003, Milinski, 2006). But, Wedekind et al. (1998) found that the size was larger (74 µm$^3$) in pairs than in singles (67 µm$^3$). They also noted that at a chosen age, the larger worms produced larger eggs, irrespective of being reared in singles or pairs. Further, the size progressively decreased with advancing worm age from 0–3 days to the 5th day, again irrespective whether reared in singles or pairs (Fig. 5.9B, C). Briefly, the egg size is a variable trait and is determined by the worm size and age as well as as the social situation namely in isolation or in pairs.

(3) Fecundity, measured either as number of eggs or cumulative of egg mass during the initial 0–3 days, was less in pairs than in singles (Wedekind et al., 1998). But it is equalized, when the mass was considered for all the five days (Fig. 5.9C). Still, relative fecundity, i.e. measured number of eggs per mg worm weight, progressively decreased with advancing age and the trend was at a lower level for pairs than that for singles (Fig. 5.9B).

(4) Hatching success is higher for the outcrossed eggs than that for the selfers: 6.7% for selfers and 24.5% for pairs (Milinski, 2006) and 28% for singles but 70% for pairs (Fig. 5.9B). It is also decreased with advancing age of the worm (Fig. 5.9B). In doubles and triplets, the outcrossed pairs received far more sperm in their seminal receptacles. With sperm competition, relatively less number of eggs remains unfertilized, resulting in less number of infertile eggs that do not hatch. It seems also that mating frequency is decreased with the advancing age of the worm, whose seminal receptacles begin to have less sperm storage and results in increased number of infertile eggs and reduced hatching success in older worms. Interestingly, the relative embryo size is larger in doubles than in singles (Fig. 5.9B).

(5) Hermaphrodites mate primarily to seek an opportunity to fertilize the partner's eggs rather than to secure their own eggs fertilized. With size-dependent sex allocation, the larger mature S. solidus is likely to be more attractive than smaller young ones. Further, a large worm has much more egg mass to offer but will get little in return, if it mates with a small worm. Mate-choice experiments have revealed a general preference for cross-fertilization over selfing. But, the large worms may not necessarily outweigh the cost of mating with small ones (Luscher and Wedekind, 2002). Inbreeding depression is considered as a major selective force for the evolution of mating systems, which favor mating among unrelated individuals. By offering simultaneous choice between an unrelated partner and a sibling in S. solidus, Schjorring and Jager (2007) found that the worm preferred its sibling. Rearing S. solidus

220    *Reproduction and Development in Platyhelminthes*

for two consecutive generations of selfing, Benesh et al. (2014) found that several fitness correlates over the whole life cycle. For example, irrespective of whether *G. aculeatus* hosted plerocercoids arising from outcrossed or selfed eggs, the cumulative growth of plerocercoid is more dependent on parasitic density rather than selfed or outcrossed plerocercoid. Hence, fitness may not totally be dependent on selfing or outcrossing alone. Finding poor correlation between larval size and fitness, Benesh and Hafter (2012) suggested that other parameters like physiological or ontogenetic variation may predict fitness more reliably.

## 5.7 Clonal Multiplication

It may be defined as the production of more than one potential adult from a single oncosphere/metacestode (Whitfield and Evans, 1983). It occurs sporadically in intermediate mammalian hosts in 19 species belonging to four cestode families namely Echinobothriidae (1 species), Dilepidae (1 species), Hymenolepididae (4 species) and Taeniidae (13 species) (Table 5.7). The claims made by many authors for its incidence in *Sparganum proliferum* (Diphyllobothriidae) and *Mesocestoides corti* (Mesocestoididae) could not be proved (e.g. Conn, 1990). Based on the topological relationship, the metacestoides were earlier classified under three categories namely (a) exogenous, (b) endogenous (Fig. 5.10A) and (c) irregularly branched scoleces (Fig. 5.10B). Based on the potency and synchronous/asynchronous multiplication, they are presently grouped into two types. In the coenurus type (Fig. 5.10A, B), the potency of germinative cells to initiate the production of scoleces is (by symmetric mitosis?) synchronous but limited to the metacestoide stage alone. In this type, the cysticerci are proliferated from the inner wall of the bladder and are not detached. At the maximum, 100 scolex buds may synchronously be generated. This type of clonal multiplication occurs in the teaniids on *Taenia crassiceps, T. endothoracicus, T. multiceps, T. mustelae, T. parva, T. pisiformis, T. selousi, T. serialis* and *T. twitchelli* (Whitfield and Evans, 1983). In hydatid type (Fig. 5.10C), while a fraction of germinative cells initiate the production of new scoleces or buds, the other fraction remains as uncommitted stem cells and thereby retain unlimited potency for proliferation of new scoleces (by asymmetric mitosis?). In this type, the inner germinal surface vesicles as capsules contain up to 30 scolices each and are soon detached into the interior of the bladder. The detached capsule, in its turn, begins to proliferate to generate daughter and granddaughter capsules both internally and externally. Hence, a single metacestoide can generate more than a million cysticerci.

Not surprisingly, considerable efforts have been made to understand this multiplication process in some of these larval cestodes. The presence of undifferentiated 'germinative cells' has been frequently described in many

## TABLE 5.7

Clonal reproduction by metacestodes of some cestodes (condensed from Whitfield and Evans, 1983, Moore and Brooks, 1987). bb = branching, budding, bu = budding

| Parasite species | Proliferating metacestodes | Host supporting proliferation | Host infected by asexual progeny |
|---|---|---|---|
| *Paricterotaenia paradoxa* | Polycerus | *Allolobophora* sp (Earthworm) | *Scolopax rusticola* (Woodcock) |
| *Staphylepis cantaniana* | Polycerus | *Didymogaster sylvatica* | *Gallus gallus* |
| *Staphylocystis pistillum* | Metacestoide, bb | *Ataenius cognatus* (beetle) | Shrews |
| *S. scalaris* | Metacestoide, bb | *Glomeris*, millipedes | Shrews |
| *Pseudodiorchis prolifer* | Parenchyma, bu | *Glomeris*, millipedes | Shrews |
| *Taenia multiceps* | Coenurus | Herbivorous mammals | Canids |
| *T. serialis* | Coenurus | Herbivorous mammals | Canids |
| *T. endothoracicus* | Coenurus, bb | Rodents | *Vulpes vulpes* |
| *T. mustelae* | Coenurus, bb | *Talpa*, rodents | Muselids |
| *T. parva* | Coenurus, bb | Rodents | Viverrids, felids |
| *T. selousi* | Coenurus, bb | *Rhabomys pumilo* | *Felis silvestris* |
| *T. twitchelli* | Coenurus, bb | Rodents | *Gulo gulo* |
| *T. crassiceps* | Coenurus, bb | Rodents | Canids |
| *T. pisiformis* | Coenurus, bb | *Lagomorphs*, rodents | Canids, felids |
| *Echinobothrium affine* | Metacestoide, bb | *Carcinus maenas* | Elasmobranch, rays |
| *Echinococcus granulosus* | Hydatid | Herbivorous mammals | Canids |
| *E. oligarthrus* | Hydatid | *Dasyprocta punctata* | Felids |
| *E. multicularis* | Alveloar hydatid | Herbivorous mammals | Canids |
| *E. vogeli* | Polycystic hydatid | *Dasyprocta punctata* | Canids |

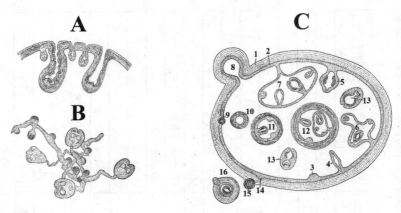

**FIGURE 5.10**

Schematic free-hand drawings to show (A) the inwardly branching coenurus cyst of *Taenia multiceps*, (B) irregular branching coenurus cyst of *Hymenolepis nana* and (C) hydatid cyst of *Echinococcus granulosus*. (1) Lamellated cyst wall, (2) Germinating layer, (3–7) Different stages of scolice formation from germinal layer, (8) Formation of daughter cyst, (9–12) Different stages of scolices formation in daughter cyst, (14–16) Stages of exogenous daughter cyst (based on Hyman, 1951).

adult and larval cestodes. Like neoblasts, these cells have a large nucleus with a prominent nucleolus surrounded by scant basophilic cytoplasm with a few mitochondria. Electron microscopic and radioactive labeled studies have revealed that the germinative cells—like their counterparts in planarians—are the only mitotically active cells that give rise to all other cells in cestodes. A breakthrough has been achieved to successfully culture the isolated cells *in vitro* from the vesicle of *Echinococcus multilocularis* metacestoide. Brehm (2010) showed that it is possible to obtain the completely regenerated metacestoides of *E. multilocularis* from the cultured dispersed cells. Amazingly, isolated cells from the cysticerci of *Taenia crassipes*, on intraperitoneal injection into mice, developed into complete cysticerci (Toledo et al., 1997). Hence, the so called asexual reproduction in some of these cestodes is proven as clonal multiplication. Incidentally, experimental studies like those of Toledo et al. and Brehm in sporocyst/redia of digeneans may also prove that the digenean asexual reproduction is also clonal multiplication and not parthenogenic or polyembryonic reproduction.

## 5.8 The Transmission

With the indirect life cycle involving two or three larval forms and one or two intermediate host(s), the cestodes encounter critical risks at each transmission event. However, these events provide excellent opportunities for dispersal. It has been shown for the first time that the life cycle of a majority (~ 30%)

*Cestoda* 223

of aquatic cestodes are of coracidium type and with three distinct larval forms involving two intermediate hosts and a definitive host. With regard to transmission mode and habitats, the following may have to be noted: (1) The digenean cycle involves penetrative and/or trophic mode(s) of transmission but the cestodes rely solely on trophic mode transmission. (2) Regarding habitats, in which transmission occurs, three types can be recognized: (i) aquatic, (ii) amphibious and (iii) terrestrial. In the aquatic type, the entire life cycle is completed by most digeneans (Fig. 4.2A, B) and cestodes (Fig. 5.4A). The cycle is commenced in the aquatic habitat but is completed in terrestrial DH in amphibious type (Fig. 4.2D, E, F). In both these types, the first development stage, the egg/miracidium in digeneans and the oncosphere/coracidium in cestodes is always aquatic and transmission can be penetrative or trophic in the former but trophic only in the latter. In the terrestrial type, the infective cysticercoids/tetrathyridium/hexacanth is developed within a terrestrial host. In this type, trophic mode of transmission is the rule (Fig. 5.4C, E, F).

In these three types, two exceptions should be noted. (a) In the completely terrestrial and less speciose digenean Brachylaimidae, the ingested egg is hatched within the amphibious/terrestrial snail (e.g. *Succinea*) and is 'bathed' in the aquatic medium of the host; the development of miracidium, sporocyst and cercaria is completed prior to the ingestion of the snail by DH. In others like *Renylaima capensis* (Fig. 4.2C), involving two intermediate slug hosts too, the larval stages are completed within the 'aquatic medium' of the slug hosts and transmission occurs by trophic mode alone. In Nematotaeniidae and *Hymenolepis nana*, the life cycle commences from aquatic habitat but with no intermediate host and transmission by trophic mode only (Fig. 5.4D). Firstly, the grouping of aquatic, amphibious and terrestrial types is valid. Secondly, the terrestrial transmission can occur only through trophic ingestion in both digeneans and cestodes. Thirdly, it is less efficient, as indicated by the less speciose brachylaimoids (228 species, Table 4.2) or as an exception among Hymenolepididae (~ 300 species).

Relevant information available for cestode transmission is limited to two publications only. Of them, the first one is an experimental study in the hexacanth-cysticercoid type of *H. nana* (cf Fig. 5.4D) with mice as DH (Ghazal and Avery, 1976). The *H. nana* eggs were offered as sprinkled or fecal pelleted eggs to starved or satiated mice. The relation for the number of eggs eaten vs exposure egg density increased asymptotically. At a given density, about 100, the infection was twofold higher for the starved mice, in comparison to the satiated mice. The transmission success was $2.9 \times 10^{-6}$ and $5.4 \times 10^{-6}$ for the sprinkled and pelleted eggs, respectively. Notably, the success was higher $2.6 \times 10^{-4}$ for the starved mice. The second publication by Jarroll (1980) involved field investigation on the coracidium type of *Bothriocephalus rarus* at transmission levels from coracidium to copepod *Macrocyclops ater* and from tadpoles to DH, the red spotted newt *Notophthalmus viridescens*. The tapeworm oviposited (6,927 eggs) at the rate of 26.5 proglottids, each containing 8.57

224   *Reproduction and Development in Platyhelminthes*

eggs/day for 30.5 days during the summer months. The efficiency values were 2.2% for coracidium to copepod and 3.8% for plerocercoid to DH, i.e. 0.8% eggs may infect DH. Incidentally, the newt's tadpole may metamorphose into an adult or become neotenic (Takahashi and Parris, 2008). However, the available value may be compared with that of 0.31% for the digenean *Haematoloechus coloradensis* with *Physa virgata* and the dragonfly nymph as intermediate hosts. In this aquatic transmission of digenean involving two penetrative intermediate hosts, the efficiency is higher than that for the cestodes involving two trophic (unlinked) intermediate hosts. Incidentally, the efficiency may be lower for the trophic mode of transmission in both digeneans (e.g. *Renylaima capensis*, Fig. 4.2C) and cestodes (e.g. *Raillietina echinobothrida*, Fig. 5.4C). In fact, the efficiency arrived from experimental investigation by Ghazal and Avery (1976) on trophic terrestrial transmission in *H. nana* stands to support this view.

## 5.9 Fishes and Losses

Cestode infection certainly causes diseases in a definitive host and losses in an intermediate host. This account, however, is limited to losses incurred by intermediate hosts alone. Surprisingly, no information is available on the loss incurred by an intermediate host, the copepods involved in the oncosphere type. In the coracidium type too, a rare publication by Scholz and Kutcha (2012) reports on up to cent percent mortality inflicted by procercoid on cyclopid copepods. Sakanari and Moser (1985) found progressive increase in mortality of the intertidal copepod *Tigriopus californicus* due to infection with *Lacistorhynchus tenuis* procercoid at different salinities. More surprisingly, except for a couple of publications (e.g. Pulkkinen and Valoten, 1999), all the available publications describe the loss incurred by the stickleback *Gasteorsteus aculeatus* due to infection with *Schistocephalus solidus* plerocercoid.

In the fields, the infected sticklebacks ate more of higher plants, nematodes, ostrocods and rotifers during the spring-summer but invertebrate eggs, chironomids and daphnids during the autumn-winter than the naïve ones (Tierney, 1994). In experiments, the stickleback infected with *S. solidus* ate more number of preferred chironomid blood worms than the naïve ones (Fig. 5.11A). Conversely, it accepted only less number (25/5–minutes) of daphnids than the naïve (33/5–minutes) ones. Nevertheless, the time required to ingest successive prey organisms was equal in all sizes of naïve and infected sticklebacks, when preys were simultaneously offered (Fig. 5.11A). Both naïve and infected sticklebacks grew equally in body length but not in weight, when fed a high ration of 8% body weight/d (Barber and Sevensson, 2003). An organ-wise analysis of growth of infected stickleback revealed that the ovarian growth was greater (25 mg) up to the size of ~ 0.7 g fish, but beyond which the liver and perivisceral fat body at the cost

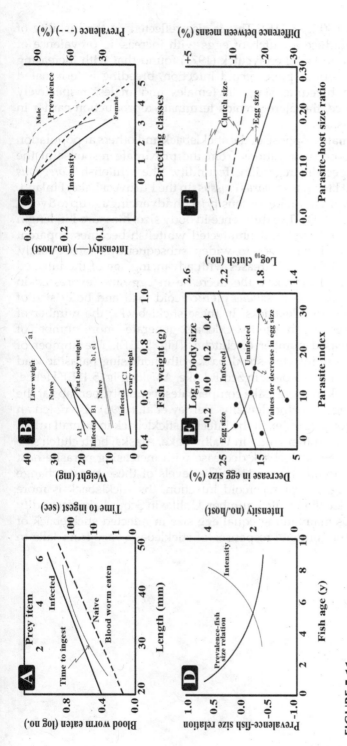

### FIGURE 5.11

(A) Number of chironomus worm eaten as a function of body size of *Gasterosteus aculeatus* infected with *Schistocephalus solidus* plerocercoid. Time required to ingest successive prey by naïve and infected sticklebacks (compiled from Cunningham et al., 1994, Barber and Ruxton, 1998). (B) Growth of ovary, liver and periviseral fat body as a function of body weight in infected and naïve sticklebacks (compiled from Barber and Svensson, 2003). (C) Effect of prevalence and intensity on termination of breeding in infected male and female sticklebacks (compiled from Pennycuick, 1971). (D) Increasing intensity of *Triaenophorus crassus* plerocercoid as a function of age in *Coregonus lavaretus*. Effect of increasing prevalence-fish size relationship as a function of fish age (compiled from Pulkkinen and Valtonen, 1999). (E) Effect of *S. solidus* plerocercoid load (index) on decrease (as %) in egg size and clutch number in uninfected sticklebacks (compiled from Heins and Baker, 2003, Heins, 2012). (F) Differences between clutch size and egg size as function of parasite-host size ratio during different years (Heins et al., 2014).

226 *Reproduction and Development in Platyhelminthes*

of the ovarian growth (Fig. 5.11B). This is also reflected on the inability of the stickleback to produce a clutch of eggs with increasing prevalence of infection (Fig. 5.11B). In fact, Pennycuick (1971) found that with increasing prevalence and intensity of plerocercoid infection, breeding is terminated at the 3rd and 5th breeding stage in females and males, respectively (see Fig. 5.11C). Hence, the plerocercoids terminate reproduction earlier in females than in males.

However, the **ultimate** response of the stickleback and others at population level differs with those observations from the **proximate** response at the laboratory level, especially regarding fecundity. The whitefish *Coregonus lavaretus* serves as SIH to *Triaenophorus crassus*. In the Puruvesi Lake, Finland, *T. crassus* plerocercoid accumulates (intensity) with advancing age up to 8 years in the whitefish (Fig. 5.11D). The difference in body size, measured in length and weight between infected and uninfected whitefish becomes apparent at the age 3 years and continues to widen subsequently. Consequently, prevalence-body size relation decreases with advancing age of the infected whitefish. In nine Alaskan Lakes, the decrease in egg size (expressed in %) increases with increasing *S. solidus* plerocercoid load and body size of infected and uninfected sticklebacks. In these sticklebacks, the number of clutches also increases; the infected stickleback generates more number of smaller eggs in increased number of clutches (Fig. 5.11E). The number of clutch increases more in infected stickleback with increasing parasitic load than that in uninfected stickleback in Scout Lake, Alaska (Fig. 5.11E). Hence, the infected stickleback in Scout Lake compensates fecundity, perhaps at the cost of reduction in egg size. Incidentally, Tierney et al. (1996) also noted an increase in gonado-somatic index in the infected stickleback in a small urban pond, Scotland. In yet another study in Walby Lake, Alaska, both clutch size and egg size in infected stickleback decrease with increasing parasite-host size ratio in parallel trends (Fig. 5.11F). The levels of these trends change from year to year. Against plerocercoid infection, ths stickleback is more sensitive to the egg size than clutch number. Unlike in Scout Lake, fecundity is decreased, perhaps maintaining equal egg size in infected stickleback of Walby Lake. Thus, the ultimate responses of stickleback vary from lake to lake and time to time.

# 6

---

## *Sexualization*

---

## Introduction

Platyhelminthes are hermaphrodites. However, some of them are protandrics (e.g. *Catenulida*, p 59, Monogenea: *Benedenia seriolae*, Lackenby et al., 2007). A few planarian species switch from clonal to sexual reproduction (Part 2.13). Sporadic incidence of gonochorism is strewn all over the major three classes of Platyhelminthes, especially in 23 schistosomatid species in four subfamilies (Table 6.1). Hence, a brief account is provided here on sexualization in flatworms.

---

## 6.1 Chromosomes and Genes

As of 2014, the number of platyhelminth species, for which karyological information is available, was limited to < 2% of flatworms. The information is reported for 115, 278 and 117 species in Turbellaria, Trematoda and Cestoda, respectively (Sofi et al., 2015). In them, the haploid chromosome number ranges from 3 to 8 for turbellarians (Birstein, 1991), 4 to 10 for monogeneans, 6 to 14 for digeneans and 3 to 14 for cestodes (Spakulova and Cassanova, 2004). The incidence of B chromosome is also known; for more details, Pongratz et al. (2003) and Spakulova and Cassanova (2004) may be consulted. Using ploidy levels, different biotypes have been identified in tricladids (Knakievicz et al., 2007). Increase in chromosome number with increasing latitude and decreasing temperature is reported for turbellarians *Phagocata vitta*, *Polycelis felina* and *Crenobia alpina*; the high chromosome numbers in these planarians are also accompanied by a switch from sexual to parthenogenic reproduction (Lorch et al., 2016). From an examination of well prepared metaphase plates with 2n = 14 chromosomes in a karyological study of the gonochoric planarian *Sabussowia dioica*, Charbagi-Barbirou and Tekaya (2009) did not detect the sex related difference in chromosome numbers

## TABLE 6.1
Gonochoric platyhelminths (compiled from Hyman, 1951, Brant, 2007, Ramm, 2017)

| | |
|---|---|
| Class **Turbellaria**, Order Tricladida,<br>Family Procerodidae, Subfamily Cercyrinae<br><br>*Sabussowia diocia*, Sexuality of *S. hastata, S. macrostoma*,<br>*S. papillosa, S. verrucosa* not yet confirmed<br>Fecampidae |  |
| Class **Trematoda**, Subclass Digenea, Family Didymozoonidae<br>*Wedlia retrovitalis* and *W. submaxillinis* |  |
| Family Schistosomatidae<br><br>    Subfamily Bilharziellinae, *Bilharziella polonica* – ducks<br>    Subfamily Denrobilharziinae, *Dendritobilharzia pulverulenta* – ducks, swans<br>    Subfamily Gigantobilharziinae, *Gigantobilharzia huronensis* – birds<br><br>    Subfamily Schistosomatinae<br>    *Allobilharzia visceralis* – birds<br>    *Austrobilharzia terrigalensis* – waterfowl<br>    *Bivitellobilharzia nairi* – elephants<br>    *Heterobilharzia americana* – raccoons<br>    *Macrobilharzia macrobilharzia* – birds<br>    *Ornithobilharzia canaliculata, O. turkestanicum* – cattle, cats<br>    *Schistosomatium douthitii* – rodents<br>    Schistosoma – mammals including human *S. haematobium*,<br>        *S. hippotami, S. incognitum, S. indicum, S. intercalatum*,<br>        *S. japanicum, S. malayensis, S. mansoni*,<br>        *S. mattheei, S. mekongi*<br><br>*Trichobilharzia franki, T. regent, T. szidati* – mainly waterfowl | |
| Class **Cestoda**, Family Taenioidae,<br>*Dioecocestus asper*<br>is a gonochore; 23 valid species in 4<br>genera are described but their sexuality<br>is not reported | 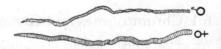 |

or in their morphology. However, the C-banded metaphase chromosomal investigation has revealed the differentiation between Z and W chromosomes resulting from (i) a partial constitutive heterochromatization of the W chromosome in African *Schistosoma mansoni* and *S. haematobium*, (ii) deletion of a part of the W in Asian *S. japonicum* and *S. mekongi* and (iii) translocation of one sex chromosome on to another in American *Schistosomatium douthitti* and *Heterobilharzia americanum* (Grossman et al., 1981). In some gonochoric adult digeneans like *Schistosomatium douthitti*, generative parthenogenesis occurs. In some others, hybridization between *S. japonicum* and *S. mansoni* results in production of viable parthenogenic eggs but not in others (e.g. *S. haematobium* and *S. mansoni*). Incidentally, some of these observations on hybridization ability, induction of generative parthenogenesis and production of viable eggs may be explained by the differences in sex chromosomal morphology (see p 157).

*Sexualization* 229

With the advent of molecular biology, investigations are switched from karyological studies to genes, especially during this decade. In testicular differentiation, the role of *DM* domain genes is known. Chong et al. (2013) showed that a sex specific gene *Smed-dmd-1* is required for testicular differentiation, spermatogenesis and male specific accessory reproductive organs in intact and regenerating turbellarians. Interestingly, a homolog of *dmt-1* exhibits male specific expression in *S. mansoni*. In male heterogamety, the functional loss incurred by Y-linked genes during evolution of sex chromosome favors global dosage compensation. The compensation increases the expression of genes on the X chromosomes, a common feature of female heterogamety. However, such dosage compensation does not occur in *S. mansoni* (Vicoso and Bachtrog, 2011). The expression and function of the *Deleted in Azoopermia* (*DAZ*) gene family is crucial for the development of germline and generation of male and female gametes (e.g. Kerr and Cheng, 2010). Three genes belonging to the *DAZ* family namely *DAZ*, *DAZL* (*DAZ*-like) and *Boule* are proved to be indispensable for gametogenesis. Demonstrating the presence of three *boule* orthologs in the flatworm *Macrostomum ligano*, Kuales et al. (2011) reported that (i) *macbol1* and *macbol2* were expressed in testes, while *macbol3* was expressed in the ovary. (ii) *Macbol3* RNAi induced aberrant egg maturation and female sterility. (iii) Similarly, *Macbol1*/RNAi blocked spermatocyte differentiation but *macbol2* RNAi showed no effect upon RNAi treatment. Clearly, *macbol1* and *macbol2* are the genes responsible for testicular and *macbol3* for ovarian differentiation. By deep sequencing, Marco et al. (2013) measured the relative expression levels of conserved and newly identified microRNAs between male and female samples of *S. mansoni*. Of 13 microRNAs that exhibited sex-biased expression, 10 were more abundant in females than in males. As sex chromosomes showed a paucity of female biased genes, Marco et al. proposed that microRNAs are likely to participate in sex differentiation. Available information reveals the role played by genes through RNA system.

## 6.2 Endocrine Differentiation

Flatworms lack endocrine glands and the circulatory system. Hence, humoral signaling has to be transmitted to cells by neuroendocrine system directly on their targets or via intercellular spaces and extracellular matrix (ECM). Bioactive peptides, i.e. neuropeptides or peptide hormones represent the largest class of cell–cell signaling molecules in metazoans and are potent regulators of neural and physiological functions.

*Morphogenesis*: Substances produced by the nervous system influence fission and subsequent regeneration through (i) regulation of mitosis by biogenic amines and neuropeptides and (ii) pattern formation through the appearance

230　*Reproduction and Development in Platyhelminthes*

**TABLE 6.2**

The role played by neuropeptides and others in regeneration and fission of flatworms (condensed from Reuter and Kreshchenko, 2004)

| Factors | Reported observations |
|---|---|
| 5-HT receptors | Melatonin inhibits fission (Morita and Best, 1984) and regeneration (Yoshizawa et al., 1991) in planarians. |
| Neuropeptides | Six native flatworm neuropeptides (NPFs) have been isolated. The NPF precursor gene *npf* has also been identified and characterized in *Moniezia expansa* (Mair et al., 2000) and *Arthurdendyus triangulatus* (Dougan et al., 2002).<br>Neuropeptides substance P and K stimulate cell proliferation during regeneration (Baguna et al., 1989) and neuropeptide Y accelerates regeneration in planarians and acoels (Hori, 1997, Hori et al., 1999).<br>A $Ca^{2+}$ transporting peptide vasopresin, dalargin, somatostatin, and hydra head activator (HHA) stimulates cephalic ganglion regeneration in *Dugesia tigrina* (Kreshchenko and Scheiman, 1994). |
| Nitric oxide (No) | No is one of the lightest molecules known to act as biological messengers. No has a role in growing genital anlagen in *D. dendriticum* (Gustafsson et al., 2002). |
| Electromagnetic fields | Weak EMF increases mitotic index by ~ 30% in the post blastema area of regenerating planarians following decapitation. |

and dominance of peptidergic immunoreactivity in the peripheral nerve net (see Reuter and Kreshchenko, 2004). Table 6.2 lists some of these neuropeptides and others playing a role in regeneration and fission.

*Reproduction*: Using biochemical and bioinformatic techniques to identify peptides in the genome of *Schmidtea mediterranea*, Collins et al. (2010) identified 51 genes encoding > 200 peptides. Analysis of these genes in both sexual and clonal strains of the flatworm has identified a neuropeptide Y superfamily member as important for the normal development and maintenance of the planarian reproductive system. One prohormone gene *npy-8* is enriched in the nervous system of sexual planarian and the gene is required for the proper development and maintenance of reproductive tissues (see p 43). A large scale RNA interference (RNAi) screen has identified that *smed-pc2* is essential for coordinated movement and normal regeneration in clonal planarians. Briefly, defective neuropeptide processing results in defects in the development of reproductive system.

# 6.3 Endocrine Disruption

In gonochorics like *Sabussowia dioica*, structural differentiation is commenced with the appearance of gonad as a cluster of large cells, the neoblasts, followed by development of ovary and vitellaria in females and testes in males. Finally, the gonoducts are developed (Charbagi-Barbirou and Tekaya, 2009). In protandrics like *Benedenia seriolae* testes appear first at the fluke's size of 1.1–2.1 mm body length followed by copulatory organs and vas deferens. Only after the completion of the male reproductive system, the ovary and vitellaria are developed, when the fluke attains a size of 2.2–2.8 mm. The fluke begins to produce eggs at 2.9–3.8 mm size (Lackenby et al., 2007).

The flatworms, either free-living or parasitic, are too fragile and small for surgical endocrine investigation. However, parasitic flatworms are capable of disrupting endocrines of their hosts. About 30% of cestodes pass through plerocercoid stage usually in a teleost host (Table 5.4). Like adults, the plerocercoid can also absorb amino acids from the host and synthesize polypeptide hormones like Growth Hormone (GH). Available information on endocrine disruption by plerocercoid is briefly summarized. It is based on an excellent review by Phares (1996) and others. (1) Experimentally infected host animals with *Spirometra mansonoides* plerocercoid do not eat more but convert the ingested food into growth more efficiently. The reasons for the increased growth are elaborated below: (2) Publications during 1963–1968 provide evidence to suggest that plerocercoids somehow substitute GH, insulin and thyroxin. (3) GH is a 22 kDa non-glycosylated polypeptide secreted by somatotrophs of the anterior pituitary. GH increases growth indirectly by stimulating synthesis and release of secondary hormones (e.g. insulin like growth factor, IGF-1). (4) In rats, [35]S-labeled studies have revealed that plerocercoids secrete a GH-like factor (PGF), which, on transportation by blood, substitutes 30% greater growth through increase in IGF-1. Studies on [3]H-thymidine uptake by cartilages also revealed the stimulation of 20–50 times greater growth. (5) Briefly, GH influences tissue sensitivity to prolactin and estrogen by increasing the number of target tissue receptors for these hormones. The PGF seems to do some of what GH can do. For example, the PGF-treated female rats gain 225% increase in weight, 93% reduction in serum GH, 37% decrease in pituitary weight and 64% reduction in the number of hepatic prolactin receptors. (6) In PGF-treated rats, the number and affinity of hepatic receptors for estrogen are reduced by > 50% and binding of estrogen is saved by 54% in the anterior pituitary (7) Purification and characterization of PGF has revealed that the GH like PGF is a 27.5 kDa cysteine proteinase. Sequencing of PGF $_c$DNA shows 40–50% homology to that of mammals.

It must, however, be pointed out that the results summarized by Phares (1996) are all drawn from experimental animals like rats, hamsters, raccoons and others. The host response to plerocercoid infection also differs from *S. mansonoides* to *S. theileri* (Opuni et al., 1974). In the natural fish host, the plerocercoids suppress reduction rather than increasing body weight. In many cyprinids, the gonads of fish infected with coracidial type *Ligula intestinalis*

232 *Reproduction and Development in Platyhelminthes*

remain immature (Arme, 1968). Further investigation has revealed that *L. intestinalis* infection is accompanied by low mRNA expression of pituitary gonadotropin subunits (e.g. Carter et al., 2005), indicating the inadequate gonadotropin underlying the depressed gametogenesis (Trubiroha et al., 2009). A more detailed study by Trubiroha et al. (2010) demonstrated that (i) the plasma levels of estradiol ($E_2$), testosterone (T) and 11-Ketotestosterone (11-KT) were significantly lower in the roach *Rutilus rutilus* infected with *L. instestinalis* plerocercoid than in the naïve roach. (ii) The differences in levels of these hormones were also sex specific, i.e. females suffered more than males (Fig. 6.1A, see also p 224–225). (iii) These sex specific depressed hormone levels were traced to the reduced levels of relative mRNA expression in *GnRH2* and *GnRH3* (*GnRH3* not shown in figure) and *Cyp19a* and *19b* (Cyp 19b not shown in figure) (Fig. 6.1B) as well as in receptors of estrogen (Esr1, Esr2a and 2b not shown in figure), *vitellogenin* (*VTG*), androgen (AR) and IGF 1 (Fig. 6.1C). Consistently, the expression of $E_2$ dependent genes like *VTG* and brain type aromatase in the brain and liver was reduced. Strikingly, the expression of *Esr 1* was also reduced in the liver of infected females but not in males. Insulin-like growth factor 1 mRNA too was increased in infected females but not in males (cf Fig. 5.11C). Despite all these severe impacts, brain mRNA levels of *GnRH* precursors encoding *GnRH2* and *GnRH3* were not affected by *L. intestinalis* plerocercoid infection. In all, the following remain to be investigated: (i) the difference in responses to plerocercoid infection by stimulation of growth in experimental host animals to *S. mansonoides* and suppression of reproduction in natural fish host to *L. intestinalis* are due to experimental vs natural host or due to difference in inherent disrupting ability between *S. mansonoides* and *L. intestinalis*, and (ii) whether the vertebrate hormones released by the cestodes have any effect on the parasites themselves.

The other oncosphere-procercoid or hexacanth-cysticercoid/tetrathyridium type involves only an arthropod host. In these types, the infection occurs mostly during the larval stages of copepods or insects. Ecdysteroids regulate molting and growth processes in arthropod larvae (for Crustacea, see Pandian, 2016). Cestodes are not capable of synthesizing steroids and must acquire them from the host (Section 1.5). Relatively less information is available on the ability of cestode to secrete ecdysteroids. However, the presence of ecdysteroids has been reported in taenioids *Echinococus granulosus*, *Hymenolepis diminuta* and *Monezia expansa*. Active secretion of 20-hydroxyecdysone by *H. diminuta in vivo* and *in vitro* suggests that the ecdysteroids are of parasitic origin (Mercer et al., 1987). Ecdysteroids have been circumstantially implicated in regulation of growth and development of cestodes. Exogenous 20-hydroxyecdysone stimulates growth and propagatory multiplication in *Mesocestoides corti* (see Gamble et al., 1995).

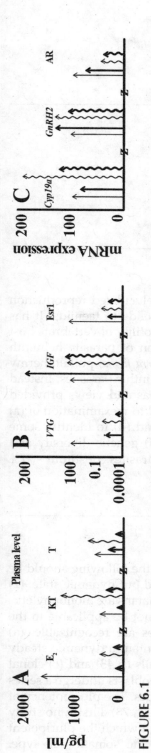

**FIGURE 6.1**

Levels of plasma estradiol (E$_2$), 11-Ketotestosterone (11-KT) and testosterone (T) hormones and relative mRNA expression of *VTG*, *IGF*, *Esr1*, *Cyp19a*, *GnRH2* and AR in uninfected (thin line arrows) and infected (thick line arrows) male (wavy lines) and females (straight lines) of *Rutilus rutilus* infected with *Ligula intestinalis* procercoid (modified and compiled from Trubiroha et al., 2010).

# 7

## Comparison and Highlights

### Introduction

Besides updating, this book has comprehensively elucidated reproduction and development of flatworms covering from acoelids to taeniids. It has brought to light that (i) of motility and body size, motility of vertebrate host is the driving force in evolution and diversification of parasite helminth lineages, (ii) despite their earlier colonization of *terra firma*, poikilothermy has limited herptiles from serving as a host to helminthic parasites. Instead of pondering over the new angle of approach, ideas and views provided throughout the book, the present chapter is devoted to reexamination of (a) neoblasts, (b) clonal selection and (c) progenesis and (d) to identify some factors responsible for (i) host distribution and (ii) genetic diversity and speciation in parasitic flatworms; incidentally, the parasites constitute ~ 77% of flatworms.

### 7.1-A Neoblast Types

Prior to considering it from a different perspective, the following should be noted: (1) Taxonomy of Platyhelminthes is in a fluid but dynamic state. (2) Whereas turbellarians are polyphyletic, the neodermatans are monophyletic. Hence, a generalization made for the former may not be applicable to the latter *in toto*. Among neoblasts, three major types are recognizable: (a) renewal and turnover of somatic cell types to maintain a dynamic steady state, (b) regeneration including Intestinal Stem Cells (p 43) and (c) clonal multiplication. In a few turbellarians, pluripotent neoblasts undergo a series of **symmetric** mitosis to generate 14 somatic cell types. The pluripotency of *nanos 2+* and its loss with *nanos 2-* is reported (p 40–41, 78). Firstly, no study has yet been made to trace the entire course, through which the pluripotent neoblasts become unipotent and generate a specific somatic cell type.

*Comparison and Highlights* 235

Secondly, it is not known whether the neodermatans also undergo a similar process. Thirdly, the digenean germinal cells in sporocyst and redia undergo 'asymmetric' mitosis and generate sporocysts, rediae and/or cercariae (p 156); the cercaria is a miniature of the adult and possesses almost all the adult tissue types. In clonally multiplying hydatid (but not in coenurus cyst [pp 221]) cyst of cestodes also, a fraction of germinative cells initiates the production of new scolices, while the other fraction remains uncommitted totipotent. With regard to mitosis in these stem cells, there is a fundamental difference between the turbellarian neoblasts and neodermatan germinal/germinative cells. At the molecular level too, Koziol et al. (2014) demonstrated this basic difference; for instance, the germinative cells of *Echinococcus multilocularis* lack *vasa* and *piwi* orthologs but they are present in the neoblasts. Fourthly, clonal and regenerative potencies may have independently evolved at different times in different turbellarian orders and species. It is not clear whether the original clonal potency is lost (Table 7.1) in some species, which can regenerate but not multiply clonally.

Thanks to Baguna and Romero (1981), free-living clonal flatworms are known to possess 14 cell types. The addition of a few more to include reproductive system and gametes, adhesive organs and associated structures may increase the number of cell types to not more than two dozen. (1) Turbellaria and Platyhelminthes include 6,376 and 27,559 species, respectively (see p 72). The number of species capable of regeneration and clonal multiplication within each turbellarian order is listed in Tables 2.9, 2.10, *albeit* no claim is made that the lists are complete. Information summarized in Table 7.1 identifies the need for researchers to address clonal multiplication in macromorphids and regeneration in polycladids and proseriates. Assuming almost all species belonging to Tricladida (1,000 species, or 15.4% of Turbellaria), Acoela (350

## TABLE 7.1

Regenerative and clonal potency of turbellarians (based on Fig. 2.1 A, Tables 2.8, 2.10)

| Taxon | Total species (no.) | Regenerating species (no.) | Clonal species (no) |
| --- | --- | --- | --- |
| Tricladida | 1,000 | 23 (2.3%) | 30 (3.0%) |
| Acoela | 350 | 7 (2.0%) | 9 (2.6%) |
| Catenulida | 120 | 1 (0.8%) | 5 (4.2%) |
| Macrostomorpha | 200 | 7 (3.5%) | 1 (0.5%) |
| Polycladida | 2,000 | 13 (1.3%) | – |
| Proseriata | 350 | 7 (2.0%) | – |
| Lecithoepitheliata | 30 | 1 (2.3%) | – |
| Bothrioplanida | – | 1 | – |
| Prolecithophora | – | 1 | – |
| Rhabdocoela | – | 4 | – |
| Total | 6,376 | 65 (1.02%) | 45 (0.71%) |

or 5.4%), Catenulida (120, 1.8%) and Macromorpha (200, 3.1%) are capable of clonal multiplication, not more than 1,670 species or 25.7% turbellarians possess the pluripotent neoblasts responsible for clonal multiplication. It is not known whether the remaining ~ 75% of flatworms do possess the neoblasts for clonal multiplication or for regeneration. Only 5 triclad species, *Dugesia gonocephala*, *D. japonica*, *D. lugubris*, *D. sicula* and *Schmidtea mediterranea*, and one species in each of catanulid (*Catanula*) and acoelid (*Convolutriloba longifissura*) are reported as capable of both regeneration and clonal multiplication (see Table 2.9). Of 65 species capable of regeneration, 59 are not known to clonally multiply. Similarly, of 46 species capable of clonal multiplication, 39 are not thus far reported to possess regenerative potency. Hence, there is a need to extend research on regeneration and clonal multiplication to other turbellarian species as well as to distinguish the pluripotent neoblasts associated with clonal potency from the multipotent neoblastss capable of regeneration of one or more organs alone as well as from the pluripotent neoblasts capable of renewing cell types.

## 7.1-B Clonal Selection

Available contribution on clonal selection is limited to a couple of publications only. Nevertheless, it is chosen to highlight it, as incorporation of SIH into the life history is a landmark step in digenean evolution. For, SIH is relatively more motile, facilitates a higher level clonal diversity, mixing and selection of the fittest metacercaria clone(s) for onward transmission as well as purging of deleterious clones from the propagatory multiplications in FIH. In this context, estimates on the number of digenean species that have incorporated SIH and **actual** or **potential** number of species serving as SIH are of great importance. From their database, Cribb et al. (2003) were the first to estimate the **actual** number of SIH from nine phyla as 556 species. It must, however, be noted that the estimate is limited to 15 of 22 superfamilies and does not include the number of digenean species in these 15 superfamilies. According to Littlewood et al. (2015), these 15 superfamilies include as many as 8,795 species. Hence, the estimate of 556 SIH species for 8,795 digenean species is certainly an underestimate. It is in this context, the need for estimation on the number of digenean species that have incorporated SIH is obvious. For the first time, it is shown that of 12,012 digeneans, 88.3% or 9,496 species may engage as many as 33,014 **potential** SIH species (Table 4.4). Hence, the digeneans engaging SIH are flexibly euryxenic or have the choice to select one among the awaiting/available 3.5 SIH species. Clearly, the incorporation of SIH by > 88% digeneans may have purged the deleterious clones from FIH and selected the fittest clone(s) for onward transmission, increased genetic diversity and lineage diversification. Of course, estimate of 33,014 SIH species may be an overestimate but still it indicates the need for a reliable database.

Comparison and Highlights 237

Further analysis of the data summarized in Tables 4.3 and 4.4 reveals that more than motility of SIH, it is the (euryxenic flexibility) wider choice to select one among the many SIH that increases the lineage diversification (cf pp 21–25). Of 16 digenean superfamilies that have incorporated SIH, two of them namely Apocreadioidea (111 species that select one of the two potential invertebrate species as SIH) and Monorchioidea (336 species that select one of the four potential invertebrate species as SIH) engage relatively less motile invertebrate hosts alone, i.e. in this Group 1, each digenean species (1,566 SIH intermediate host species ÷ 447 digenean species) has a choice of one among 3.5 potential invertebrate SIH species awaiting/available to be selected. All digenean species belonging to Group 2 engage relatively more motile vertebrate hosts as SIH. Within it, Group 2A comprising four superfamilies Bivesiculoidea (28 species), Azygioidea (40 species), Bucephaloidea (416 species) and Opisthorchioidea (839 species) has a limited choice of one vertebrate host species per digenean species (see Table 4.4). In the remaining 10 superfamilies, i.e. each of 7,499 species in Group 2B has the choice to select one among a vertebrate SIH species + 1 to 5 invertebrate species as potential SIH, i.e. each digenean species in Group 2B has the widest choice to select one among 4.0 SIH species potentially awaiting/available. In these groups, the mean digenean species number is 225, 331 and 536/superfamily, respectively. Remarkably, Group 1 has relatively less motile invertebrate host with a choice for host species, Group 2A has relatively a more motile vertebrate host but with no choice for digenean to select SIH and Group 2B has the widest choice among a vertebrate + 1–5 invertebrate species as potential SIH to be selected by a digenean species. More than motility of SIH, it is the widest choice (euryxenic flexibility) for SIH species for digenean selection that seems to have increased lineage diversification. Notably, motility of DH and euryxenic flexibility or choice to select SIH has been the driving force for lineage diversification in digeneans.

## 7.1-C Progenesis

Not surprisingly, some fluke species that are subjected to one or other intense stress, have adapted a strategy to eliminate one or more larval stage(s) and/or intermediate host(s) to enable them to complete the life cycle. Progenesis is an adaptive response to the uncertainty of transmission toward a penultimate or definitive host and a reproductive insurance strategy to a complete life cycle (see Lefebvre and Poulin, 2005). In a few anurans and urodeles, progenesis is reported. Its incidence is also known from a few monogeneans and cestodes and from many digeneans. In some of them, the life cycle is truncated by abbreviation, mostly involving neoteny. Gould's (1977) definition of progenesis, as heterochronic development, in which the first reproduction commences in the juvenile stage, may not directly be applicable to flatworms,

238   *Reproduction and Development in Platyhelminthes*

as progenesis occurs in them mostly during the larval/developmental stage. All viviparous gyrodactylids are progenetic; their life cycle is direct, as oncomiracidium larva is eliminated. In them, the stress of low competitive ability seems to have triggered progenesis. The obligate and facultative truncation occurs, the latter involving neoteny, in four monogenean species too (Table 7.2). In Cestoda, the obligatory progenesis occurs in *Bothriocephalus* and *Archigetes limnodrili* and facultative progenesis in *A. iowensis* and *A. sieboldi*. It must be noted that the truncation is limited to polystomatids alone in Monogenea and Bothriocephalidae and Caryophyllidae alone in Cestoda—all of them living in freshwater. Evidently, unpredictable water level has been the single most critical environmental factor that triggers the truncation. These incidences in Monogenea and Cestoda as well as the incoming description of progenesis in Digenea indicate that progenesis may be a common feature of neodermatans.

Contrastingly, the truncation occurs in digeneans that are residents in Marine (M), Brackishwater (B), Freshwater (F) and Terrestrial (T) habitats. The first list by Poulin and Cribb (2002) revealed the incidences of truncation in 20 genera in 17 families as facultative and 25 genera in 13 families as obligatory. The truncation is further divided into four groups: (i) use of the snail as the only intermediate host, (ii) use of FIH also as SIH, (iii) progensis in SIH and (iv) use of SIH also as a Definitive Host (DH). In them, clonal selection (a comprehensive term to include clonal diversity, mixing and selection) is limited to a few clones in FIH but it is nearly 5-times more in SIH. In the second list, which is limited to species involving metacercariae in SIH alone, Lefebvre and Poulin (2005) included 13 and 48 species that were characterized by facultative and obligate in truncation, respectively. Hence,

## TABLE 7.2

Incidence of truncation in life cycle of Monogenean and Cestoda

| Obligatory | Facultative |
|---|---|
| **Monogenea: Polystomidae in freshwater** | |
| *Polystomoidella oblonga* (p 116)—Hatched after completion of development up to 6-sucker stage, while still in the parent | *Polystoma gallieni* (Fig. 3.6C) and *P. integerrimum* (Fig. 3.6D): neotenic branchial phenotype and normal bladder Type |
| *Scapiopus couchi*—Miracidium eliminated- Releasing completely developed young one | |
| Viviparity and hyperviviparity enforces progenesis in gyrodactylids | |
| **Cestoda in freshwater** | |
| *Bothriocephalus?* - Neotenic procercoid develops into plerocercoid in cyclops itself | *Archigetes iowensis* –3% progenetic eggs in FIH tubificids |
| *A. limnodrili* with 100% progenetic eggs in FIH *Limnodrilus* | *A. sieboldi* –56% progenetic eggs in FIH *Limnodrilus* |

*Comparison and Highlights* 239

the choice for digeneans is facultative truncation, i.e. if a population or a fraction of it in a species undergoes facultative progenetic truncation, the others may pass through a regular life cycle.

If all the species included in the genera listed in the first list are valid, then the cycle may be truncated in as many as 866 digenean species (see Table 4.4). Of them, clonal selection (see pp 170–173) may occur at a low level in 65 snail species or 7.5% of 866 digenean species, which have incorporated the snail host alone. In the remaining 92.5% of truncated digeneans, the cycle involves a minimum of two intermediate hosts. In them, the probability for a higher level of clonal selection is expected. Hence, a vast majority of the truncated digeneans have ensured a certain level of clonal selection to purge the inbreeding depression arising from propagatory multiplication in FIH. In the truncated digeneans, species richness is 21.7, 23.9, 29.7 and 58.5 species/genus in fluke species with a snail host alone, FIH serving also as SIH, progenesis in SIH and SIH serving also as DH, respectively. Evidently, clonal selection in SIH ensures the fittest metacercarial clone(s) for onward transmission and genetic diversity leading to lineage diversification.

But, this conclusion may be questionable, as there are many discrepancies between the first and second lists. The following are some examples: (i) the second list adds 10 families (from Heterophyidae to Psilostomidae), (ii) in the family Macroderoididae *Glypthelmins* and *Haplometra* mentioned in the first list are not found in the second one, (iii) *Microphallus* is listed as FIH used also as SIH in the first list, but it is progenetic SIH in the second list and so on. Hence, the need for a new reliable database to be prepared by leading experts like Drs. T.H. Cribb, T. Lefebvre and R. Poulin is obvious.

However, in view of elaborate details summarized for 79 species in the second list, other features of progenesis may be discussed. To the list of 79 species, *Neochasmus* sp with the choice of seven fish species as SIH may be added (McLaughlin et al., 2006). The factors responsible for switching to progenesis can be traced to the instability of environmental factors like (a) unpredictable water levels (e.g. *Alloglossidium anomorphosis*), (b) water temperature (e.g. *Stegodexamene anguillae*), (c) salinity (e.g. *Bunocotyle meriodionalis*) and/or (d) unavailability of SIH and/or DH. The second list provides more details for 56 progenetic digenean species from five families. For them, there are 99 species belonging to crustaceans, insects, leeches, molluscs, polychaete, fish and amphibians to serve as SIH. With reference to unavailability of SIH, the second list indicates the incidence of 7/13 obligatory and 18/48 of facultative digenean species with a single species to serve as SIH; this may imply that unavailability of SIH seems to trigger progenesis in (7 out of 13) 41% of progenetic digeneans. For, with increasing abundance of SIH fish, for example, the proportion of progenetic *S. anguillae* decreases from 0.3% with no host to 0% at the availability of 225 host individuals (Herrmann and Poulin, 2011). Temporal unavailability of DH may be associated with seasonal migration of birds and fish, and aestivation

# TABLE 7.3

Estimated number of digenean species undergoing facultative or obligatory truncation in life cycle (reorganized from Poulin and Cribb, 2002, and updated with species number and habitat). F = Freshwater, M = Marine, B = Brackish water, T = Terrestrial; ITIS indicates Integrated Taxonomic Informational System; WORMS–World Register of Marine Species

| Family | Genus | Snail host only | FIH as SIH | Progenesis in SIH | SIH as DH | Remarks |
|---|---|---|---|---|---|---|
| **Facultative truncation** | | | | | | |
| Brachycoeliidae | *Brachycoelium* | | | + | | 3, F, M (ITIS) |
| Bucephalidae | *Bucephaloides* | | | + | | 23, M (WORMS) |
| Eumegacetidae | *Orthetrotrema* | | | + | | |
| Lecithodendriidae | *Pleurogenes* | | | + | | 1, F (WORMS) |
| | *Pleurogenoides* | | | + | | 5, F, T (WORMS) |
| | *Prosotocus* | | | + | | |
| Leprocreadiidae | *Stegodexamene* | | | + | | 3, M, B, F (WORMS) |
| Lissorchiidae | *Asymphylodera* | | | + | | 16, M, B, F (WORMS) |
| Opecoelidae | *Coitocaecum* | | | + | | 56, M, B, F (WORMS) |
| | *Plagioporus* | | | | + | 159, B, F (WORMS) |
| Opisthorchiidae | *Ratzia* | | | + | | |
| Zoogonidae | *Deretrema* | | | + | | 43, M (WORMS) |
| **Plagiorchiidae** | ***Opisthoglyphe*** | | | | **+** | |
| Cephalogonomidae | *Cephalogonimus* | | | | + | 3, F (WORMS) |
| Diplostomidae | *Alaria* | | | | + | 43, F, M (WORMS) |
| Fellodistomidae | *Protoeces* | + | + | | | |
| Gorgoderidae | *Phyllodistomum* | | + | + | | 134, M, B, F (WORMS) |
| **Gymnophallidae** | ***Gymnophallus*** | | **+** | | | **38, M (WORMS)** |

| Cyathocotylidae | *Mesostephanus* | ⊕ | | | | 6, M, B, F (WORMS) |
|---|---|---|---|---|---|---|
| Derogenidae | *Genarchella* | ⊕ | | | | 14, F (WORMS) |
| **Obligatory truncation** | | | | | | |
| Cyclocoeliidae | All genera | | + | | | 22, 8 g, M, F, T (WORMS) |
| Echinostomatidae | *Echinostoma* | | + | | | 11, M, F, T (WORMS) |
| Eucotylidae | All genera | | + | | | 7, 3 g (WORMS) |
| **Gymnophallidae** | ***Parvatrema*** | | + | | | **28, M (WORMS)** |
| Hasstilessidae | *Hasstilesia* | | + | | | 1, M (ITIS) |
| Leucochloridiidae | *Leucochloridium* | | + | | | 1, M, T (WORMS) |
| Microphallidae | *Microphallus* | | + | | | 62, M, F, T (WORMS) |
| | *Maritrema* | | + | | | 41, M (WORMS) |
| Cryptogonomidae | *Aphalloides* | ⊕ | | | | 2, M (WORMS) |
| Macroderoididae | *Alloglossidium* | | | + | | 10, F (WORMS) |
| | *Glypthelmins* | | | + | + | 29, F, T (WORMS) |
| | *Haplometra* | | | + | + | |
| Allocreadiidae | *Allocreadium* | + | + | + | | 56, F, B (WORMS) |
| Bunocotylidae | *Bunocotyle* | + | + | + | | 6, M (WORMS) |
| Hemiuridae | *Parahemiurus* | ⊕ | | | | 43, M (WORMS) |
| **Plagiorchiidae** | ***Paralepoderma*** | | | | + | |
| **Total** | | | | | | **866 species** |

⊕ = families involving snail host only.

242 *Reproduction and Development in Platyhelminthes*

in frogs (e.g. *Plerogenoides ovatus*, Janardanan et al., 1987). In *S. anguillae*, temperature elevation of ~ 4°C increases the proportion of progenesis from 0.1% at 12°C to ~ 0.25% at 16°C (Herrman and Poulin, 2011b).

There are contradictory reports on the viability of progenetic eggs. Many studies have shown that both male and female reproductive systems are completely developed and functional in progenetic digeneans. But the eggs produced by progenetic metacercariae of *Crepidostomum cornutum* failed to hatch or when hatched, the miracidia were incapable of infecting the snail host. However, experimental studies have shown that the obligate progenetic leptophallid *Paralepoderma brumpti* was able to maintain 10 progenetic generations. The opisthorchiid *Ratzia joyeuxi* successfully passed through three successive cycles from progenetic eggs. This area merits further studies (see Lefebvre and Poulin, 2005).

The autoradiographic study measuring the level of incorporation of trace-labeled substrate has suggested that thickness and permeability of the cyst wall is important for normal development of metacercaria. The contradictory reports on viability of progenetic eggs can be resolved in the context of the location of the encysted metacercariae. Metacercariae may encyst in the muscle, head, body cavity and/or gonads. As they provide resource-rich nutrients at high levels, metacercariae encysted in the body cavity and gonads are more likely to become progenetic (Herrman and Poulin, 2011a). A few metacercariae that encyst in muscles may not be able to acquire adequate or desired nutrients. They may not become progenetic, and if they become, they may produce sterile eggs. In fact, an experimental study by Herman and Poulin (2011a) showed that on encystation in the gonads of fish, the progenetic *S. anguillae* metacercaria grows to the largest size of 0.8 mm$^2$. With increasing body size, the metacercaria not only produces increased number of eggs but also the largest eggs. The progenetic eggs are released to the exterior along the host's eggs (e.g. *S. anguillae*), excretory products (e.g. *Alloglossidium macrobdellensis*), host predation (e.g. *Prostocus confusus*), host skin rupture (e.g. *Ratzia joyeuxi*) or on host death (e.g. *A. lobatum*, *Neochasmus* sp).

## 7.2 Habitat Distribution

Reorganization of information listed in Figs. 16.3 and 16.4 by Littlewood et al. (2015) provides an opportunity for further analysis of taxonomic distribution of digeneans and cestodes on hosts. The restricted distribution of six digenean taxa in aquatic teleosts alone has limited the lineage diversification; for example, the number of species in these six families ranges from 28 in Bivesiculoidea to 932 Opecoeloidea with an average of 258 species/superfamily (Table 7.4, left panel). Similar restricted distribution of two fluke families within terrestrial tetrapod hosts has not

# TABLE 7.4

Vertebrate host parasitized by major digenean (left panel) and cestode (right panel) groups. ● = marine, ○ = freshwater, ◐ = brakishwater, △ = terrestrial, ▲ = marine cum terrestrial, ⊘ = freshwater cum terrestrial, ◭ = brakishwater cum terrestrial, Ⓐ = coastal, H = Holocephali, E = Elasmobranchii, C = Chondrostei, I = Holostei, T = Teleostei, A = Amphibia, R = Reptilia, B = Aves, M = Mammalia (compiled from Littlewood et al., 2015)

| Major group | Sp (no.) | H | E | C | I | T | A | R | B | M |
|---|---|---|---|---|---|---|---|---|---|---|
| Aspidogastrea | | ● | ● | | | ◐ | | ● | | |
| Heronimoidea | 1 | | | | | | ○ | | | |
| Bivesiculoidea | 28 | | | | | ◐ | | | | |
| Transversotrematoidea | 30 | | | | | ◐ | | | | |
| Haplosplanchoidea | 50 | | | | | ◐ | | | | |
| Haploporoidea | 170 | | | | | ◐ | | | | |
| Opecoeloidea | 932 | | | | | ◐ | | | | |
| Monorchioidea | 336 | | | | | ◐ | | | | |
| Aporcreadioidea | 111 | | | | | ◐ | | △ | | |
| Bucephaloidea | 416 | | | | | ◐ | ○ | | | |
| Gymnophalloidea | 231 | | | | | ◐ | | | ⊘ | △ |
| Azygioidea | 40 | | ● | | ○ | ○ | | | | |
| Hemiuroidea | 1334 | | ● | | | ◐ | ○ | ● | | |
| Gorgoderoidea | 1084 | | ● | ○ | | ◐ | ○ | △ | △ | △ |
| Microphalloidea | 1335 | | ● | | | ◐ | ○ | △ | △ | △ |
| Lepocreadioidea | 542 | | | | | ● | | | Ⓐ | |
| Opisthorchioidea | 839 | | | | | ● | ○ | △ | △ | △ |
| Schistosomatoidea | 453 | ● | ● | | | ◐ | ○ | ◐ | ● | ▲ |
| Paramphistomoidea | 431 | | | | | ◐ | ○ | ▲ | | △ |
| Pronocephaloidea | 293 | | | | | ◐ | | ● | Ⓐ | ▲ |
| Echinostomatoidea | 1098 | | | | | ◐ | | △ | △ | ▲ |
| Brachylaimoidea | 227 | | | | | | ○ | △ | ◐ | △ |
| Diplostomoidea | 797 | | | | | | ○ | ⊘ | △ | △ |
| Plagiorchioidea | 953 | | | | | | ○ | ○ | △ | △ | △ |
| **Total** | **12,012** | | | | | | | | | |

| Major group | Sp (no.) | H | E | C | I | T | A | R | B | M |
|---|---|---|---|---|---|---|---|---|---|---|
| Diphyllidea | 58 | | ● | | | | | | | |
| Trypanorhyncha | 389 | | ● | | | | | | | |
| Litobothriidea | 9 | | ● | | | | | | | |
| Lecanicephalidea | 161 | | ● | | | | | | | |
| Cathetocephalidea | 10 | | ● | | | | | | | |
| Tetraphyllidea | 141 | | ● | | | | | | | |
| Phyllobothriidea | 164 | ● | ● | | | | | | | |
| Rhinebothriidea | 114 | | ◐ | | | | | | | |
| Gyrocotylidea | 10 | ● | | | | | | | | |
| Caryophyllidea | 167 | | | | | ○ | | | | |
| Spathebothriidea | 6 | | | ● | | ● | | | | |
| Tetrabothriidea | 73 | | | | | | | | ● | ● |
| Diphyllobothriidea | 129 | | | | | | ○ | △ | ○ | ◐ |
| Nippotaeniidea | 6 | | | | | ○ | | | | |
| Onchoproteocephalidea | 658 | | ● | | | ○ | △ | △ | | △ |
| Taenioidea | 2264 | | | | | | ⊘ | ⊘ | ▲ | ▲ |
| Amphilinidea | 15 | | | ● | | ◐ | | △ | | |
| Haplobothriidea | 2 | | | | ○ | | | | | |
| Bothriocephalidea | 263 | | | | ○ | | | | | |
| Mesocestoididae | 32 | | | | | | | | | △ |
| **Total** | **4,761** | | | | | | | | | |

244  *Reproduction and Development in Platyhelminthes*

facilitated lineage diversification; the species number falls between 227 in Brachylaimoidea and 797 in Diplostomoidea. Conversely, the unrestricted distribution of four families over a wider range of aquatic and terrestrial hosts has facilitated lineage diversification; their species number ranges from 453 in Schistosomatoidea to 1335 in Microphalloidea with an average of 1052 species/superfamily. As the exact number of species distributed over aquatic and terrestrial hosts is not known for each of the six families in the third group, no further analysis can be made.

However, cestodes allow further analysis, as only a few families overlap in distribution; for example, diphyllobothriids are distributed on all the taxa among tetrapods (Table 7.4, right panel); but the number of species for each of the host taxa is not provided. Of 4,671 cestode species, 27.7% (1,294+ species) and 17.0% (794+ species see p 195) are hosted by elasmobranchs and teleosts, respectively. The dominance of terrestrial hosts is due to a single group, the taenioides comprising 2,264 species. Intertwining of one or another life stage into different food chains opens new avenues for diversification of lineages. For example, by attaching cercarial cysts on aquatic vegetation and grasses, digeneans have gained access to a range of herbivorous ruminants and other hosts (see Shoop 1988). Aquatic cestodes engage carnivorous food chains namely copepods and carnivorous fish to complete a generation in their life cycle. The species number in eight families from Caryophyllidea to Tetraphyllidea engaging in carnivorous food chains totals to only 1,266 species (Table 5.4). But a single family Taenioidea is explosively diversified (2,264 species), as they involve insectivorous and herbivorous (taeniids) food chains for dissemination. No invertebrate taxa are as much speciose as insects are. Herbivorous mammals may outnumber the carnivorous mammals. Within taenioids, members of families Devaineidea, Diphyllidae, Hymenolepidae and Anoplocephalidae (anoplocephaline and thysanosomine groups only) engage in insectivorous food chains to infect the definitive hosts such as the non-ruminant herbivorous mammals (horses, rhinos), rodents and insectivorous birds. In Taeniidae, the members engage in herbivorous food chains involving rodents, ruminants and pigs.

In aquatic habitats, elasmobranchs host cestodes more than teleosts. At this juncture, a hitherto unrecognized aspect of urea on taxonomic distribution of parasitic flukes and flatworms in aquatic hosts must be highlighted. **Cestoda**: A compilation of information reported in Figs. 16.2 and 16.4 of Littlewood et al. (2015) indicates that of ~ 2,000 cestode species hosted by marine fish species, 85% of worms occur exclusively in elasmobranchs, while the others in primitive fish. **Digenea**: The analysis of Gibson and Bray (1994) on a relatively smaller sample of 5,350 digenean species indicates that fishes serve as a definitive host for 53% of the sampled fluke species; unfortunately, the authors have not divided the data pertaining to teleosts and elasmobranchs. Teleosts serve as a host exclusively for only three species in two genera and a family the ptychogoninid (within the superfamily Hemiuroidea) digeneans (see Hyman, 1951). **Monogenea**: For monogeneans too, the relevant information is not readily available. However, of 360 host marine fish species recorded from China, only 17 (or 5%) elasmobranch species serve as hosts for 17 fluke species (Table 3.1). **Microhabitats**: Of known 4,671 cestode species, 1,294 (or 27.7%) and 794 (or 17%) tapeworm species are localized in the intestine of elasmobranchs and teleosts, respectively (see Littlewood et al., 2015). Hyman (1951) hinted that digenean flukes occur in the digestive tract and many other organs. Being suctorial feeders on body fluids, blood and soft tissues, the digeneans have indeed made an extensive exploration into many internal organs within a definitive host as the final niche (see p 18). Selection of the niche may have an implication for the exit of their eggs to the exterior (see p 173). However, a majority of digeneans have selected the intestine as their destination. As the hard skin substratum of elasmobranchs is unsuitable of deployment of hooks, for example, *Leptocotyle minor* on the dogfish *Styliorhinus canalicula*, even the limited incidences of elasmobranchs on monogenean flukes are relegated to the gills. Surprisingly, a computer search reveals that most of the 17 elasmobranch species, that host 17 fluke species of China are estuarine or visitors to brackishwaters (e.g. *Dasyatis kuhli*, *Rhina ancylostoma*, *Triakis scyllium*). **Food and acquisition**: Essentially, the trematodes are suctorial feeders (see p 35), whereas cestodes osmotrophically acquire low molecular nutrients through the tegumental surface from the surrounding digested nutrients in the intestine.

In parasitic flatworms, a pair (or more) of protonephridia are provided with terminal flame cells; the nephridia open to the exterior through nephridial pore(s) (Hyman, 1951). However, no information is available whether the protonephridia can excrete ammonia alone or ammonia and urea together. In teleosts, urea constitutes ~ 30% of total nitrogenous end products and is excreted through the kidney, while the remaining ~ 67% ammonia is lost diffusively through the gills (see Pandian, 1975). Presumably, the parasitic flatworms may also excrete ammonia alone but not urea. For, they do not accept or acquire blood, body fluids and other cells loaded with high concentration of urea in elasmobranchs but they thrive in their intestine.

Ammonia and urea, the nitrogenous excretory end products, are soluble and diffusible. In comparison to ammonia, urea is less toxic. Urea is a dipole like water, and has low oil-water partition coefficient and penetrates through aqueous pores more readily than in the lipid-protein component cell membrane. Despite similarities in diffusion coefficient of ammonia and urea, ammonia passes through most biological membranes faster than urea. The principal excretory end product is urea in marine elasmobranchs and mammals. But elasmobranchs retain a major fraction of urea and trimethylamine oxide (TMAO, ~ 70 mM/l) in their blood to balance osmoregulation. For example, blood urea level is high in marine elasmobranchs and varies between 260 and 764 mM/l (see Pandian, 1975). Being inhabitants of both marine and estuarine habitats around Australian

246 *Reproduction and Development in Platyhelminthes*

coastal waters, the bull shark *Carcharhinus leucas* has blood urea levels at 192 and 370–873 mM/l in brackish water and seawater, respectively (Pillans and Franklin, 2004). In freshwater sharks like *Dasyatis uarnak* and *Pristis microdon*, Smith (1931) reported even lower levels of blood urea. The high blood urea level in the gills and other organs (other than the intestine) of elamobranchs has eliminated successful invasion and colonization by parasitic trematodes.

## 7.3 Diversification and Speciation

The life cycle of all the major parasitic taxa is indirect and involves one or more larval stage(s). As of now, the confirmed species number is 6,500 for Monogenea, 12,012 for Digenea and 4,761 for Cestoda. Table 7.5 summarizes a comparative account on the life history traits of parasitic flatworms. The following factors may be considered as responsible for lineage diversification: (i) fecundity, (ii) incidence of propagatory multiplication in intermediate host(s), (iii) incorporation of more and more intermediate hosts, (iv) genetic diversity, (v) host specificity (vi) penetrative or trophic mode of transmission and (vii) intensity of infection, mating type. In ~ 27,700 speciose platyhelminths, the share is 23.5, 16.2, 43.4% and 16.9% for turbellarians, monogeneans, digeneans and cestodes, respectively. The digeneans are 2-times more speciose than other taxa within platyhelminths, as (i) their relatively higher fecundity (4,088 eggs/day, see Table 1.5) is supplemented by propagatory multiplication in one or more larval stage(s). (ii) 88% digeneans engage more mobile SIH; they are highly flexible and completely euryxenic to select one among 3.5 species awaiting/available to serve as SIH and (iii) higher level of clonal selection in SIH and the consequent onward transmission of the fittest metacercaria clone(s).

Despite the inclusion of an indirect life cycle involving penetrative oncomiracidium, the monogeneans are not diversified due to (i) the lowest fecundity (0.123 eggs/days), which is not supplemented by propagatory multiplication, (ii) incorporation of no intermediate host and (iii) the highest level of oioxenic host specificity. Irrespective of the (i) highest fecundity (223,124 eggs/day), (ii) indirect life cycle involving two to three larval stages and one to two intermediate hosts and (iii) the lowest level of host specificity, cestodes are not diversified due to the ubiquitous selfing and inbreeding.

Incidentally, a notable fact is that the first intermediate hosts of digeneans are always a hemocoelomic mollusc (rarely hemocoelomic leech) and so are the cestodes on a hemocoelomic arthropod (except in taeniids). Both digenean sporocyst and cestodean procercoid in aquatic habitats as well as cysticercoids in terrestrial habitats have to osmotrophically abstract the host body fluids as their nutrients. In them, the presence of hemocyanin in the hemocoel is a common feature. Excretion of insoluble hematin poses a

**TABLE 7.5**

Comparison of life history traits in parasitic flatworms. FW = freshwater

| Selected traits | Remarks | | |
|---|---|---|---|
| | Monogenea | Digenea | Cestoda |
| Species number | 4,500 (Table 1.2) | 12,012 (Table 4.2) | 4,761 (Table 5.1) |
| Life style | Ectoparasitic on teleost's skin, gills & urinary bladder of amphibian | Endoparasitic in different organs (see Table 7.2, p 18) | Bereft of mouth & gut, endoparasites relegated to intestine |
| Larval form | Only 1; penetrative oncomiracidium | Miracidium, sporocyst, redia, cercaria | Coracidium, oncosphere, hexacanth types (see pp 198–200) |
| Intermediate host | Nil | 1 to 3 hosts; 74% flukes involve 2 hosts (see Table 4.4) | 30% in coracidium types involves 2 hosts; remaining 70% involves 1 host |
| Propagatory multiplication | Nil | In sporocyst and/or redia; when it occurs in a bivalve metacercarial multiplication in second intermediate gastropod host (e.g. *Parvatrema margaritense*) | Clonal multiplication limited to 22 species (Table 5.7) |
| Host habitats | 93% in marine and FW teleosts, 4% in FW elasmobranch & 3% in amphibians; 80% marine & 16.7% FW teleosts (Young, 1970) | 48% in fishes and 52% in tetrapods | 49% hosted by tetrapods, 28% by elasmobranchs, 17% by most FW teleosts (see p 195) |
| Host specificity | 71% oioxenics, 14% stenoxenics, 15% euryxenics | Euryxenic to all the hosts (see pp 150–151, Fig 4.3) | Euryxenic to all the hosts |
| Intensity, mating, genetic diversity | The presence of 2–5 flukes at 10% prevalence provide ample scope for cross breeding to increase genetic diversity | Inbreeding depression due to selfing and clonal multiplication are neutralized by clonal mixing in second intermediate host (p 150–151) | Selfing is ubiquitous. Cross breeding yields a fraction of selfed eggs |
| Transmission | No information | By penetrative alone or trophic alone or by penetrative and trophic (see Table 4.6, Fig. 4.7) or both by trophic. At an efficiency of 0.03% (p 34) | Rely solely on trophic transmission at an efficiency of 0.08% (see p 34) |

serious problem to blood sucking trematode adults (Section 1.5). Firstly, can these osmotrophic larvae excrete copper? Secondly, is it easier for the larvae to abstract oxygen from hemocoel carrying the less efficient hemocyanin? (see pp 54–55). Thirdly, is there a specific factor like schistomin present in the hemocoel of molluscs (De Jong-Brink et al., 1991) that induces propagatory multiplication? Is the schistomin like factor not present in arthropod hosts? In these virgin areas, research is required.

# 8

# References

Abebe, R., Abunna, F., Berhane, M. et al. 2010. Fasciolosis: Prevalence, financial losses due to liver condemnation and evaluation of a simple, sedimentation diagnostic technique in cattle slaughtered at Hawassa municipal abattoir, Southern Ethiopia. Ethiop Vet J, 14: 39–51.

Aboobaker, A.A. 2011. Planarian stem cells: A simple paradigm for regeneration. Trends Cell Biol, 21: 304–311.

Abrous, A., Rondelaud, D. and Dreyfuss, G. 1999. *Paramphistomum daubneyi* and *Fasciola hepatica*: influence of temperature changes on the shedding of cercariae from dually infected *Lymnaea truncatula*. Parasitol Res, 85: 765–769.

Adell, T., Cebria, F. and Salo, E. 2010. Gradients in planarian regeneration and homeostasis. Cold Spring Harb Perspect Biol, 2: a000505.

Adell, T., Martin-Duran, J.M., Salo, E. and Cebria, F. 2015. Platyhelminthes. In: *Evolutionary Developmental Biology of Invertebrates*. (ed) Wanninger, A., Springer Verlag, Wien, 2: 21–40.

Adou, Y.E., Blahoua, K.G., Kamelan, T.M. and N'Douba, V. 2017. Prevalence and intensity of gill monogenean parasites of *Tilapia guineensis* (Bleeker, 1862) in man-made Lake Ayame 2, Cote d'Ivoire according to season, host size and sex. Int J Biol Chem Sci, 11: 1559–1576.

Aguirre-Macedo, M.L., May-Tec, A.L., Martinez-Aquino, A. et al. 2016. Diversity of helminth parasites in aquatic invertebrate hosts in Latin America: how much do we know? J Helminthol, 91: 1–13.

Akesson, B., Gschwentner, R., Hendelberg, J. et al. 2001. Fission in *Convolutriloba longifissura*: asexual reproduction in acoelous turbellarians revisited. Acta Zool, 82: 231–239.

Akoll, P., Konecny, R., Mwanja, W.W. et al. 2011. Parasite fauna of farmed Nile tilapia (*Oreochromis niloticus*) and African catfish (*Clarias gariepinus*) in Uganda. Parasitol Res, 110: 315–323.

Allen, J.D., Klompen, M.L., Alpert, E.J. and Reft, A.J. 2017. Obligate planktotrophy in the Gotte's larva of *Stylochus ellipticus* (Platyhelminthes). Invert Reprod Dev, 61: 110–118.

Alvarado, A.S. and Newmark, P.A. 1999. Double-stranded RNA specifically disrupts gene expression during planarian regeneration. Proc Natl Acad Sci USA, 96: 5049–5054.

Alvarez-Presas, M. and Riutort, M. 2014. Review planarian (Platyhelminthes, Tricladida) diversity and molecular markers: A new view of an old group. Diversity, 6: 323–338.

Alvite, G. and Esteves, A. 2011. Lipidic metabolism in parasitic platyhelminthes. In: *Research in Helminths*. (ed) Esteves, A. Transworld Res Network, Tiruvandrum, India, pp 1–12.

Anderson, G.A., Schell, S.C. and Prattt, I. 1965. The life cycle of *Bunoderella metteri* (Allocreadiidae: Bunoderinae), a trematode parasite of *Ascaphus truei*. J Parasitol, 51: 579–582.

Anderson, R.M., Mercer, J.G., Wilson, R.A. and Carter, N.P. 1982. Transmission of *Schistosoma mansoni* from man to snail: experimental studies on miracidial survival and infectivity in relation to larval age, water temperature, host size and host age. Parasitology, 85: 339–360.

Antonelli, L., Quilichini, Y. and Marchand, B. 2010. *Sparicotyle chrysophrii* (Van Beneden and Hesse 1863) (Monogenea: Polyopisthocotylea) parasite of cultured gilthead sea bream *Sparus aurata* (Linnaeus 1758) (Pisces: teleostei) from Corsica: ecological and morphological study. Parasitol Res, 107: 389–398.

250 *Reproduction and Development in Platyhelminthes*

Arme, C. 1968. Effect of plerocercoid larva of a pseudophyllidean cestode *Ligula intestinalis* on pituitary gland and gonads of its host. Biol Bull, 134: 15–25.

Armstrong, J.C. 1965. Mating behaviour and development of schistosomes in the mouse. J Parasitol, 51: 605–616.

Auld, S.K.J.R. and Tinsley, M.C. 2015. The evolutionary ecology of complex life cycle parasites: linking phenomena with mechanisms. Heredity, 114: 125–132.

Ax, P. and Schulz, E. 1959. Ungeschlechtliche Fortpflanzung durch Paratomie bei acoelen Turbellarien. Biol Zentbl, 78: 613–622.

Ayalneh, B., Bogale, B. and Dagnachew, S. 2018. Review on ovine fasciolosis in Ethiopia. Acta Parasitol Global, 9: 7–14.

Badets, M. and Verneau, O. 2009. Origin and evolution of alternative developmental strategies in amphibious sarcopterygian parasites (Platyhelminthes, Monogenea, Polystomatidae). Org Diver Evol, 9: 155–164.

Badets, M., Morrison, C. and Verneau, O. 2010. Alternative parasite development in transmission strategies: how time flies! J Evol Biol, 23: 2151–2162.

Bagge, A.M. and Valtonen, E.T. 1999. Development of monogenean communities on the gills of roach fry (*Rutilus rutilus*). Parasitology, 118: 479–489.

Baguna, J. 1974. Dramatic mitotic response in planarians after feeding and a hypothesis for the control mechanism. J Exp Zool, 190: 117–122.

Baguna, J. 1976a. Mitosis in the intact and regenerating planarian *Dugesia mediterranea* n. sp. I. mitotic studies during growth, feeding and starvation. J Exp Zool, 195: 53–64.

Baguna, J. 1976b. Mitosis in the intact and regenerating planarian *Dugesia mediterranea* n. sp. II. Mitotic studies during regeneration and a possible mechanism of blastema formation. J Exp Zool, 195: 65–80.

Baguna, J. 1981. Planarian neoblasts. Nature, 290: 14–15.

Baguna, J. and Romero, R. 1981. Quantitative analysis of cell types during growth, degrowth, and regeneration in the planarians *Dugesia (S) mediterranea* and *Dugesia (G) tigrina*. Hydrobiologia, 84: 181–194.

Baguna, J., Salo, E. and Romero, R. 1989. Effects of activators and antagonists of the neuropeptides substance P and substance K on cell proliferation in planarians. Int J Dev Biol, 33: 261–264.

Baguna, J., Romero, R., Salo, E. et al. 1990. Growth, degrowth and regeneration as developmental phenomena in adult freshwater planarians. In: *Experimental Embryology in Aquatic Plant and Animal Organisms*. (ed) Marthy, H.J., NATO-ASI Series, Plenum Press, New York, pp 129–162.

Baguna, J. 1998. Planarians. In: *Cellular and Molecular Basis of Regeneration: From Invertebrates to Humans*. (eds) Ferretti, P. and Geraudie, J. John Wiley, Chichester, pp 135–165.

Baguna, J., Salo, E., Collet, J. et al. 1998. Cellular, molecular, and genetic approaches to regeneration and pattern formation in planarians. Fortschr Zool, 36: 65–78.

Baguna, J., Carranza, S, Pala, M. et al. 1999. From morphology and karyology to molecules. New methods for taxonomical identification of asexual populations of freshwater planarians. A tribute to Professor Mario Benazzi. Ital J Zool, 66: 207–214.

Baguna, J. 2012. The planarian neoblast: The rambling history of its origin and some current black boxes. Int J Dev Biol, 56: 19–37.

Bakhraibah, A.O. 2018. Effect of locality, host species and sex on the metazoan parasitic infestation of two species of *Scarus* fish from the Red Sea Coast at Jeddah and Rabigh in Saudi Arabia. Sci Res Pub, 8: 252–258.

Bakke, T.A., Jansen, P.A. and Hansen, L.P. 1990. Differences in the host resistance of Atlantic salmon, *Salmo salar* strains to the monogenean *Gyrodactylus salaris* Malmberg, 1957. J Fish Biol, 37: 577–587.

Bakke, T.A., Harris, P.D. and Jansen, P.A. 1992a. The susceptibility of *Salvelinus fontinalis* (Mitchell) to *Gyrodactylus salaris* Malmberg (Platyhelminthes; Monogenea) under experimental conditions. J Fish Biol, 41: 499–507.

Bakke, T.A., Harris, P.D., Jansen, P.A. and Hansen, L.P. 1992b. Host specificity and dispersal strategy in gyrodactylid monogeneans, with particular reference to *Gyrodactylus salaris* (Platyhelminthes, Monogenea). Dis Aquat Org, 13: 63–74.

References 251

Bakke, T.A., Soleng, A. and Harris, P.D. 1999. The susceptibility of Atlantic salmon (*Salmo salar* L.) x brown trout (*Salmo trutta* L.) hybrids to *Gyrodactylus salaris* Malmberg and *Gyrodactylus derjavini* Mikailov. Parasitology, 119: 467–481.

Bakke, T.A., Soleng, A., Lunde, H. and Harris, P.D. 2000. Resistance mechanisms in *Salmo salar* stocks infected with *Gyrodactylus salaris*. Acta Parasitol, 45: 272.

Bakke, T.A., Harris, P.D. and Cable, J. 2002. Host specificity dynamics: observations on gyrodactylid monogeneans. Int J Parasitol, 32: 281–308.

Bakke, T.A., Cable, J. and Harris, P.D. 2007. The biology of gyrodactylid monogeneans: The "Russian–Doll Killers". Adv Parasitol, 64: 161–376.

Ballabeni, P. 1995. Parasite-induced gigantism in snails: a host adaptation? Funct Ecol, 9: 887–893.

Barber, I. and Ruxton, G.D. 1998. Temporal prey distribution affects the competitive ability of parasitized sticklebacks. Anim Behav, 56: 1477–1483.

Barber, I. and Svensson, P.A. 2003. Effects of experimental *Schistocephalus solidus* infections on growth, morphology and sexual development of female three-spined sticklebacks, *Gasterosteus aculeatus*. Parasitology, 126: 359–367.

Bardhan, A., Kumarb, R.R., Nigamc, S. et al. 2014. Estimation of milk losses due to fasciolosis in Uttarakhand. Agri Econo Res Rev, 3: 32–37.

Barger, M.A. 2012. Life span of cercariae of *Postharmostomum helicis* (Trematoda Brachylaimidae) under different temperatures and relative humidity. Comp Parasitol, 79: 169–172.

Barkar, S.C. and Cribb, T.H. 1993. Sporocysts of *Mesostephanus haliasturis* (Digenea) produce miracidia. Int J Parasitol, 23: 137–139.

Barnes, R. 1974. *Invertebrate Zoology*. Saunders, International Student Edition, p 870.

Bartel, M.H. 1965. The life cycle of *Raillietina* (R.) *loeweni* Bartel and Hansen, 1964 (Cestoda) from the black-tailed jackrabbit, *Lepus californicus melanotis* Mearns. J Parasitol, 51: 800–806.

Basch, P.F. and Basch, N. 1984. Intergeneric reproductive stimulation and parthenogenesis in *Schistosoma mansoni*. Parasitology, 89: 369–376.

Beck, M.A., Goater, C.P. and Colwell, D.D. 2015. Comparative recruitment, morphology and reproduction of a generalist trematode *Dicrocoelium dendrictum* in three species of host. Parasitology, 142: 1297–1305.

Bednarz, S. 1962. The developmental cycle of germ cells in *Fasciola hepatica* L. 1758 (Trematodes, Digenea). Zool Polon, 12: 439–466.

Behensky, C., Shurmann, W. and Peter, R. 2001. Quantitative analysis of turbellarians cell suspensions by fluorescent staining with acridin orange and videomicroscopy. Belg J Zool, 131: 131–136.

Beisner, B.E., Mccauley, E. and Wrona, F.J. 1997. Predator-prey instability: individual-level mechanisms for population-level results. Funct Ecol, 11: 112–120.

Benazzi–Lentati, G. 1966. Amphimixis and pseudogamy in freshwater triclads: experimental recombination of polyploidy pseudogamic biotypes. Chromosoma, 20: 1–14.

Benazzi–Lentati, G. 1970. Gametogenesis and egg fertilization in planarians. Int Rev Cytol, 27: 101–179.

Benazzi, L.G. and Deri, P. 1981. On the production of diploid offspring from specimens of the triplo–hexaploid biotype of the planarian *Dugesia benazzii*. Accad Naz Lincei Rend, 69: 445–449.

Benazzi–Lentati, G., Deri, P. and Benazzi, M. 1988. Does the genetic constitution of the EGG influence fission frequency of the offspring in planarians? Accad Naz Lincei Rend, 82: 131–136.

Benazzi, M., Baguna, J. and Ballester, R. 1970. First report on an asexual form of the planarian *Dugesia lugubris*. s.l. Accad Naz Lincei Rend, 48: 42–44.

Benazzi, M. and Ball, I.R. 1972. The reproductive apparatus of sexual specimens from fissiparous populations of *Fonticola morgani* (Tricladida, Paludicola). Can J Zool, 50: 703–704.

Benazzi, M. 1974. Fissioning in planarians from a genetic standpoint. In: *Biology of the Turbellaria*. (eds) Riger, N.W. and Morse, M.P., McGraw-Hill, New York, pp 476–492.

Benazzi, M. and Grasso, M. 1977. Comparative research on the sexualization of fissiparous planarians treated with substances contained in sexual planarians. Monit Zool Ital, 11: 9–19.

252  *Reproduction and Development in Platyhelminthes*

Benazzi, M. 1981. Reproductive biology of *Dugesia sanchezi*, a freshwater planarian from Chile. Hydrobiologia, 84: 163–165.

Benazzi, M. 1993. Occurrence of sexual population of *Dugesia* (*Girardia*) *tigrina*, a freshwater planarian native to America in a lake of southern Italy. Ital J Zool, 60: 129–130.

Benesh, D.P. and Hafer, N. 2012. Growth and ontogeny of the tapeworm *Schistocephalus solidus* in its copepod first host affects performance in its stickleback second intermediate host. Parasite Vect, 5: 90.

Benesh, D.P., Weinreich, F., Kalbe, M. and Millinski, M. 2014. Life time inbreeding depression, purging, and mating system evolution in a simultaneous hermaphrodite tapeworm. Evolution, 68: 1762–1774.

Berger, J. and Mettrick, D.F. 1971. Microtrichial polymorphism among hymenolepid tapeworms as seen by scanning electron microscopy. Trans Am Microsc Soc, 90: 393–403.

Best, J.B., Goodman, A.B. and Pigeon, A. 1969. Fissioning in planarians: Control by the brain. Science, 164: 565–566.

Beukeboom, L.W., Seif, M., Mettenmeyer, T. et al. 1996a. Paternal inheritance of B chromosomes in a parthenogenetic hermaphrodite. Heredity, 77: 646–654.

Beukeboom, L.W., Weinzierl, R.P., Reed, K.M. and Michiels, N.K. 1996b. Distribution and origin of chromosomal races in the freshwater planarian *Dugesia polychroa* (Turbellaria: Tricladida). Hereditas, 124: 7–15.

Beukeboom, L.W., Sharbel, T.F. and Michiels, N.K. 1998. Reproductive modes, ploidy distribution, and supernumerary chromosome frequencies of the flatworm *Polycelis nigra* (Platyhelminthes; Tricladida). Hydrobiologia, 383: 277–285.

Birstein, V.J. 1991. On the karyotypes of the *Neorhabdocoela* species and karyological evolution of Turbellaria. Genetica, 83: 107–120.

Blackshaw, R.P. 1992. The effect of starvation on size and survival of the terrestrial planarian *Artiposthiatri angulata* (Dendy) (Tricladida, Terricola). Ann Appl Biol, 120: 573–578.

Blackwelder, R.E. and Shepherd, B.A. 1981. *The Diversity of Animal Reproduction*. CRC Press, USA, p 272.

Blahoua, G.K., Adou, E.Y., Etile N.R. et al. 2018. Occurrence of gill monogenean parasites in redbelly tilapia, *Tilapia zillii* (Teleostei: Cichlidae) from Lobo River, Cote d'Ivoire. J Anim Plant Sci, 35: 5674–5688.

Blahoua, K.G., Yao, S.S., Etile, R.N. and N'Douba, V. 2016. Distribution of gill monogenean parasites from *Oreochromis niloticus* (Linne, 1758) in man-made Lake Ayame I, Coted'Ivoire. Afr J Agri Res, 11: 117–129.

Blasco-Costa, I. and Poulin, R. 2017. Parasite life-cycle studies: a plea to resurrect an old parasitological tradition. J Helminthol, 91: 647–656.

Blythe, M.J., Kao, D., Malla, S. et al. 2010. A dual platform approach to transcript discovery for the planarian *Schmidtea mediterranea* to establish RNAseq for stem cell and regeneration biology. PLoS One, 5: e15617.

Boag, B., Neilson, R. and Scrimgeour, C.M. 2006. The effect of starvation on the planarian *Arthurdendyus triangulatus* (Tricladida: Terriola) as measured by stable isotopes. Biol Fert Soil, 43: 267–270.

Boeger, W.A., Kritsky, D.C., Pie, M.R. and Engers, K.B. 2005. Mode of transmission, host switching, and escape from the red queen by viviparous gyrodactylids (Monogenoidea). J Parasitol, 91: 1000–1007.

Bondad-Reantaso, M.G., Ogawa, K., Fukudome, M. and Wakabayashi, H. 1995a. Reproduction and growth of *Neobenedenia girellae* (Monogenea: Capsalidae), a skin parasite of cultured marine fishes of Japan. Fish Pathol, 30: 227–231.

Bondad-Reantaso, M.G., Ozawa, K., Yoshinaga, T. and Wakabayashi, H. 1995b. Acquired protection against *Neobenedenia girellae* in Japenese flounder. Fish Pathol, 30: 233–238.

Boray, J.C. 2007. Liver fluke disease in sheep and cattle. Primefacts, 446: 1–10.

Boufana, B., Zibrat, U., Jehle, R. et al. 2011. Differential diagnosis of *Triaenophorus crassus* and *T. nodulosus* experimental infection in *Cyclops abyssorum praealpinus* (Copepoda) from the Alpine Lake Grundlsee (Austria) using PCR–RFLP. Parasitol Res, 109: 745–750.

## References 253

Brant, S.V. 2007. The occurrence of the avain schistosome *Allobilharzia visceralis* Kolarova, Rudolfova, Hampl et Skirnisson, 2006 (Schistosomatidae) in the Tundra swan, *Cygnus columbianus* (Anatidae) from North America. Folia Parasitol, 54: 99–104.

Braun, F. 1996. Beitrage zur mikroskopischen Anatomie und Fortpfanzungsbiologie von *Gyrodactylus wageneri*. Z Parasitenkd, 17: 31–63.

Brehm, K. 2010. *Echinococcus multilocularis* as an experimental model in stem cell research and molecular host–parasite interaction. Parasitology, 137: 537–555.

Brondsted, A. and Bronsted, H.V. 1961. Influence of temperature on rate of regeneration in the time–graded regeneration field in planarians. J Embryol Exp Morph, 9: 159–166.

Brondsted, H.V. 1969. *Planarian Regeneration*. Pergamon Press, Oxford, p 298.

Brumpt, E. 1936. *Precis de Parasitologie*. Paris: Masson and Cie, p 1082.

Bryant, C. and Williams, J.P.G. 1962. Some aspects of the metabolism of the liver fluke, *Fasciola hepatica* L. Exp Parasitol, 12: 373–376.

Buchmann, K. 1997a. Infection biology of gill parasitic monogeneans with special reference to the congeners *Pseudodactylogyrus bini* and *P. anguillae* (Monognea: Platyhelminthes) from European eel. Thesis, Royal Veterinary and Agricultural University, Frederiksberg, Denmark.

Buchmann, K. 1997b. Population increase of *Gyrodactylus derjavini* on rainbow trout induced by testosterone treatment of the host. Dis Aquat Org, 30: 145–150.

Buchmann, K. 1998a. Binding and lethal effect of complement from *Oncorhynchus mykiss* on *Gyrodactylus derjavini* (Platyhelminthes, Monogenea). Dis Aquat Org, 32: 195–200.

Buchmann, K. 1998b. Some histochemical characteristics of the mucus microenvironment in four salmonids with different susceptibilities to gyrodactylid infections. J Helminthol, 72: 101–107.

Buchmann, K. and Bresciani, J. 1999. Rainbow trout leucocyte activity: influence on the ectoparasitic monogenean *Gyrodactylus derjavini*. Dis Aquat Org, 35: 13–22.

Buchmann, K. and Lindenstrom, T. 2001. Interactions between monogenean parasites and their fish hosts. ISM 4, 9–13, July, Brisbane, Australia, 51.

Buchmann, K. and Lindenstrom, T. 2002. Interactions between monogenean parasites and their fish hosts. Int J Parasitol, 32: 309–319.

Bullock, T.H. and Horridge, G.A. 1965. *Structure and Function in the Nervous System of Invertebrates*. Freeman, San Francisco.

Bush, A.O., Aho, J.M. and Kennedy, C.R. 1990. Ecological versus phylogenetic determinants of helminth parasite community richness. Evol Ecol, 4: 1–20.

Cable, J. and Harris, P.D. 2002. Gyrodactylid developmental biology: historical review, current status and future trends. Int J Parasitol, 32: 255–280.

Caira, J. and Littlewood, D.T.J. 2013. Worms, platyhelminthes. Encyclo Biodiv, 5: 863–899.

Caira, J.N. and Jensen, K. 2014. A digest of elasmobranch tapeworms. J Parasitol, 100: 373–391.

Caira, J.N., Jensen, K., Waeschenbach, A. and Littlewood, D.T.J. 2014. An enigmatic new tapeworm, *Litobothrium aenigmaticum*, sp. nov. (Platyhelminthes: Cestoda: Litobothriidea), from the pelagic thresher shark with comments on development of known *Litobothrium* species. Invert Syst, 28: 231–243.

Calentine, R.L. 1963. The life cycle of *Archigetes iowensis* (Cestoda: Caryophyllidea). Thesis, Iowa State University.

Calow, P. 1987. Platyhelminthes and Rhynchocoela, with special reference to the triclad turbellarians. In: *Animal Energetics*. (eds) Pandian, T.J. and Vernberg, F.J. Academic Press, San Diego, 1: 121–158.

Calow, P., Beveridge, M. and Sibly, R. 1979. Heads and tails: adaptational aspects of asexual reproduction in freshwater triclads. Integ Comp Biol, 19: 715–727.

Campbell, R.A. 1983. Parasitism in the deep sea. In: *The Sea*. (ed) Rowe, G.T., John Wiley, New York, 8: 473–552.

Cannon, L.R.G. and Lester, R.J.G. 1988. Two turbellarians parasitic in fish. Dis Aquat Org, 5: 15–22.

Cardona, A., Haritenstein, V. and Romero, R. 2005. The embryonic development of the triclads *Schmidtea polychroa*. Dev Genes Evol, 215: 109–131.

254  *Reproduction and Development in Platyhelminthes*

Carter, V., Pierce, R., Dufour, S. et al. 2005. The tapeworm *Ligula intestinalis* (Cestoda: Pseudophyllidea) inhibits LH expression and pituitary in its host *Rutilus rutilus*. Reproduction, 130: 939–945.

Cebria, F. 2007. Regenerating the central nervous system: how easy for planarians? Dev Genes Evol, 217: 733–748.

Cebria, F. and Newmark, P.A. 2005. Planarian homologs of netrin and netrin receptor are required for proper regeneration of the central nervous system and the maintenance of nervous system architecture. Development, 132: 3691–3703.

Cebria, F. Jopek, G.T. and Newmark, P.A. 2007. Regeneration and maintenance of the planarian midline is regulated by a slit orthologue. Dev Biol, 14: 394–406.

Cebria, F., Salo, E. and Adell, T. 2015. Regeneration and growth as modes of adult development: The Platyhelminthes as a case study. In: *Evolutionary Developmental Biology of Invertebrates*. (ed) Wanninger, A., Springer-Verlag, Vienna, 2: 41–78.

Cecchini, S., Sarglia, M., Berni, P. and Cognetti-Varriale, A.M. 1998. Influence of temperature on the life cycle of *Diplectanum aequans* (Monogenea, Diplectanidae), parasitic on sea bass, *Dicentarchus labrax* (L.). J Fish Dis, 21: 73–75.

Cerda, J.R., Buttke, D.E. and Ballweber, L.R. 2018. *Echinococcus* spp. tapeworms in North America. Emerg Infect Dis, 24: 230–235.

Chambers, C.B. and Ernst, I. 2005. Dispersal of the skin fluke *Benedenia seriolae* (Monogenea: Capsalidae) by tidal currents and implications for sea-cage farming of *Seriola* spp. Aquaculture, 250: 60–69.

Chan, B. and Wu, B. 1984. Studies on the pathogenicity, biology and treatment of *Pseudodactylogyrus* for eels in fishfarms. Acta Zool Sin, 30: 173–180.

Chandebois, R. 1980. The dynamics of wound closure and its role in the programming of planarian regeneration. II. Destalization, Dev Growth Diff, 22: 693–704.

Chapman, A.D. 2009. Number of living species in Australian and world. Rep Austr Biol Study, Dept of Environment, Govt of Australia, p 80.

Chapuis, E. 2009. Correlation between parasite prevalence and adult size in a Trematode–Mollusc system: Evidence for evolutionary gigantism in the freshwater snail, *Galba truncatula*? J Moll Stud, 75: 391–396.

Charbagi-Barbirou, K. and Tekaya, S. 2009. Sexual differentiation and karyological study in the gonochoric planarian *Sabussowia dioica* (Platyhelminthes: Tricladida). Cah Biol Mar, 50: 303–309.

Charnov, E.L. 1979. Simultaneous hermaphroditism and sexual selection. Proc Natl Acad Sci USA, 76: 2480–2484.

Child, C.M. 1911. Studies on the dynamics of morphogenesis and inheritance in experimental reproduction. I. The axial gradient in *Planaria dorotocephala* as a limiting factor in regulation. J Exp Zool, 10: 265– 320.

Child, C.M. 1914. Studies on the dynamics of morphogenesis and inheritance in experimental reproduction. VII. The stimulation of pieces by section in *Planaria dorotocephala*. J Exp Zool, 16: 413–443.

Child, C.M. 1941. *Patterns and Problems of Development*. University of Chicago Press, Chicago, p 820.

Ching, H.L. 1995. Evaluation of characters of the digenean family Gymnophallidae Morozov, 1955. Can J Fish Aquat Sci, 52: 78–83.

Chintala, M.M. and Kennedy, V.S. 1993. Reproduction of *Stylochus ellipticus* (Platyhelminthes: Polycladida) in response to temperature, food and presence or absence of a partner. Biol Bull, 185: 373–387.

Choisy, M., Brown, S.P., Lafferty, K.D. and Thomas, F. 2003. Evolution of trophic transmission in parasites: Why add intermediate hosts? Am Nat, 162: 172–181.

Chong, T., Collins, J.J., Brubacher, J.L. et al. 2013. A sex–specific transcription factor controls male identity in a simultaneous hermaphrodite. Nat Commun, 4: 1814.

Christen, M. and Milinski, M. 2003. The consequences of self-fertilization and outcrossing of the cestode *Schistocephalus solidus* in its second intermediate host. Parasitology, 126: 369–378.

Christensen, A.M. and Kanneworff, B. 1964. *Kronborgia amphipodicola* Gen, et Sp. Nov., A dioecious turbellarian parasitizing ampeliscid amphipods. Ophelia, 1: 147–166.

Christensen, A.M. and Kanneworff, B. 1965. Life history and biology of *Kronborgia amphipodicola* Christensen and Kanneworff (Turbellaria, Neorhabdocoela). Opehlia, 2: 237–251.

Christensen, B. 1984. Asexual propagation and reproductive strategies of aquatic Oligocheata. Hydrobiologia, 115: 91–95.

Christensen, N.O. 1979. *Schistosoma mansoni*: interference with cercarial host-finding by various aquatic organisms. J Helminthol, 53: 7–14.

Christensen, N.O. 1980. A review of the influence of host and parasite-related factors and environmental conditions on the host-finding capacity of the trematode miracidium. Acta Tropica, 37: 303–318.

Christensen, N.O. and Nansen, P. 1976. The influence of temperature on the infectivity of *Fasciola hepatica* miracidia to *Lymnaea truncatula*. J Parasitol, 62: 698–701.

Christensen, N.O., Nansen, P. and Fradsen, F. 1978. The influence of some physio-chemical factors on the host finding capacity of *Fasciola hepatica* miracidia. J Helminthol, 52: 61–67.

Christensen, N.O. Frandsen, F. and Nansen, P. 1980. The interaction of some environmental factors influencing *Schistosoma mansoni* cercarial host-finding. J Helminthol, 54: 203–205.

Chu, K.Y. and Dawood, I.K. 1970. Cercarial production from *Biomphalaria alexandrina* infected with *Schistosoma mansoni*. Bull World Health Organ, 42: 569–574.

Ciordia-Davila, H. 1956. Cytological studies of the germ cell cycle of the trematode family Bucephalidae. Trans Am Microsc Soc, 75: 103–116.

Clark, W.C. 1974. Interpretation of life history pattern in the Digenea. Int J Parasitol, 4: 115–123.

Clausen, K.T., Larsen, M.H., Iversen, N.K. and Mouritsen, K.N. 2008. The influence of trematodes on the macroalgae consumption by the common periwinkle *Littorina littorea*. J Mar Biol Ass UK, 88: 1481–1483.

Collins, J.J. III, Hou, X., Romanova, E.V. et al. 2010. Genome-wide analyses reveal a role for peptide hormones in planarian germline development. PLoS Biol, 8(10): e1000509.

Collins, J.J. and Newmark, P.A. 2013. It's no fluke: The planarian as a model for understanding schistosomes. PLoS Pathogens, 9 (7): e1003396.

Collins, J.J., Wang, B., Lambrus, B.G. et al. 2013. Adult somatic stem cells in the human parasite, *Schistosoma mansoni*. Nature, 494: 476– 479.

Collins, J.J. III. 2017. Platyhelminthes. Curr Biol, 27: 252–256.

Comai, L. 2005. The advantages and disadvantages of being polyploidy. Nature Rev Genet, 6: 836–846.

Combes, C. 1995. *Ecologie et Evolution du Parasitism*. Masson Paris, Milan, Barcelone, p 485.

Cone, D.K., Beverley-Burton, M., Wiles, M. and McDonald, T.E. 1983. The taxonomy of *Gyrodactylus* (Monogenea) parasitizing certain salmonid fishes of North America, with a description of *Gyrodactylus nerkae* n. sp. Can J Zool, 61: 2587–2597.

Conn, D.B. 1990. The rarity of asexual reproduction among *Mesocestoides* tetrathyridia (Cestoda). J Parasitol, 76: 453–455.

Corkum, K.C. and Beckerdite, F.W. 1975. Observations on the life history of *Alloglossidium macrobdellensis* (Trematoda: Macroderoididae), from *Macrobdella ditetra* (Hirudinea: Hirudinidae). Am Midland Natl, 93: 484–491.

Cowles, M.W., Brown, D.D.R., Niosperos, S.V. et al. 2013. Genome-wide analysis of the bHLH gene family in planarians identifies factors required for adult neurogenesis and neuronal regeneration. Development, 140: 4691–4702.

Crews, A.E. and Esch, G.W. 1986. Seasonal dynamics of *Halipegus occidualis* (Trematoda: Hemuridae) in *Helisoma anceps* and its impact on fecundity of the snail host. J Parasitol, 72: 646–651.

Cribb, T.H., Bray, R.A. and Littlewood, D.T.J. 2001. The nature and evolution of the association among digeneans, molluscs and fishes. Int J Parasitol, 31: 997–1011.

Cribb, T.H., Chisholm, L.A. and Bray, R.A. 2002. Diversity in the Monogenea and Digenea: does lifestyle matter? Int J Parasitol, 32: 321–328.

Cribb, T.H., Bray, R.A., Olson, P.D. and Littlewood, D.T.J. 2003. Life cycle evolution in the Digenea: a new perspective from phylogeny. Adv Parasitol, 54: 198–254.

256 *Reproduction and Development in Platyhelminthes*

Cribb, T.H., Bott, N.J., Bray, R.A. et al. 2014. Trematodes of the Great Barrier Reef, Australia: emerging patterns of diversity and richness in coral reef fishes. Int J Parasitol, 44: 929–939.

Cribb, T.H., Bray, R.A., Diaz, P.E. et al. 2016. Trematodes of fishes of the Indo-West Pacific: told and untold richness. Syst Parasitol, 93: 237–247.

Criscione, C.D. and Blouin, M.S. 2006. Minimal selfing, few clones and no among-host genetic structure in a hermaphroditic parasite with asexual larval propagation. Evolution, 60: 553–562.

Cunningham, E., Tierney, J.F. and Huntingford, F.A. 1994. Effects of the cestode *Schistocephalus solidus* on food intake and foraging decisions in the three-spined stickleback *Gasterosteus aculeatus*. Ethology, 97: 65–75.

Curtis, W.C. 1902. *The Life History, the Normal Fission and Reproductive Organs of Planaria maculata*. The Society, Boston, p. 92.

Davison, J. 1973. Population growth in planaria *Dugesia tigrina* (Gerard). J Gen Physiol, 61: 767–785.

De Jong-Brink, M., Elsaadany, M. and Soto, M.S. 1991. The occurrence of schistosomin, an antagonist of female gonadotropic hormones, is a general phenomenon in haemolymph of schistosome-infected freshwater snails. Parasitology, 103: 371–378.

De Montaudouin, X., Blanchet, H., Deoclaux-Marchand, C. et al. 2015. Cockle infection by *Himasthla quissetensis*–I. From cercaria emergence to metacercariae infection. J Sea Res, 113: 99–107.

De Mulder, K., Kulaes, G., Pfister, D. et al. 2009. Characterization of the stem cell system of the acoel *Isodiametra pulchra*. BMC Dev Biol, 9: 69 DOI: 10.1186/1471-213X-9-69.

De Robertis, E.M. 2010. *Wnt* signaling in axial patterning and regeneration: lessons from planaria. Sci Sig, 3: pe21.

Deblock, S. 1980. Inventaire des trematodes larvaires parasites des mollusques *Hydrobia* (Prosobranches) des cotes de France. Parasitologia, 22: 1–105.

Delogu, V. and Galletti, M.C. 2011. *Sabussowia ronaldi* sp. nov. (Platyhelminthes: Tricladida: Maricola), a new Mediterranean species and its life cycle. Meiofauna Marina, 19: 41–47.

Desdevises, Y., Morand, S., Jousson, O. and Legendre, P. 2002. Coevolution between *Lamellodiscus* (Monogenea: Diplectanidae) and Sparidae (Teleostei): the study of a complex host-parasitesystem. Evolution, 56: 2459–2471.

Dhakal, S., Buss, S.M., Cassidy, E.J. et al. 2018. Establishment success of the beetle tapeworm *Hymenolepis diminuta* depends on dose and host body condition. Insects, 9: 14–39.

Diaz Briz, L.M., Martorelli, S.R. and Genzano, G.N. 2016. The parasite *Monascus filiformis* (Trematoda, Digenean, Fellodistomidae) on *Stromateus brasiliensis* (Pisces, Perciformes, Stromateidae): possible routes of transmission involving jellyfish. J Mar Biol Ass UK, 96: 1483–1489.

Diggles, B.K., Roubal, F.R. and Lester, R.J.G. 1993. The influence of formaline, benzocaine and hyposalinity on the fecundity and viability of *Polylabroides multispinosus* (Monogenea: Microcotylidae) parasitic on the gills of *Acanthopagrus australis* (Pisces: Sparidae). Int J Parasitol, 23: 877–884.

Dmitrieva, E.V. and Gerasev, P.I. 2000. Two new species of *Gyrodactylus* (Gyrodactylidae, Monogenea) from Black Sea fishes. Vest Zool, 34: 98.

Dmitrieva, E.V. 2003. Tranmission triggers and pathways in *Gyrodactylus sphinx* (Monogenea, Gyrodactylidae). Vest Zool, 32: 67–72.

Domenici, L. and Gremigni, V. 1977. Fine structure and functional role of the coverings of the eggs in *Mesostoma ehrenbergii* (Focke) (Turbellaria, Neorhabdocoela). Zoomorphologie, 88: 247–257.

Donges, J. 1971. The potential number of redial generations in echinostomatids (Trematoda). Int J Parasitol, 1: 51–59.

Dorchies, P.H. 2007. Comparison of methods for the veterinary diagnosis of liver flukes (*Fasciola hepatica*) in cattle. Bulletin USAMV–CN, 64: 1–2.

Dorovskikh, G.N. and Matrokhina, S.N. 1987. Distribution of some species of parasites on the gills of ruff. Parazitologiya, 21: 64–68.

Dougan, P.M., Mair, G.R., Halton, D.W. et al. 2002. Gene organization and expression of a neuropeptide Y homolog from the land planarian *Arthurdendyus triangulatus*. J Comp Neurol, 454: 189–193.

Downie, A.J. and Cribb, T.H. 2011. Phylogenetic studies explain the discrepant host distribution of *Allopodocotyle heronensis* sp. nov. (Digenea, Opecoelidae) in Great Barrier Reef serranids. Acta Parasitol, 56: 296–300.

Dronen, N.O. Jr. 1978. Host–parasite population dynamics of *Haematoloechus coloradensis* Cort, 1915 (Digenea: Plagiorchiidae). Am Midland Natl, 99: 330–349.

Dubois, F. 1949. Contribution a l'etude de la migration des cellules de regeneration chez les planarians dulcicoles. Biol Bull Fr Belg, 83: 213–283.

Dubois, S., Savoye, N., Sauriau, P.-G. et al. 2009. Digenean tramtodes–marine mollusc relationships: a stable isotope study. Dis Aquat Org, 84: 65–77.

Dumont, H.J., Rietzler, A.C. and Han, B.-P. 2014. A review of typhloplanid flatworm ecology, with emphasis on pelagic species. Inland Waters, 4: 257–270.

Dzik, J.M. 2006. Molecules released by helminth parasites involved in host colonization. Acta Biochim Polon, 53: 33–64.

Eckert, J. and Deplazes, P. 2004. Biological, epidemiological, and clinical aspects of Echinococcosis, a zoonosis of increasing concern. Clinic Microbiol Rev, 17: 107–135.

Egger, B., Gschwentner, R. and Rieger, R. 2006. Free-living flatworms under the knife: past and present. PMCID, DOI: 10.1007/s00427-006-0120-5.

Egger, B., Gschwenter, R., Hess, M.W. et al. 2009. The caudal regeneration blastema is an accumulation of rapidly proliferating stem cells in the flatworm *Macrostomum lignano*. BMC Dev Biol, 9: 14, DOI: 10.1186/1471-213X-9-41.

Ehlers, U. 1985. *Das Phylogenetische System der Platyhelminthes*. Gustav Fischer, Stuttgart.

Eisenhoffer, G.T., Kang, H. and Alvarado, A.S. 2008. Molecular analysis of stem cells and their descendants during cell turnover and regeneration in the planarian *Schmidtea mediterranea*. Cell Stem Cell, 11: 327–339.

Erazo-Pagador, G. and Cruz-Lacierda, E.R. 2010. The morphology and life cycle of the gill monogenean (*Pseudorhabdosynochus lantauensis*) on orange-spotted grouper (*Epinephelus coioides*) cultured in the Philippines. Bull Eur Ass Fish Pathol, 30: 55–64.

Ernst, I., Fletcher, A. and Hayward, C. 2000. *Gyrodactylus anguillae* (Monogenea: Gyrodactylidae) from anguillid eels (*Anguilla australis* and *Anguilla reinhardtii*) in Australia: a native or an exotic? J Parasitol, 86: 1152–1156.

Ernst, I., Whittington, I.D., Corenillie, S. and Talbot, C. 2005. Effects of temperature, salinity, desiccation and chemical treatments on egg embryonation and hatching success of *Benedenia seriolae* (Monogenea: Capsalidae), a parasite of farmed *Seriola* spp. J Fish Dis, 28: 157–164.

Esch, G.W., Barger, M.A. and Fellis, K.J. 2002. The transmission of digenetic trematodes: style, elegance, complexity. Integ Comp Biol, 42: 304–312.

Evans, N.A. 1985. The influence of environmental temperature upon transmission of the cercariae of *Echinostoma liei* (Digenea: Echinostomatidae). Parasitology, 90: 269–275.

Faliex, E., Da Silva, C., Simon, G. and Sasal, P. 2008. Dynamic expression of immune response genes in the sea bass *Dicentrarchus labrax* experimentally infected with monogenean *Diplectanum aequans*. Fish Shellfish Immunol, 24: 759–769.

Faltynkova, A., Sures, B. and Kostadinova, A. 2016. Biodiversity of trematodes in their intermediate mollusk and fish hosts in the freshwater ecosystems of Europe. Syst Parasitol, 93: 282–293.

FAO. 2014. The State of World Fisheries and Aquaculture: Opportunities and Challenges. Food and Agricultural Organization, Rome, p 221.

Fernandez, J. and Esch, G.W. 1991. Effect of parasitism on the growth rate of the pulmonate snail *Helisoma anceps*. J Parasitol, 77: 937–944.

Ferrari-Hoeinghaus, A.P., Takemoto, R.M., Oliveira, L.C. et al. 2006. Host-parasite relationship of monogeneans in gills of *Astyanax altiparanae* and *Rhamdia quelen* of the Sao Francisco Verdadeiro River, Brazil. Parasite, 13: 315–320.

258 *Reproduction and Development in Platyhelminthes*

Ferreira, S.M., Jensen, K.T., Martins, P.A. et al. 2005. Impact of microphallid trematodes on the survivorship, growth, and reproduction of an isopod (*Cyathura carinata*). J Exp Mar Biol Ecol, 318: 191–199.

Fiore, L. 1971. A mechanism for self-inhibition of population growth in the flatworm *Mesostoma ehrenbergii* (Focke). Oecologia, 7: 356–360.

Fredensborg, B.L., Mouritson, K.N. and Poulin, R. 2005. Impact of trematodes on host survival and population density in the intertidal gastropod *Zeacumantus subcarinatus*. Mar Ecol Pro Ser, 290: 109–117.

Fredensborg, B.L. and Poulin, R. 2006. Parasitism shaping host life–history evolution: adaptive responses in a marine gastropod to infection by trematodes. J Anim Ecol, 75: 44–53.

Fried, B. 1994. Matacercarial excystment of trematodes. Adv Parasitol, 33: 92–120.

Friedlander, M.R., Adamidi, C., Han, T. et al. 2009. High resolution profiling and discovery of planarian small RNAs. Proc Natl Acad Sci USA, 106: 11546–11551.

Galaktionov, K.V. and Dobrovolskij, A.A. 2003. *The Biology and Evolution of Trematodes*, Springer, Dordrecht, p 592.

Galaktionov, K.V. 2006. Phenomenon of parthenogenetic metacercariae in gymnophallids and aspects of trematode evolution. Proc Zool Inst Russ Acad Sci, 310: 51–58.

Galaktionov, K.V., Irwin, S.W.B. and Saville, D.H. 2006. One of the most complex life cycles among trematodes: a description of *Parvatrema margaritense* (Ching, 1982) n. comb. (Gymnophallidae) possessing parthenogenetic metacercariae. Parasitology, 132: 733–746.

Gamble, H.R., Fetterer, R.J. and Urban, J.F. 1995. Reproduction and development in helminthes. In: *Biochemisry and Molecular Biology of Parasites*. (ed) Marr, J.J. and Muller, M. Academic Press, New York, pp 289–306.

Gannicott, A.M. and Tinsley, R.C. 1997. Egg hatching in the monogenean gill parasite *Discocotyle sagittata* from the rainbow trout (*Oncorhynchus mykiss*). Parasitology, 114: 569–579.

Gannicott, A.M. and Tinsley, R.C. 1998. Environmental effects on transmission of *Discocotyle sagittata* (Monogenea): egg production and development. Parasitology, 117: 499–504.

Gardner, S.L. 2002. Book review: Interrelationships of the Platyhelminthes. Syst Biol, 51: 192–194.

Gaston, K.J. and Blackburn, T.M. 1995. The frequency of distribution of bird body weights: aquatic and terrestrial species. Ibis, 137: 237–240.

Gerard, C., Mone, H. and Theron, A. 1993. *Schistosoma mansoni-Biomphalaria glabrata*: dynamics of the sporocyst population in relation to the miracidial dose and the host size. Can J Zool, 71: 1880–1885.

Gerard, C. and Theron, A. 1997. Age/size- and time-specific effects of *Schistosoma mansoni* on energy allocation patterns of its snail host *Biomphalaria glabrata*. Oecologia, 112: 447–452.

Gercken, J. and Renwrantz, L. 1994. A new mannan-binding lectin from the serum of the eel (*Anguilla anguilla* L.): Isolation, characterization and comparison with fucose-specific serum lectin. Comp Biochem Physiol, 108: 449–461.

Ghazal, A.M. and Avery, R.A. 1976. Observations on coprophagy and the transition of *Hymenolepis nana* infections in mice. Parasitology, 73: 39–45.

Gibson, D.I. and Bray, R.A. 1994. The evolutionary expansion and host-parasite relationships of the Digenea. Int J Parasitol, 24: 1213–1226.

Girstmair, J., Schnegg, R., Telford, M.J. and Egger, B. 2014. Cellular dynamics during regeneration of the flatworm *Monocelis* sp. (Proseriata, Platyhelminthes). Evol Dev, 5: 37.

Gonchar, A. and Galaktionov, K.V. 2017. Life cycle and biology of *Tristriata anatis* (Digenea: Notocotylidae): morphological and molecular approaches. Parasitol Res, 116: 45–59.

Gonzales-Lanza, C., Alvarez-Pellitero, P. and Sitja-Bobbadilla, A. 1991. Dipletanidae (Monogenea) infestations of sea bass *Dicentrarchus labrax* (L.) from the Spanish Mediterranean area. Parasitol Res, 77: 307–314.

Gonzalez-Moreno, O. and Gracenea, M. 2006. Life cycle and description of a new species of brachylaimid (Trematoda: Digenea) in Spain. J Parasitol, 92: 1305–1312.

Gorbushin, A.M. 1997. Field evidence of trematode-induced gigantism in *Hydrobia* spp. (Gastropoda: Prosobranchi). J Mar Biol Ass UK, 77: 785–800.

Gould, S.J. 1977. *Ontogeny and Phylogeny*. Belknap Press of Harvard University Press, Cambridge. p 503.

Grabda-Kazubska, B. 1976. Abbreviation of the life cycles in plagiorchid trematodes: general remarks. Acta Parasitol Polon, 24: 125–141.

Granovitch, A.I. and Sergievsky, S.O. 1990. Reproductive structure of the White Sea populations of mollusk *Littorina saxatilis* (Olivi) (Gastropoda: Prosobranchia). Zool Zh, 69: 32–41.

Granovitch, A.I., Yagunova, E.B., Maximovich, A.N. and Sokolova, I.M. 2009. Elevated female fecundity as a possible compensatory mechanism in response to trematode infestation in populations of *Littorina saxatilis* (Olivi). Int J Parasitol, 39: 1011–1019.

Grasso, M. and Benazzi, M. 1973. Genetic and physiologic control of fissioning and sexuality in planarians. J Embryol Exp Morph, 30: 317–328.

Gremigni, V. and Miceli, C. 1980. Cytophotometric evidence for cell 'transdifferentiation' in planarian regeneration. Arch Dev Biol, 188: 107–113.

Gremigni, V., Miceli, C. and Picano, E. 1980. On the role of germ cells in planarian regeneration. J Embryol Exp Morph, 55: 65–76.

Grobler, N.J., Christison, K.W., Olivier, P.A.S. and Van As, J.G. 2003. Observations on the development of *Udonella caligorum* Johnston, 1835 (Monogenea: Polyonchoinea) on a parasitic copepod species of *Caligus* (Copepoda: Caligidae), collected from Lake St Lucia, South Africa. Afr Zool, 38: 393–396.

Grossman, A.I., Short, R.B. and Cain, G.D. 1981. Karyotype evolution and sex chromosome differentiation in schistosomes (Trematoda, Schistosomatidae). Chromosoma, 84: 413–430.

Guegan, J.F. and Hugueny, B. 1994. A nested parasite species subset pattern in tropical fish: host as major determinant of parasite infracommunity structure. Oecologia, 100: 184–189.

Guegan, J.F. and Morand, S. 1996. Polyploid hosts: strange attractors for parasites! Oikos, 7: 366–370.

Guegan, J.-F., Lambert, A., Leveque, C. et al. 1992. Can host body size explain the parasite species richness in tropical freshwater fishes? Oecologia, 90: 197–204.

Guilford, H.G. 1961. Gametogenesis, egg-capsule formation, and early miracidial development in the digenetic trematode *Halipegus eccentricus* Thomas. J Parasitol, 47: 757–764.

Guneydag, S., Ozkan, H. and Ozer, A. 2017. *Scolex pleuronectis* (Cestoda) infections in several bony fish species collected from sinop coasts of the Black Sea. Sinop Uni J Nat Sci, 2: 150–158.

Gupta, S.C. and Singh, B.P. 2002. Fasciolosis in cattle and buffaloes in India. J Vet Parasitol, 16: 139–145.

Gurley, K.A., Rink, J.C. and Sanchez-Alvarado, A. 2008. B-catenin defines head versus tail identity during planarian regeneration and homeostasis. Science, 319: 323–327.

Gurley, K.A., Elliott, S.A., Simkov, O. et al. 2010. Expression of secreted *Wnt* pathway components reveals unexpected complexity of the planarian amputation response. Dev Biol, 347: 24–39.

Gururajan, R., Perry-O'Keefe, H., Melton, D.A. and Weeks, D.L. 1991. The *Xenopus* localized messenger RNA An3 may encode an ATP–dependent RNA helicase. Nature, 349: 717–719.

Gustafsson, M.K.S., Halton, D.W., Maule, A.G. et al. 2002. Neuropeptides in flatworms. Peptides, 23: 2053–2061.

Gutierrez, P.A. and Martorelli, S.R. 1999. The structure of the monogenean community on the gills of *Pimelodus maculates* from Riode la Plata (Argentina). Parasitology, 119: 177–182.

Haas, W., Haberl, B., Kalbe, M. and Korner, M. 1995. Snail host-finding behavior by miracidia and cercariae: chemical host cues. Parasitol Today, 11: 468–472.

Hager, A. 1941. Effects of dietary modification of host rats on the tapeworm *Hymenolepis diminuta*. Iowa State Coll J Sci, 15: 127–153.

Haight, M., Davidson, D. and Paternak, J. 1977. Relationship between nuclear morphology and the phases of the cell cycle during cercarial development of the digenetic trematode *Trichobilharzia ocellata*. J Parasitol, 63: 267–273.

Hall, S.R., Backer, C. and Caceres, C.E. 2007. Parasitic castration: a perspective form a model of dynamic energy budgets. Integ Comp Biol, 47: 295–309.

Halton, D.W. 1975. Intracellular digestion and cellular defecation in a monogenean, *Diclidophora merlangi*. Parasitology, 70: 331–340.

260 *Reproduction and Development in Platyhelminthes*

Halton, D.W. 1979. The surface topography of monogenean *Diclidophora merlangi* revealed by scanning electron microscopy. Z Parasitenkd, 61: 1–12.

Halton, D.W. 1982. An unusual structural organization to the gut of a digenetic trematode, *Fellodistomum fellis*. Parasitology, 85: 633–647.

Halton, D.W. 1997. Nutritional adaptations to parasitism within the Platyhelminthes. Int J Parasitol, 27: 693–704.

Handberg-Thorsager, M. and Salo, E. 2007. The planarian *nanos*-like gene *smednos* is expressed in germline and eye precursor cells during development and regeneration. Dev Genes Evol, 217: 403–411.

Hansen, H., Bakke, T.A. and Buchmann, L. 2007. DNA taxonomy and barcoding of monogenean parasites: lessons from *Gyrodactylus*. Trends Parasitol, 23: 363–367.

Hanson, E.D. 1960. Asexual reproduction in acoelous Turbellaria. Yale J Bull Med, 33: 107–111.

Harkema, R. 1943. The cestodes of North Carolina poultry with remakrs on the life history of *Raillietina tetragona*. J Elisha Mitchell Sci Soc, 59: 127.

Harrington, W.C. Bearse, H.M. and Firth, F.E. 1939. Observations on 748 the life history, occurrence and distribution of the redfish parasite *Sphyrion lumpi*. US Bur Fish Sp Rep, 5: 1–18.

Harris, A.H., Harkema, R. and Miller, G.C. 1967. Maternal transmission of *Pharyngostomoides procyonis* Harkema, 1942 (Trematoda: Diplostomatidae). J Parasitol, 53: 1114–1115.

Harris, P.D. 1983. The morphology and life cycle of the oviparous *Oogyrodactylus farlowellae* gen. et sp. n. (Monogenea, Gyrodactylidae). Parasitology, 87: 405–420.

Harris, P.D. 1985. Observations on the development of the male reproductive system in *Gyrodactylylus gasterostei* Glaser, 1974 (Monogenea, Gyrodactylidae). Parasitology, 91: 519–529.

Harris, P.D. and Tinsley, R.C. 1987. The biology of *Gyrodactylus gallieni* (Gyrodactylidea), an unusual viviparous monogenean from the African clawed toad, *Xenopus laevis*. J Zool, 212: 57–64.

Harris, P.D. 1993. Interactions between reproduction and population biology in gyrodactylid monogeneans—A review. Bull Fr Peche Piscic, 1: 47–65.

Harris, P.D. 1998. Ecological and genetic evidences for clonal reproduction in *Gyrodactylus gasterostei* Glaser, 1974. Int J Parasitol, 28: 1595–1607.

Harris, P.D., Soleng, A. and Bakke, T.A. 2000. Increased susceptibility of salmonids to the monogenean *Gyrodactylus salaris* following administration of hydrocortisone acetate. Parasitology, 120: 57–64.

Harris, P.D., Shinn, A.P., Cable, J. and Bakke, T.A. 2004. Nominal species of the genus *Gyrodactylus* von Nordmann 1832 (Monogenea: Gyrodactylidae), with a list of principal host species. Syst Parasitol, 59: 1–27.

Hartmann, M. 1922. Uber den dauernden Ersatz der ungeschlechtlichen Fortpflanzung durch fortgesetzte Regenerationen. Biol Zentrabl, 42: 364–381.

Hase, S., Kobayashi, K. Koyanagi, R. et al. 2003. Transcriptional pattern of a novel gene, expressed specifically after the point of no return during sexualization in planaria. Dev Genes Evol, 212: 585–592.

Hass, W., Haberl, B., Kalbe, M. and Korner, M. 1995. Snail host–finding by miracidia and cercariae: chemical host cues. Parasitol Today, 11: 468–472.

Hassanine, R.M.EL–S. 2006. The life-cycle of *Diploproctodaeum arothroni* Bray and Nahhas, 1998 (Digenea: Lepocreadiidae), with a comment on the parasitic castration of its molluscan intermediate host. J Natl Hist, 40: 1211–1223.

Hauser, J. 1987. Sexualization of *Dugesia anderiani* by feeding. Acta Buiol Leopoldensia, 9: 111–128.

Heins, D.C., Baker, J.A. and Martin, H.C. 2002. The "crowding effect" in the cestode *Schistocephalus solidus*: density-dependent effects on plerocercoid size and infectivity. J Parasitol, 88: 302–307.

Heins, D.C. and Baker, J.A. 2003. Reducton of egg size in natural populations of three spined stickleback infected with a cestode macroparasite. J Parasitol, 89: 1–6.

Heins, D.C. 2012. Fecundity compensation in the three-spined stickleback *Gasterosteus aculeatus* infected by the diphyllobothriidean cestode *Schistocephalus solidus*, Biol J Linn Soc, 106: 807–819.

Heins, D.C., Barry, K.A. and Petrauskas, L.A. 2014. Consistency of host responses to parasitic infection in the three-spined stickleback fish infected by the diphyllobothriidean cestode *Schistocephalus solidus*. Biol J Linn Soc, 113: 958–968.

Heitkamp, U. 1977. Zur Fortpflanzungsbiologie von *Mesostoma ehrenbergii* (Focke, 1836) (Turbellaria). Hydrobiologia, 55: 21–31.

Henderson, D.J. and Hanna, R.E.B. 1988. *Hymenolepis nana* (Cestoda: Cyclophyllidea): DNA, RNA and protein synthesis in 5-day old juveniles. Int J Parasitol, 18: 963–972.

Herrmann, H.K. and Poulin, R. 2011a. Encystment site affects the reproductive strategy of a progenetic trematode in its fish intermediate host: is host spawning an exit for parasite eggs? Parasitology, 138: 1183–1192.

Herrmann, H.K. and Poulin, R. 2011b. Life cycle truncation in a trematode: Does higher temperature indicate shorter host longevity? Int J Parasitol, 41: 697–704.

Hirazawa, N., Takano, R., Hagiwara, H. et al. 2010. The influence of different water temperatures on *Neobenedenia girellae* (Monogenea) infection, parasite growth, egg production and emerging second generation on amberjack *Seriola dumerili* (Carangidae) and the histopathological effect of this parasite on fish skin. Aquaculture, 299: 2–7.

Hoai, T.D. and Hutson, K.S. 2014. Reproductive strategies of the insidious fish ectoparasite, *Neobenedenia* sp. (Capsalidae: Monogenea). PLoS ONE, 9: e108801.

Hoberg, E.P., Brooks, D.R. and Siegel-Causey, D. 1997. Host-parasite cospeciation: history, principles and prospects. In: *Host-Parasite Evolution: General Principles and Avian Models*. (eds) Clayton, D.H. and Moore, J., Oxford University Press, Oxford, pp 212–235.

Hodasi, J.K.M. 1972. The effects of *Fasciola hepatica* on *Lymnaea truncatula*. Parasitology, 65: 359–369.

Hori, I. 1982. An ultrastructural study of the chromatoid body in planarian regenerative cells. J Electron Microsc, 31: 63–72.

Hori, I. 1992. Cytological approach to morphogenesis in the planarian blastema. I. Cell behavior during blastema formation. J Submicrosc Cytol Pathol, 24: 75–84.

Hori, I. 1997. Cytological approach to morphogenesis in the planarian blastema. II. The effect of neuropeptides. J Submicrosc Cytol pathol, 29: 91–97.

Hori, I., Hikosaka-Katayama, T. and kishida, Y. 1999. Cytological approach to morphogenesis in the planarian blastema. III. Ultra-structure and regeneration of the acoel turbellarian *Convoluta naikaisensis*. J Submicrosc Cytol pathol, 31: 247–258.

Horsfall, M.W. 1938. Observations on the life history of *Raillietina echinobothrida* and *R. tetragona* (Cestoda). J Parasitol, 24: 409–421.

Hoshi, M., Kobayashi, K., Ariyoka, S. et al. 2003. Switch from asexual to sexual reproduction in the planarian *Dugesia ryukyuensis*. Integ Comp Biol, 43: 542–246.

Hosier, D.W. and Goodchild, C.G. 1970. Suppressed egg-laying by snails infected with *Spirorchis scripta* (Trematoda: Spirorchidae). J Parasitol, 56: 302–304.

Huston, D.C., Cutmore, S.C. and Cribb, T.H. 2016. The life-cycle of *Gorgocephalus yaagi* Bray & Cribb, 2005 (Digenea: Gorgocephalidae) with a review of the first intermediate hosts for the superfamily Lepocreadioidea Odhner, 1905. Syst Parasitol, 93: 653–665.

Huxham, M., Raffaelli, D. and Pike, A. 1993. The influence of *Cryptocotyle lingua* (Digenea: Platyhelminthes) infections on the survival and fecundity of *Littorina littorea* (Gastropoda: Prosobranchia); an ecological approach. J Exp Mar Biol Ecol, 168: 223–238.

Huyse, T., Van den Broeck, F., Hellemans, B. et al. 2013. Hybridisation between the two major African schistosome species of humans. Int J Parasitol, 43: 687–689.

Hyman, L.H. 1951. *The Invertebrates: Platyhelminthes and Rhynhocoela*. McGraw-Hill Book Company, New York, p 550.

Iglesias, M., Gomez-Skarmeta, J.L., Salo, E. and Adell, T. 2008. Silencing of *Smed-bcatenin1* generates radial-like hypercephalized planarians. Development, 135: 1215–1221.

Imbert-Establet, D., Xia, M. and Jourdane, J. 1994. Parthenogenesis in the genus *Schistosoma*: electrophoretic evidence for this reproduction systein in *S. japonicum* and *S. mansoni*. Parasitol Res, 80: 186–191.

262  *Reproduction and Development in Platyhelminthes*

Irwin, S.W.B., Galaktionov, K.V., Malkova, I.I. et al. 2003. An ultrastructural study of reproduction in the parthenogenetic metacercariae of *Cercaria margaritensis* Ching, 1982 (Digenea: Gymnophallidae). Parasitology, 126: 261–271.

Ishii, Y. 1934. Studies on the development of *Fasciolopis buski*. Part I. Development of the egg outside the host. J Med As Formosa, 33: 1–30.

Ishikawa, K. and Yamasu, T. 1992. An acoel flatworm species related closely to species *Convolutriloba retrogemma* Hendelberg & Akesson occurs in Okinawa Island, Ryuku Archipelago. Zool Sci, 9: 1281.

Ishizuka, H., Maezawa, T., Kawauchi, J. et al. 2007. The *Dugesia ryukyuensis* database as a molecular resource for studying switching of the reproductive system. Zool Sci, 24: 31–37.

Iyaji, F.O., Etim, L. and Eyo, J.E. 2009. Parasite assemblages in fish hosts. Biol Res, 7: 561–570.

Jagersten, G. 1972. *Evolution of the Metazoan Life Cycle: A Comprehensive Theory*. Academic Press, London, p 282.

James, B.L. and Bowers, E.A. 1967. Reproduction in the daughter sporocyst of *Cercaria bucephalopsis haimeana* (Lacaze–Duthiers, 1854) (Bucephalidae) and *Cercaria dichotoma* Lebour, 1911 (non Muller) (Gymnophallidae). Parasitology, 57: 607–625.

Janardanan, K.P., Ramanandan, S.K. and Usha, N.v. 1987. On the progenetic metacercaria of *Pleurogenoides ovatus* Rao, 1977 (Trematoda: Pleurogenitinae) from the freshwater crab, *Paratelphusa hydrodromous* (Herbst), with observations on its *in vitro* excystment. Zool Anz, 219: 313–320.

Jansen, P.A. and Bakke, T.A. 1995. Susceptibility of brown trout to *Gyrodactylus salaris* (Monogenea) under experimental conditions. J Fish Biol, 46: 415–422.

Jarecka, L. 1961. Morphological adaption of tapeworm eggs and their importance in the life cycles. Acta Parasitol Polon, 23: 93–114.

Jarroll, E.L. Jr. 1980. Population dynamics of *Bothriocephalus rarus* (Cestoda) in *Notophthalmus viridiscens*. Am Midland Natl, 103: 360–366.

Jennings, J.B. 1968. Platyhelminthes: Nutrition. In: *Chemical Zoology*. (eds) Florkin, M. and Scheer, B.T. Academic Press, New York, pp 303–326.

Jennings, J.B. 1971. Parasitism and commensalism in the turbellaria. Adv Parasitol, 9: 1–32.

Jennings, J.B. and Calow, P. 1975. The relationship between high fecundity and the evolution of entoparasitism. Oecologia, 21: 109–115.

Jennings, J.B. 1997. Nutritional and respiratory pathways to parasitism exemplified in the Turbellaria. Int J Parasitol, 27: 679–691.

Jensen, K. and Bullard, S.A. 2010. Characterization of diversity of tetraphyllidean and rhinebothriidean cestode larval types, with comments on host associations and life-cycles. Int J Parasitol, 40: 889–910.

Jensen, K., Caira, J.N. Cielocha, J.J. et al. 2016. When proglottids and scoleces conflict: phylogenetic relationships and a family-level classification of the Lecanicephalidea (Platyhelminthes: Cestoda). Int J Parasitol, http://dx.doi.org/10.1016/j.ijpara.2016.02.002.

Jensen, T., Jensen, K.T. and Mouritsen, K.N. 1998. The influence of the trematode *Microphallus claviformis* on two congeneric intermediate host species (*Corophium*): infection characteristics and host survival. J Exp Mar Biol Ecol, 227: 35–48.

Jianying, Z., Tingbao, Y., Lin, L. and Xuejuan, D. 2003. A list of monogeneans from Chinese marine fishes. Syst Parasitol, 54: 111–130.

Jocelyn, M.C. 2009. *Australapatemon* sp. (Trematoda) infection in *Valvata macrostoma* and in two leech species, *Helobdella stagnalis* and *Erpobdella octoculata*. MS Thesis, University of Jyvaskyla.

Joffe, B.I. and Reuter, M. 1993. The nervous system of *Bothriomolus balticus* (Proseriata)–a contribution to the knowledge of the orthogon in the Platyhelminthes. Zoomorphology, 113: 113–127.

Joffe, B.I., Solovei, I.V. and MacGreger, H.C. 1996. Ends of chromosomes in *Polycelis tenuis* (Platyhelinthes) have telomere repeat TTAGGG. Chromosome Res, 4: 323–324.

Johansson, L.C. and Norberg, U.M.L. 2000. Biomechanics—Asymmetric toes aid underwater swimming. Nature, 406: 582–583.

Johnson, M.B., Lafferty, K.D., Oosterhout, C.V. and Cable, J. 2011. Parasite transmission in social interacting hosts: Monogenean epidemics in guppies. PLoS ONE, 6(8): e22634. DOI: 10.1371/journal.pone.0022634.

Johnson, T.J., Lunde, K.B., Ritchie, E.G. and Launer, A.E. 1999. The effect of trematode infection on amphibian limb development and survivorship. Science, 284: 802–804.

Johnstone, T.H. and Simpson, R.E. 1940. The anatomy and life history of *Cyclocoelum jaenschi*, n. sp. Trans R Soc South Austr, 64: 273–278.

Jokela, J. and Lively, C. 1995. Parasites, sex and early reproduction in a mixed population of freshwater snails. Evolution, 49: 1268–1271.

Jokela, J., Uotila, L. and Taskinen, J. 1993. Effects of castrating trematode parasite *Rhipidocotyle fennica* on energy allocation of freshwater clam *Anodonta piscinalis*. Funct Ecol, 7: 332–338.

Jokiel, P.L. and Townsley, S.J. 1974. Biology of the polyclad *Prosthiostomum* (*Prosthiostomum*) sp., a new coral parasite from Hawaii. Pac Sci, 28: 361–373.

Jordaens, K., Dillen, L. and Backeljau, T. 2007. Effects of mating, breeding system and parasites on reproduction in hermaphrodites: pulmonate gastropods (Mollusca). Anim Biol, 57: 137–195.

Jourdane, J. and Theron, A. 1980. *Schistosoma mansoni*: cloning by microsurgical transplantation of sporocysts. Exp Parasitol, 50: 349–57.

Jurberg, A.D. and Brindley, P.J. 2015. Gene function in schistosomes: recent advances toward a cure. Front Genet, 6: 144. DOI: 10.3389/fgene.2015.00144.

Kanneworff, B. and Christensen, A.M. 1966. *Kronborgia caridicola* sp. nov., an endoparasitic turbellarians from North Atlantic shrimp. Ophelia, 3: 65–80.

Kavana, N.J., Lim, L.H.S. and Ambu, S. 2014. The life cycle of *Siprometra* species from Peninsular Malaysia. Trop Biomet, 31: 487–495.

Kearn, G.C. 1963. Feeding in some monogenean skin parasites: *Entobdella soleae* on *Solea solea* and *Acanhocotyle* sp. on *Raia clavata*. J Mar Biol Ass UK, 43: 749–766.

Kearn, G.C. 1967. Experiments on host-finding and host-specificity in the monogenean skin parasite *Entobdella soleae*. Parasitology, 57: 585–605.

Kearn, G.C. 1974. The effect of fish skin mucus on hatching in the monogenea *Entobdella soleae* from the skin of common sole *Solea solea*. Parasitology, 66: 173–183.

Kearn, G.C. 1985. Observations on egg production in the monogenean *Entobdella soleae*. Int J Parasitol, 15: 187–194.

Kearn, G.C. 1986a. Role of chemical substences from fish hosts in hatching and host-finding in monogeneans. J Chem Ecol, 12: 1651–1658.

Kearn, G.C. 1986b. The eggs of monogeneans. Adv Parasitol, 25: 175–273.

Kearn, G.C., Ogawa, K. and Maeno, Y. 1992. Egg production, the oncomiracidium and larval development of *Benedenia seriolae*, a skin parasite of the yellowtail *Seriola quinqueradiata* in Japan. Publ Seto Mar Biol Lab, 35: 351–362.

Kearn, G.C. 1999. The survival of monogenean (platyhelminth) parasites on fish skin. Parasitology, 119S: 57–88.

Kearn, G.C. 2014. Some aspects of the biology of monogenean (platyhelminth) parasites of marine and freshwater fishes. J Oceanogr Mar Res, 2: 1–8.

Keas, B.E. and Esch, G.W. 1997. The effect of diet and reproductive maturity on the growth and reproduction of *Helisoma anceps* (Pulmonata) infected *by Halipegus occidualis* (Trematoda). J Parasitol, 83: 96–104.

Keeney, D.B., Waters, J.M. and Poulin, R. 2007. Clonal diversity of the marine trematode *Maritrema novaezealandensis* within intermediate hosts: the molecular ecology of parasites life cycles. Mol Ecol, 16: 431–439.

Kenk, R. 1937. Sexual and asexual reproduction in *Euplanaria tigrina* (Girard). Biol Bull, 73: 280–294.

Kenk, R. 1941. Induction of sexuality in the asexual form of *Dugesia tigrina*. J Exp Zool, 87: 55–69.

Kerr, C.L. and Cheng, L. 2010. The dazzle in germ cell differentiation. J Mol Cell Biol, 2: 26–29.

Keymer, A.E. and Anderson, R.M. 1979. The dynamics of infection of *Tribolium confusum* by *Hymenolepis diminuta*: the influence of infective-stage density and spatial distribution. Parasitology, 79: 195–207.

264  *Reproduction and Development in Platyhelminthes*

Khalil, G.M. and Cable, R.M. 1968. Germinal development in *Philophthalmus megalurus* (Cort, 1911) (Trematoda: Digenea). Z Parasitenkd, 31: 211–231.

Khidr, A.A., Said, A.E., Abu Samak, O.A. and Abu Sheref, S.E. 2012. The impacts of ecological factors on prevalence, mean intensity and seasonal changes of the monogenean gill parasite, *Microcotylides* sp., infesting the *Terapon puta* fish inhabiting coastal region of Meditterranean Sea at Damietta region. J Basic Appl Zool, 65: 109–115.

Knakievicz, T., Lau, A.H., Pra, D. and Erdtmann, B. 2007. Biogeography and karyotypes of freshwater planarians (Platyhelminthes, Tricladida, Paludicola) in Southern Brazil. Zool Sci, 24: 123–129.

Kobayashi, C., Saito, Y., Ogawa, K. and Agata, K. 2007. *Wnt* signaling is required for antero-posterior patterning of the planarianbrain. Dev Biol, 306: 714–724.

Kobayashi, K., Koyanagi, R., Matsumoto, M. et al. 1999. Switching from asexual to sexual reproduction in the planarian *Dugesia ryukyuensis*: Bioassay system and basic description of sexualizing process. Zool Sci, 16: 291–298.

Kobayashi, K. and Hoshi, M. 2002. Switching from asexual to sexual reproduction in the planarian *Dugesia ryukyuensis*: Change of the fissiparous capacity along with the sexualizing process. Zool Sci, 19: 661–666.

Kobayashi, K., Arioka, S., Hase, S. and Hoshi, M. 2002a. Signification of the sexualizing substance produced by the sexualized planarians. Zool Sci, 19: 667–672.

Kobayashi, K., Arioka, S. and Hoshi, M. 2002b. Seasonal changes in the sexualization of the planarian *Dugesia ryukyuensis*. Zool Sci, 19: 1267–1278.

Kobayashi, K., Ishizu, H., Arioka, S. et al. 2008. Production of diploid and triploid offspring by inbreeding of the triploid planarian *Dugesia ryukyuensis*. Chromosoma, 117: 289–296.

Kobayashi, K., Arioka, S., Hoshi, M. and Matsumoto, M. 2009. Production of asexual and sexual offspring in the triploid sexual planarian *Dugesia rhykyuensis*. Integ Zool, 4: 265–271.

Kobayashi, K., Meazawa, T., Nakagawa, H. and Hoshi, M. 2012. Existence of two sexual races in the planarian species switching between asexual and sexual reproduction. Zool Sci, 29: 265–272.

Kocur, R. 2011. "*Cotylurus flabelliformis*" (On-line), Animal Diversity Web. Museum of Zoology, University of Michigan.

Koie, M. 1995. The life-cycle and biology of *Hemiurus communis* Odhner, 1905 (Digenea: Hemiuridae). Parasite, 2: 195–202.

Koie, M., Karlbakk, E. and Nylund, A. 2010. A cystophorous cercaria and metacercaria in *Antalis entails* (L.) (Mollusca, Scaphopoda) in Norwegian waters, the larval stage of *Lecithophyllum botryophorum* (Olsson, 1868) (Digenea, Lecithasteridae). Sarsia, 87: 302–311.

Kok, D.J. and Du Preez, L.H. 1987. *Polystoma australis* (Monogenea): Life cycle studies in experimental and natural infections of normal and substitute hosts. J Zool Lond, 212: 235–243.

Komuniecki, R. and Harris, B.G. 1995. Carbohydrate and energy metabolism in helminthes. In: *Biochemisry and Molecular Biology of Parasites*. (ed) Marr, J.J. and Muller, M., Academic Press, New York, pp 49–66.

Koziol, U., Dominguez, M.F., Marin, M. et al. 2010. Stem cell proliferation during *in vitro* development of the model cestode *Mesocestoides corti* from larva to adult worm. Front Zool, 7, http://www.frontiersinzoology.com/content/7/1/22.

Koziol, U., Rauschendorfer, T., Rodriguez, L.Z. et al. 2014. The unique stem cell system of the immortal larva of the human parasite *Echinococcus multilocularis*. Evol Dev, 5, http://www.evodevojournal.com/content/5/1/10.

Kreshchenko, N.D. and Sheiman, I.M. 1994. Pharynx regeneration in *Planaria*: The effects of neuropeptides. Russ J Dev Biol, 25: 350–356.

Krichinskaya, E.B. 1986. Asexual reproduction, regeneration and somatic embryogenesis in the planaria *Dugesia tigrina*. Hydrobiologia, 132: 195–200.

Krist, A.C. and Lively, C.M. 1998. Experimental exposure of juvenile snails (*Potamopyrgus antipodarum*) to infection by trematode larvae (*Microphallus* sp.): infectivity, fecundity compensation and growth. Oecologia, 116: 575–582.

## References 265

Kuales, G., De Mulder, K., Glashauser, J. et al. 2011. *Boule*-like genes regulate male and female gametogenesis in the flatworm *Macrostomum lignano*. Dev Biol, 357: 117–132.

Kuchta, R., Brabec, J., Kubackova, P. and Scholz, T. 2013. Tapeworm *Diphyllobothrium dendriticum* (Cestoda)–neglected or emerging human parasite? PLoS ONE, 7: e2535.

Lackenby, J.A., Chambers, C.B., Ernst, I. and Whittington, I.A. 2007. Effect of water temperature on reproductive development of *Benedenia seriolae* (Monogenea: Capsalidae) from *Seriola lalandi* in Australia. Dis Aquat Org, 74: 235–242.

Ladurner, P., Rieger, R. and Baguna, J. 2000. Spatial distribution pattern and differentiateion potential of stem cells in hatchlings and adults in the marine platyhelminth *Macrostomum* sp.: a bromodeoxyuridin analysis. Dev Biol, 226: 231–241.

Lagrue, C. and Poulin, R. 2009. Life cycle abbreviation in tramatode parasites and the developmental time hypothesis: is the clock ticking? J Evol Biol, 22: 1727–1738.

Lambert, A. and Gharbi, S.E. 1995. Monogenean host specificity as a biological and taxonomic indicator for fish. Biol Conserv, 72: 227–235.

Lange, C.S. 1967. A quantitative study of the number and distribution of neoblast in *Dugesia lugubris* (Planaria) with reference to size and ploidy. J Embryol Exp Morph, 18: 199–213.

Lange, C.S. 1968. A possible explanation in cellular terms of the physiological ageing of the planarians. Exp Geront, 3: 219–230.

Laumer, C.E. and Giribet, G. 2014. Inclusive taxon sampling suggests a single stepwise origin of ectolecithality in Platyhelminthes. Biol J Linn Soc, 111: 570–588.

Lawson, J.R. and Wilson, R.A. 1980. The survival of the cercariae of *Schistosoma mansoni* in relation to water temperature and glycogen utilization. Parasitology, 81 : 337–348.

Lazaro, E.M., Sluys, R., Pala, M. et al. 2009. Molecular barcoding and phylogeography of sexual and asexual freshwater planarians of the genus *Dugesia* in the Western Mediterranean (Platyhelminthes, Tricladida, Dugesiidae). Mol Phylogenet Evol, 52: 835–845.

Lee, B. and Mann, B.Q. 2017. Age and growth of narrow-barred Spanish mackerel *Scomberomorus commersoni* in the coastal waters of southern Mozambique and KwaZulu-Natal, South Africa. Afr J Mar Sci, 39: 397–407.

Lefebvre, F. and Poulin, R. 2005. Progenesis in digenean trematodes: a taxonomic and synthetic overview of species reproducing in their second intermediate hosts. Parasitology, 130: 587–605.

Lefebvre, F., Georgiev, B.B., Bray, R.A. and Littlewood, D.T.J. 2009. Developing a dedicated cestode life cycle database: lessons from the hymenolepidids. Helminthologia, 46: 21–27.

Lefevre, T., Roche, B., Poulin, R. et al. 2008. Exploiting host compensatory responses: the "must" of manipulation? Trends Parasitol, 24: 435–439.

Lender, T.H. 1956. L'inhibition de la regeneration du cervau des planaries *Polycelis nigra* (Ehr.) et *Dugesia lugubris* (O. Schm.) en presence de broyats de tetes ou de queues. Bull Soc Zool Fr, 81: 192.

Lester, R.J.G. and Adams, J.R. 1974. *Gyrodactylus alexanderi*: reproduction, mortality and effect on the host. Can J Zool, 52: 827–833.

Lim, J.H. 2011. Liver flukes: the malady neglected. Korean J Radiol, 12: 269–279.

Lim, L.H.S. 1995. *Neocalcestoma* Tripathi, 1957 and Neocalcestomatidaen fam. (Monogenea) from ariid fishes of Peninsular Malaysia. Syst Parasitol, 30: 141–151.

Lindenstrom, T. and Buchmann, K. 1998. Dexamethasone treatment increases susceptibility of rainbow trout, *Oncorynchus mykisss* (Walbaum), to infections with *Gyrodactylus derjavini* Mikailov. J Fish Dis, 1: 29–38.

Lindenstrom, T. and Buchmann, K. 2000. Acquired resistance of rainbow trout against *Gyrodactylus derjavini*. J Helminthol, 74: 155–160.

Littlewood, D.T. and Bray, R.A. 2001. In: *Interrelationships of the Platyhelminthes*. Taylor and Francis Publishing, London, p 356.

Littlewood, D.T.J., Rohde, K. and Clough, K.A. 1999. The interrelationships of all major groups of Platyhelminthes: phylogenetic evidence from morphology and molecules. Biol J Linn Soc, 66: 75–114.

Littlewood, D.T.J. 2003. *The Evolution of Parasitism: Phylogenetic Perspective*. Elsevier, Amsterdam, p 404.

266 *Reproduction and Development in Platyhelminthes*

Littlewood, D.T.J., Bray, R.A. and Waeschenbach, A. 2015. Phylogenetic patterns of diversity in cestodes and trematodes. In: *Parasite Diversity and Diversification: Evolutionary Ecology Meets Phylogenetics.* (eds) Morand, S., Krasnov, B.R. and Littlewood, D.T.J., Cambridge University Press, pp 304–319.

Liu, S.Y., Selck, C., Friedrich, B. et al. 2013. Reactivating head regrowth in a regeneration–deficient planarian species. Nature, 500: 81–84.

Llewellyn, J. 1956. The host-specificity, micro-ecology, adhesive attitudes, and comparative morphology of some trematode gill parasites. J Mar Biol Ass UK, 35: 113–127.

Llewellyn, J. 1962. The life histories and population dynamics of monogenean gill parasites of *Trachurus trachurus* (L.). J Mar Biol Ass UK, 42: 587–600.

Llewellyn, J. 1970. Taxonomy, genetics and evolution of parasites: Monogenea. J Parasitol, 56: 493–504.

Lo, C.M., Morand, S. and Galzin, R. 1998. Parasite diversity/host age and size relationship in three coral–reef fishes from French Polynesia. Int J Parasitol, 28: 1695–1708.

Loker, E.S. 1978. Normal development of *Schistosomatium douthitti* in the snail *Lymnaea catascopium.* J Parasitol, 64: 977–985.

Loot, G., Francisco, P., Santoul, F. et al. 2001. The three hosts of the *Ligula intestinalis* (Cestoda) life cycle in Lavernose–Lacasse gravel pit, France. Arch Hydrobiol, 152: 511–525.

Lorch, S., Zeuss, D., Brandl, R. and Branlde, M. 2016. Chromosome numbers in three species groups of freshwater flatworms increase with increasing latitude. Ecol Evol, 6: 1420–1429.

Love, S. 2017. Liver fluke. A review. Parasitology–Sheep Unit, Armidale, Primefact, 813: 1–202.

Lowenberger, C.A. and Rau, M.E. 1994. *Plagiorchis elegans*: emergence, longevity and infectivity of cercariae, and host behavioural modifications during cercarial emergence. Parasitology, 109: 65–72.

Loy, C. and Hass, W. 2001. Prevalence of cercariae from *Lymnaea stagnalis* snails in a pond system in Southern Germany. Parasitol Res, 87: 878–882.

Luscher, A. and Wedekind, C. 2002. Size–dependent discrimination of mating partners in the simultaneous hermaphroditic cestode *Schistocephalus solidus.* Behav Ecol, 15: 254–259.

Lynch, J.E. 1945. Redescription of the species of *Gyrocotyle* from the rat-fish, *Hydrolagus collici* (Lay and Bennet), with notes on the morphology and taxanomy of the genus. J Parasitol, 31: 418–446.

MacArthur, R.H. and Wilson, E.O. 1967. *The Theory of Island Biogeography.* Princeton University Press, Princeton (NJ), p 203.

MacDonald, S. 1974. Host skin mucus as a hatching stimulant in *Acanthocotyle lobianchi,* a monogenean from the skin of *Raja* spp. Parasitology, 68: 331–338.

MacDonald, S. and Llewellyn, J. 1980. Reproduction in *Acanthocotyle greeni* n. sp. (Monogenea) from the skin of *Raja* spp. at Plymouth. J Mar Biol Ass UK, 60: 81–88.

Maciel, P.O., Muniz, C.R. and Alves, R.R. 2017. Eggs hatching and oncomiracidia lifespan of *Dawestrema cycloancistrium,* a monogenean parasitic on *Arapaima gigas.* Vet Parasitol, 247: 57–63.

MacInnis, A.J., Bethel, W.M. and Cornford, E.M. 1974. Identification of chemicals of snail origin that attract *Schistosoma mansoni* miracidia. Nature, 248: 361–363.

MacInnis, A.J., Graff, D.J., Kilejian, A. and Read, C.P. 1976. Specificity of amino acid transport in the tapeworm *Hymenolepis diminuta* and its rat host. Rice Institute Pamphlet–Rice University Studies, 62: 183–204, Rice University: http://hdl.handle.net/1911/63248.

Mackiewicz, J.S. 1981. Caryophyllidea (Cestoidea): Evolution and classification. Adv Parasitol, 19: 139–206.

Mackiewicz, J.S. 1988. Cestode transmission patterns. J Parasitol, 74: 60–71.

Madanire-Moyo, G.N., Matla, M.M., Olivier, P.A. and Luus-Powell, W.J. 2011. Population dynamics and spatial distribution of monogeneans on the gills of *Oreochromis mossambicus* (Peters, 1852) from two lakes of the Limpopo River System, SouthAfrica. J Helminthol, 85: 146–152.

Madhavi, R. 1978. Life history of *Allocreadium fasciatusi* Kakaji, 1969 (Trematoda: Allocreadiidae) from the freshwater fish *Aplocheilus melastigma* McClelland. J Helminthol, 52: 51–59.

Madhavi, R. 1980. Life history of *Allocreadium handiai* Pande, 1937 (Trematoda: Allocreadiidae) from the freshwater fish *Channa punctata* Bloch. Z Parasitenkd, 63: 89–97.

Madhavi, R. and Anderson, R.M. 1985. Variability in the susceptibility of the fish host, *Poecilia reticulata*, to infection with *Gyrodactylus bullatarudis* (Monogenea). Parasitology, 91: 531–544.

Madhavi, R. and Jhansilakshmibai, K. 1994. The miracidium of *Transversotrema patialense* (Soparkar, 1924). J Helminthol, 68: 49–51.

Madhavi, R. and Swarnakumari, V.G.M. 1995. The morphology life cycle and systematic position of *Orthotrotrema monostomium* Macy & Basch, 1972, a progenetic trematode. Syst Parasitol, 32: 225–232.

Mair, G.R., Halton, D.W., Shaw, C. and Maule, A.G. 2000. The neuropeptide F (NPF) encoding gene from the cestode *Moniezia expansa*. Parasitology, 120: 71–77.

Makanga, B. 1981. The effect of varying the number of *Schistosoma mansoni* miracidia on the reproduction and survival of *Biomphalaria pfeifferi*. J Invert Pathol, 37: 7–10.

Malinowski, P.T., Cochet–Escartin, O., Kaj, K.J. et al. 2017. Mechanics dictate where and how freshwater planarians fission. Proc Natl Acad Sci USA, 114: 10888–10893.

Malmberg, G. 1970. The excretory systems and marginal hooks as a basis for the systematics of *Gyrodactylus* (Trematoda, Monogenea). Ark Zool, 23: 1–235.

Malone, J.B. 1986. Fasciolosis and cestodiasis in cattle. Food Anim Pract, 2: 261–275.

Marchiondo, A.A., Weinstein, P.P. and Mueller, J.F. 1989. Significant of the distribution of [57]Co-Vitamin $B_{12}$ in *Spirometra mansonoides* (Cestoidea) during growth and differentiation in mammalian intermediate and definitive hosts. Int J Parasitol, 19: 119–124.

Marco, A., Kozomara, A., Hui, J.H.L. et al. 2013. Sex-biased expression of microRNAs in *Schistosoma mansoni*. PLoS Negl Trop Dis, 7: e2402.

Marcogliese, D.J. 1995. The role of zooplankton in the transmission of helminth parasites to fish. Rev Fish Biol Fisher, 5: 336–371.

Marie-Orleach, L., Janicke, T., Vizoso, D.B. et al. 2014. Fluorescent sperm in a transparent worm: Validation of a GFP marker to study sexual selection. BMC Evol Biol 14: 148.

Mason, P.R. and Fripp, P.J. 1976. Analysis of the movements of *Schistosoma mansoni* miracidia using dark-ground photography. J Parasitol, 62: 721–727.

Mathavan, S. and Pandian, T.J. 1977. Patterns of emergence, import of egg energy and energy export via emerging dragonfly populations in a tropical pond. Hydrobiologia, 54: 257–272.

Mazeri, S., Rydevik, G., Handel, I. et al. 2017. Estimation of the impact of *Fasciola hepatica* infection on time taken for UK beef cattle to reach slaughter weight. Sci Rep, www.nature.com/scientificreports.

Mazzanti, C., Monni, G. and Varriale, A.M.C. 1999. Observations on antigenic activity *Pseudodactylogyrus anguillae* (Monogenea) on the European *Anguilla anguilla*. Bull Eur Ass Fish Pathol, 19: 57–59.

McCaig, M.L.O. and Hopkins, C.A. 1963. Studies on *Schistocephalus solidus*. II. Establishment and longevity in the definitive host. Exp Parasitol, 13: 273–283.

McClelland, G. and Bourns, T.K.R. 1969. Effects of *Trichobilharzia ocellata* on growth, reproduction, and survival of *Lymnaea stagnalis*. Exp Parasitol, 24: 137–146.

McCusker, P., McVeigh, P., Rathinasamy, V. et al. 2016. Stimulating neoblast-like cell proliferation in juvenile *Fasciola hepatica* supports growth and progression towards the adult phenotype *in vitro*. PLoS Neg Trop Dis, 10: e0004994. DOI: 10.1371/journal.pntd.0004994.

McLaughlin, J.D., Marcogliese, D.J. and Kelly, J. 2006. Morphological, developmental and ecological evidence for a progenetic life cycle in *Neochasmus* (Digenea). Folia Parasitol, 53: 44–52.

Mehlhorn, H. 2016. *Echinostoma revoltum*. In: *Encyclopedia of Parasitology*, (ed) Mehlhorn, H., Springer, Berlin.

Meier, M. and Meier-Brook, C. 1981. *Schistosoma mansoni*: effect on growth, fertility and development of distal male organs in *Biomphalaria glabrata* exposed to miracidia at different ages. Z Parasitekd, 66: 121–131.

Meinkoth, N.A. 1947. Notes on the life cycle and taxonomic position of *Haplobothrium globuliforme* Cooper, a tapeworm of *Amia calva* L. Trans Am Microsc Soc, 65: 256–261.

Melvin, D.M. 1952. Studies on the life cycle and biology of *Monoecocestus sigmodontis* (Cestoda: Anoplocephalidae) from the cotton rat, *Sigmodon hispidus*. J Parasitol, 38: 346–355.

Mercer, J.G., Munn, A.E., Arme, C. and Rees, H.H. 1987. Analysis of ecdysteroids in different developmental stages of *Hymenolepis diminuta*. Mol Biochem Parasitol, 25: 61–71.

268 *Reproduction and Development in Platyhelminthes*

Meyer, M.C. and Vik, R. 1963. The life cycle of *Diphyllobothrium Sebago* (Ward, 1910). J Parasitol, 49: 962–968.

Meyrowitsch, D., Christensen, N.O. and Hindsbo, O. 1991. Effects of temperature and host density on the snail-finding capacity of cercariae of *Echinostoma caproni* (Digenea: Echinostomatidae). Parasitology, 102: 391–395.

Michiels, N. and Newman, L. 1998. Sex and violence in hermaphrodites. Nature, 391: 647.

Milinski, M. 2006. Fitness consequences of selfing and outcrossing in the cestode *Schistocephalus solidus*. Integ Comp Biol, 46: 373–380.

Minchella, D.J. and Loverde, P.T. 1981. A cost of increased early reproductive effort in the snail *Biomphalaria glabrata*. Am Nat, 118: 876–881.

Minchella, D.J., Sollenberger, K.M. and De Souza, C.P. 1995. Distribution of schistosome genetic diversity within molluscan intermediate hosts. Parasitology, 111: 217–220.

Mo, T.A. 1992. Seasonal variations in the prevalence and infestation intensity of *Gyrodactylus salaris* Malmberg, 1957 (Monogenea: Gyrodactylidae) on Atlantic salmon parr, *Salmo salar* L., in the river Batnfjordselva, Norway. J Fish Biol, 41: 697–707.

Mohandas, A. 1975. Further studies on the sporocyst capable of producing miracidia. J Helminthol, 49: 167–171.

Molina, M.D., Salo, E. and Cebria, F. 2007. The BMP pathway is essential for re-specification and maintenance of the dorsoventral axis in regenerating and intact planarians. Dev Biol, 311: 79–94.

Mondal, M., Kundu, J.K. and Misra, K.K. 2016. Variation in lipid and fatty acid uptake among nematode and cestode parasites and their host, domestic fowl: host-parasite interaction. J Parasitol Dis, DOI: 10.1007/s12639-015-0718-5.

Monteiro, C.M. and Brasil–Sato, M.C. 2010. Habitat selection and maturation of *Saccocoelioides nanii* (Digenea: Haploporidae) in *Prochilodus argenteus* (Actinopterygii: Prochilodontidae) from the São Francisco River, Brazil. Zoologia, 27: 757–760.

Mooney, A.J., Ernst, I. and Whittington, I.D. 2006. An egg-laying rhythm in *Zeuxapta seriolae* (Monogenea: Heteraxinidae), a gill parasite of yellowtail kingfish (*Seriola lalandi*). Aquaculture, 253: 10–16.

Mooney, A.J., Ernst, I. and Whittington, I.D. 2008. Egg-laying patterns and *in vivo* egg production in the monogenean parasites diseases caused by Platyhelminthes *Heteraxine heterocerca* and *Benedenia seriolae* from Japanese yellowtail *Seriola quinqueradiata*. Parasitology, 135: 1295–1302.

Moore, J. and Brooks, D.R. 1987. Asexual reproduction in cestodes (Cyclophyllidea: Taeniidae): ecological and phylogenetic influences. Evolution, 41: 882–891.

Moore, M.M., Kaattari, S.L. and Olson, R.E. 1994. Biologically active factors against the monogenetic trematode *Gyrodactylus stellatus* in the serum and mucus of infected juvenile English soles. J Aquat Anim Health, 6: 93–100.

Moraczewski, J. 1977. Asexual reproduction and regeneration of *Catenula* (Turbellaria, Archoophora). Zoomorphologie, 88: 65–80.

Morag, L., McCaig, O. and Hopkins, C.A. 1963. Studies on *Schistocephalus solidus*. 2. Establishment and longevity in the definitive host. Exp Parasitol, 13: 273–283.

Morag, L., McCaig, O. and Hopkins, C.A. 1965. Studies on *Schistocephalus solidus*. 3. The *in vitro* cultivation of the plerocercoid. Parasitology, 55: 257–268.

Morgan, T.H. 1898. Experimental studies of the regeneration of *Planaria maculata*. Arch Entwm, 7: 364–397.

Morgan, T.H. 1901. *Regeneration*. MacMillan, New York, p 316.

Morgan, T.H. 1905. "Polarity" considered as a phenomenon of gradation of materials. J Exp Zool, 2: 495–506.

Morita, M. and Best, J.B. 1984. Electron microscopic studies of planarian regeneration. IV. Cell division of neoblasts in *Dugesia dorotocephala*. J Exp Zool, 229: 425–436.

Morley, N.J., Irwin, S.W.B. and Lewis, J.W. 2003. Pollution toxicity to the transmission of larval digeneans through their molluscan hosts. Parasitology, 126S: 5–26.

Mouritsen, K.M. and Jensen, K.T. 1994. The enigma of gigantism: effect of larval trematodes on growth, fecundity, egestion and locomotion in *Hydrobia ulvae* (Pennant) (Gastropoda: Prosobranchia). J Exp Mar Biol Ecol, 181: 53–66.

*References* **269**

Mouton, S., Willems, M., Houthoofd, W. et al. 2011. Lack of metabolic ageing in the long–lived flatworm *Schmidtea polychroa*. Exp Geront, 46: 755–761.

Mouton, S., Wudarski, J., Gudniewska, M. and Berezikov, E. 2018. The regenerative flatworm *Macrostomum lignano*, a model organism with high experimental potential. Int J Dev Biol, 62: 551–558.

Mukhin, V.A., Smirnova, E.B. and Novikov, Vu. 2007. Characteristics of proteinase digestive function in invertebrates-inhabitants of cold seas. Zh Evol Biokhim Fiziol. 43: 398–403.

Muller, R. 2001. *Worm and Human Disease.* CABI Publishing, p 320.

Nagwa, E.A., Loubna, M.A., El-Madawy, R.S. and Toulan, E.I. 2013. Studies on helminthes of poultry in Gharbia Governorate. Benha Vet Med J, 25: 139–144.

Nakagawa, H., Ishizu, H. and Chinone, A. 2012a. The *Dr-nanos* gene is essential for germ cell specification in the planarian *Dugesia ryukyuensis*. Int J Dev Biol, 56: 165–171.

Nakagawa, H., Ishizu, H., Hasegawa, R. et al. 2012b. *Drpiwi-1* is essential for germline cell formation during sexualization of the planarian *Dugesia ryukyuensis*. Dev Biol, 361: 167–176.

Navarette-Perea, J., Moguel, B., Mendoza-Herandez, G. et al. 2014. Identification and quantification of host proteins in the vesicular fluid of porcine *Taenia solium* cysticerci. Exp Parasitol, 143: 11–17.

Navarro, B.S. and Jokela, J. 2013. Population genetic structure of parthenogenetic flatworm populations with occasional sex. Freshwater Biol, 58: 416–429.

Negovetich, N.J. and Esch, G.W. 2007. Long–term analysis of Charlie's pond: Fecundity and trematode communites of *Helisoma anceps*. J Parasitol, 63: 1311–1318.

Newmark, P.A. and Alvarado, A.S. 2000. Bromodexoyuridine specifically labels the regenerative stem cells of planarians. Dev Biol, 220: 142–153.

Newmark, P.A. and Alvarado, A.S. 2001. Regeneration in planaria. Encyclopedia of life sciences, Nat Pub Group/www.els.net.

Newmark, P.A. and Alvarado, A.S. 2002. Not your father's planarian: a classic model enters the era of functional genomics. Nat Rev Genet, 3: 210–219.

Newmark, P.A., Reddien, P.W., Cebria, F. and Alvarado, A.S. 2003. Ingestion of bacterially expressed double-stranded RNA inhibits gene expression in planarians. Proc Natl Acad Sci USA, 100: 11861–11865.

Newmark, P.A., Wang, Y. and Chong, T. 2008. Germ cell specification and regeneration in planarian. Gold Spring Harbor Symp Quant Biol, 75: 573–581.

Niewiadomska, K. and Pojmanska, T. 2011. Multiple strategies of digenean trematodes to complete their life cycles. Wladomoscl Parazytologlczne, 57: 233–241.

Nimeth, K.T., Egger, B., Rieger, R. et al. 2007. Regeneration in *Macrostomum lignano* (Platyhelminthes): cellular dynamics in the neoblast stem cell system. Cell Tissue Res, 327: 637–646.

Nodono, H. and Matsumoto, M. 2012. Reproductive mode and ovarian morphology regulation in chimeric planarians composed of asexual and sexual neoblasts. Mol Reprod Dev, 79: 451–460.

Nodono, H., Ishino, Y., Hoshi, M. and Matsumoto, M. 2012. Stem cells from innate sexual but not acquired sexual planarians have the capability to form a sexual individual. Mol Reprod Dev, 79: 757–766.

Nollen, P.M. 1975. Studies on the reproductive system of *Hymenolepis diminuta* using autoradiography and transplantation. J Parasitol, 61: 100–104.

Norena, C., Damborenea, C. and Brusa, F. 2015. Phylum Platyhelminthes. In: *Freshwater Invertebrates: Ecology and General Biology.* (eds) Thorp, J.H. and Rogers, D.C., Elsevier, Amsterdam, pp 181–204.

Nuttycombe, J.W. and Waters, A.J. 1935. The American species of *Stenostomum*. Proc Am Phil Soc Bull, 69: 213–301.

Ogawa, K. 2015. Diseases of cultured marine fishes caused by Platyhelminthes (Monogenea, Digenea, Cestoda). Parasitology, 142: 178–195.

Oglesby, L.C. 1961. Ovoviviparity in the monogenetic trematode *Polystomoidella oblonga*. J Parasitol, 47: 237–243.

270  *Reproduction and Development in Platyhelminthes*

Ohashi, H., Umeda, N., Hirazawa, N. et al. 2007. Expression of *vasa (vas)*-related genes in germ cells and specific interference with gene functions by double-stranded RNA in the monogenean, *Neobenedenia girellae*. Int J Parasitol, 37: 515–23.

Okano, D., Ishida, S. and Kobayashi, K. 2015. Light and electron microscopic studies of the intestinal epithelium in *Notoplana humilis* (Platyhelminthes, Polycladida): the contribution of mesodermal/gastrodermal neoblasts to intestinal regeneration. Cell Tissue Res, 362: 569–540.

Okino, T., Ushirogawa, H., Matoba, K. et al. 2017. Establishment of the complete life cycle of *Spirometra* (Cestoda: Diphyllobothriidae) in the laboratory using a newly isolated triploid clone. Parasitol Int, 66: 116–118.

Opuni, E.K., Muller, R. and Mueller, J.F. 1974. Absence of sparganum growth factor in African *Spirometra* spp. J Parasitol, 60: 375–376.

Orii, H., Sakurai, T. and Watanabe, K. 2005. Distribution of the stem cells (neoblatsts) in the planarian *Dugesia japonica*. Dev Genes Evol, 215: 143–157.

Ozer, A., Ozturk, T. and Ozturk, M.O. 2004. Prevalence and intensity of *Gyrodactylus arcuatus* Bychowsky, 1933 (Monogenea) infestations on the three-spined stickleback, *Gasterosteus aculeatus* L., 1758. Turk J Vet Anim Sci, 28: 807–812.

Ozer, A. and Ozturk, T. 2005. *Dactylogyrus cornu* Linstow, 1878 (Monogenea) infestations on Vimba (*Vimba vimba tenella* (Nordmann, 1840)) caught in the Sinop region of Turkey in relation to the host factors. Turk J Vet Anim Sci, 29: 1119–1123.

Paling, J.E. 1965. The population dynamics of the monogenean gill parasite *Discocotyle sagittata* Leuckart on Windermere trout, *Salmo trutta*, L. Parasitol, 55: 667–694.

Pallas, P.S. 1774. Specilegia zoological quibas novae imprimus et obscurae animallium species inconibus descriptionibus atque commentariis illustrantur Fasc 10. Berolini, p 41.

Palmberg, I. 1986. Cell migration and differentiation during wound healing and regeneration in *Microstomum lineare* (Turbellaria). Hydrobiologia, 132: 181–188.

Palmberg, I. 1990. Stem cells in microturbellarians. An autoradiographic and immunocytochemical study. Protoplasma, 158: 109–120.

Pampoulie, C., Morand, S., Lambert, A. et al. 1999. Influence of the Trematode *Aphalloidescoelomicola* Dollfus, Chabaud & Golvan, 1957 on the fecundity and survival of *Pomatoschistusmicrops* (Kroyer, 1838) (Teleostei, Gobiidae). Parasitology, 119: 61–67.

Pampoulie, C., Bouchereau, J.L., Rosecchi, E. and Poizat, G. 2000. Annual variations in the reproductive traits of *Pomatoschistus microps* in a Mediterranean lagoon undergoing environmental changes: Evidence of phenotypic plasticity. J Fish Biol, 57: 1441–1452.

Pandian, T.J. 1967. Changes in the chemical composition and caloric content of developing eggs of the shrimp *Crangon crangon* Helgolander wiss. Meeresunters, 16: 216–224.

Pandian, T.J. and Fluchter, J. 1968. Rate and efficiency of yolk utilization in developing eggs of the sole *Solea solea*. Helgoland Mar Res, 18: 53–60.

Pandian, T.J. 1970a. Ecophysiological studies on the developing eggs and embryos of the European lobster *Homarus gammarus*. Mar Biol, 5: 154–167.

Pandian, T.J. 1970b. Yolk utilization and hatching time in the Canadian lobster *Homarus americanus*. Mar Biol, 7: 249–254.

Pandian, T.J. 1975. Mechanism of heterotrophy. In: *Marine Ecology*. (ed) Kinne, O., John Wiley, London, 3 Part 1: 61–249.

Pandian, T.J. 1987. Fish. In: *Animal Energetics*. (eds) Pandian, T.J. and Vernberg, F.J., Academic Press, San Diego, 2: 357–465.

Pandian, T.J. 2010. *Sexuality in Fishes*. Science Publishers/CRC Press, USA, p 208.

Pandian, T.J. 2011. *Sex Determination in Fish*. Science Publishers/CRC Press, USA, p 270.

Pandian, T.J. 2012. *Genetic Sex Differentiation in Fish*. CRC Press, USA, p 214.

Pandian, T.J. 2013. *Endocrine Sex Differentiation in Fish*. CRC Press, USA, p 303.

Pandian, T.J. 2015. *Environmental Sex Determination in Fish*. CRC Press, USA, p 299.

Pandian, T.J. 2016. *Reproduction and Development in Crustacea*. CRC Press, USA, p 301.

Pandian, T.J. 2017. *Reproduction and Development in Mollusca*. CRC Press, USA, p 299.

Pandian, T.J. 2018. *Reproduction and Development in Echinodermata and Prochordata*. CRC Press, USA, p 270.

Pandian, T.J. 2019. *Reproduction and Development in Annelida*. CRC Press, USA, p 276.

## References 271

Paperna, I. 1964. Competitive exclusion of *Dactylogyrus extensus* by *Dactylogyrus vastator* (Trematoda, Monogenea) on the gills of reared carp. J Parasitol, 50: 94–98.

Pappas, P.W. and Read, C.P. 1975. Membrane transport in helminth parasites: A review. Exp Parasitol, 37: 469–530.

Park, J.K., Kim, K.H. and Kang, S. et al. 2007. A common origin of complex life cycles in parasitic flatworms: evidence from the complete mitochondrial genome of *Microcotyle sebastis* (Monogenea: Platyhelminthes). BMC Evol Biol, 7: 11. DOI: https://doi.org/10.1186/1471-2148-7-11.

Parker, G.A., Ball, M.A. and Chubb, J.C. 2015a. Evolution of complex life cycles in trophically transmitted helminthes. I. Host incorporation and trophic ascent. J Evol Biol, 28: 267–291.

Parker, G.A., Ball, M.A. and Chubb, J.C. 2015b. Evolution of complex life cycles in trophically transmitted helminthes. II. How do life–history stages adapt to their hosts? J Evol Biol, 28: 292–304.

Paterson, A.M. Gray, R.D. and Wallis, G.P. 1993. Parasites, petrels and penguins: does louse presence reflect seabird phylogeny? Int J Parasitol, 23: 515–526.

Payne, A. *Taenia saginata*. Facebook.

Pearre, S. 1979. Niche modification in Chaetognatha infected with larval trematode (Digenea). Int Rev Gesamten Hydrobiol, 64: 193–206.

Pearson, J.C. 1972. A phylogeny of life-cycle patterns of the Digenea. Adv Parasitol, 10: 153–189.

Pechenik, J.A. and Fried, B. 1995. Effect of temperature on survival and infectivity of *Echinostoma trivolvis* cercariae: a test of the energy limitation hypothesis. Parasitology, 111: 373–378.

Pechenik, J.A., Fried, B. and Simpkins, H.L. 2001. *Crepidula fornicata* is not a first intermediate host for trematodes: Who is? J Exp Mar Biol Ecol, 261: 211–224.

Pennycuick, L. 1971. Quantitative effects of three species of parasites on a population of three-spined stickleback *Gasterosteus aculeatus*. J Zool Lond, 165: 143–162.

Peoples, R.C. 2013. A review of the helminth parasites using polychaetes as hosts. Parasitol Res, 112: 3409–3412.

Peter, R. 2001. Experimental systems for studying regeneration processes: Turbellarians as model organisms with a stem cell system (In German). Ber Nat Med Verein Innsbruck, 88: 287–350.

Peter, R., Gschwentner, R., Schurmann, W. et al. 2004. The significance of stem cells in free-living flatworms: one common source for all cells in the adult. J Appl Biomed, 2: 21–35.

Peters, A., Streng, A. and Michaels, N.K. 1996. Mating behaviour in a hermaphrodite flatworm with reciprocal insemination: Do they assess their mates during copulations. Ethology, 102: 236–251.

Petersen, C.P. and Reddien, P.W. 2008. *Smed*-betacatenin is required for anteroposterior blastema polarity in planarian regeneration. Science, 319: 327–330.

Pfister, D. and Ladurner, P. 2005. The totipotent stem cell system as a source for germ line cells during development and regeneration in the flatworm *Macrostomum lignano*. Dev Biol, 283: 643.

Phalee, A., Wongsawad, C., Rojanapaibul, A. and Chai, J.Y. 2015. Experimental life history and biological characteristics of *Fasciola gigantica* (Digenea: Fasciolidae). Korean J Parasitol, 53: 59–64.

Phares, K. 1996. An unusual host-parasite relationship: the growth hormone-like factor from plerocercoids of spirometrid tapeworms. Int J Parasitol, 26: 575–588.

Pickering, A.D. and Pottinger, T.G. 1989. Stress responses and disease resistance in salmonid fish: Effects of chronic elevation of plasma cortisol. Fish Physiol Biochem, 7: 253–258.

Pillans, R.D. and Franklin, C.E. 2004. Plasma osmolyte concentrations and rectal glands mass of bull sharks *Carcharhinus leucas*, captured along a salinity gradient. Comp Biochem Physiol, 138 A: 363–371.

Planetary Biodiversity Inventory: A Survey of Tapeworms from Vertebrate Bowels of the Earth.

Pongratz, N., Storhas, M., Carranza, S. and Michiels, N.K. 2003. Phylogeography of competing sexual and parthenogenetic forms of a freshwater flatworm: patterns and explanations. BMC Evol Biol, 3: 23.

Poppe, T. 1999. *Fiskehelse og fiskesykdommer*. Universitestsforlaget, Oslo (in Norwegian).

## 272 Reproduction and Development in Platyhelminthes

Poulin, R. 1992. Determinants of host-specificity in parasites of freshwater fishes. Int J Parasitol, 22: 753–758.

Poulin, R. 1996a. How many parasites species are there: Are we close to answers? Int J Parasitol, 26: 1127–1129.

Poulin, R. 1996b. The evolution of body size in the Monogenea: the role of host size and latitude. Can J Zool, 74: 726–732.

Poulin, R. 1998. *Evolutionary Ecology of Parasites from Individuals to Communities*. Chapman and Hall, London, p 343.

Poulin, R. 1999. Speciation and diversification of parasitic lineages: an analysis of congeneric parasite species in vertebrates. Evol Ecol, 13: 455–467.

Poulin, R. and Morand, S. 2000. The diversity of parasites. Q Rev Biol, 75: 277–293.

Poulin, R. 2001. Progenesis and reduced virulence as an alternative transmission strategy in a parasitic trematode. Parasitology, 123: 623–630.

Poulin, R. 2002. The evolution of monogenean diversity. Int J Parasitol, 32: 245–254.

Poulin, R. and Cribb, T.H. 2002. Trematode life cycles: short is sweet? Trends Parasitol, 18: 176–183.

Poulin, R. 2005. Evolutionary trends in body size of parasitic flatworms. Biol J Linn Soc, 85: 181–189.

Poulin, R. and Presswell, B. 2016. Taxonomic quality of species descriptions varies over time and with the number of authors, but unevenly among parasite taxa. Syst Biol, 65: 1107–1116.

Premvati, G. 1955. *Cercaria multiplicata* n. sp. from the snail *Melanoides tuberculatus* (Muller). J Zool Soc Ind, 7: 13–24.

Presswell, B., Blasco-Costa, I. and Kostadinova, A. 2014. Two new species of *Maritrema* Nicoll, 1907 (Digenea: Microphallidae) from New Zealand: morphological and molecular characterization. Parasitol Res, 113: 1641–1656.

Presswell, R., Poulin, R. and Randhwa, H.S. 2012. First report of a gyporhynchid tapeworm (Cestoda: Cyclophyllidea) from New Zealand and from an eleotrid fish described from metcestoides and *in vitro* grown worms. J Helminthol, 86: 453–464.

Price, P.W. 1974. Strategies of egg production. Evolution, 28: 76–84.

Probst, S. and Kube, J. 1999. Histopathological effects of larval trematode infections in mudsnails and their impact on host growth: what causes gigantism in *Hydrobia ventrosa* (Gastropoda: Prosobranchia). J Exp Mar Biol Ecol, 238: 49–68.

Pulkkinen, K. and Valtonen, E.T. 1999. Accumulation of plerocercoids of *Triaenophorus crassus* in the second intermediate host *Coregonus lavaretus* and their effect on growth of the host. J Fish Biol, 55: 115–126.

Radha, E. 1971. Some notes on the population dynamics of the monogenean gill parasite *Gastrocotyle indica*. Mar Biol, 8: 213–219.

Ramm, S.A. 2017. Exploring the sexual diversity of flatworms: Ecology, evolution and the molecular biology of reproduction. Mol Reprod Dev, 84: 120–131.

Randolph, H. 1891. The regeneration of the tail in *Lumbriculus*. Zool Anz, 14: 154–156.

Randolph, H. 1897. Observations and experiments on regeneration in planarians. Arch Entwick/ Mech Org, 5: 352–372.

Rantanen, J.T., Valtonen, E.T. and Holopainen, I.J. 1998. Digenean parasites of the bivalve mollusk *Pisidium amnicum* in a small river in eastern Finland. Dis Aquat Org, 33: 201–208.

Rauch, G., Kalbe, M. and Reusch, T.B.H. 2005. How a complex life cycle can improve a parasite's sex life. J Evol Biol, 18: 1069–1075.

Rawlinson, K.A., Bolanos, D.M., Liana, M.K. and Litvaitis, M.K. 2008. Reproduction, development and parental care in two direct-developing flatworms (Platyhelminthes: Polycladida: Acotylea). J Natl Hist, 42: 2173–2192.

Rawlinson, K.A. 2014. The diversity, development and evolution of polyclad flatworm larvae. Evol Dev, 5: 9.

Reddien, P.W. and Alvarado, A.S. 2004. Fundamentals of planarian regeneration. Annu Rev Cell Biol, 20: 725–757.

Reddien, P.W., Bermange, A.L., Murfitt, K.J. et al. 2005. Identification of genes needed for regeneration, stem cell function, and tissue homeostasis by systematic gene perturbation in planaria. Dev Cell, 8: 635–649.

*References* 273

Reddien, P.W., Bermange, A.L., Kicza, A.M. and Alvarado, A.S. 2007. BMP signaling regulates the dorsal planarian midline and is needed for asymmetric regeneration. Development, 134: 4043–4051.

Reddy, A., Frazer, B.A. and Fried, B. 1997. Low molecular weight hydrophilic chemicals that attracts Echinostoma trivolvis and E. caproni cercariae. Int J Parasitol, 27: 283–287.

Redmond, M.D., Hartson, R.B., Hoverman, J.T. et al. 2011. Experimental exposure of Helisoma trivolvis and Biomphalaria glabrata (Gastropoda) to Ribeiroia ondatrae (Trematoda). J Parasitol, 97: 1055–1061.

Reed, P., Francis–Floyd, R., Klinger, R. and Petty, D. 2012. Monogenean parasites of fish. University of Florida, EDIS website at http://edis.ifas.ufl.edu.

Rees, G. 1940. Studies on the germ cell cycle of the digenetic trematode Parorchis acanthus Nicoll: Part II. Structure of the miracidium and germinal development in the larval stages. Parasitology, 32: 372–391.

Reuter, M. and Kreshchenko, N. 2004. Flatworm asexual multiplication implicates stem cells and regeneration. Can J Zool, 82: 334–356.

Reynoldson, T.B., Young, J.O. and Taylor, M.C. 1965. Effect of temperature on the life cycle of four species of lake-dwelling triclads. J Anim Ecol, 34: 23–43.

Rieger, R., Legniti, A., Ladurner, P. et al. 1999. Ultrastructure of neoblasts in microturbellaria: significance for understading stem cells in free-living Platyhelminthes. Invert Reprod Dev, 35: 127–140.

Rietzler, A.C., Dumont, H.J., Rocha, O. and Ribeiro, M.M. 2018. Predation and reproductive performance in two pelagic typhoplanid turbellarians. PLoS ONE, 13: e0193472.

Rinaldi, G., Eckert, S.E., Tsai, I.J. et al. 2012. Germline transgenesis and insertional mutagenesis in Schistosoma mansoni mediated by murine leukemia virus. PLoS Pathol, 8: e1002820.

Rink, J.C. 2013. Stem cell systems and regeneration in planaria. Dev Genes Evol, 223: 67–84.

Rivera, V.R. and Perich, M.J. 1994. Effects of water quality on survival and reproduction of four species of planaria (Turbellaria: Tricladida). Invert Reprod Dev, 25: 1–7.

Roberts, T., Murrell, K.D. and Marks, S. 1994. Economic losses caused by foodborne parasitic diseases. Parasitol Today, 10: 419–423.

Roger, A.J. and Hug, L.A. 2006. The origin and diversification of eukaryotes: problems with molecular phylogenetics and molecular clock estimation. Philos Trans R Soc, 361B: 1039–1054.

Rohde, K. 1976. Monogenean gill parasites of Scomberomorus commersoni Lacepede and other mackerel on the Australian east coast. Z Parasitenkd, 51: 49–69.

Rohde, K. 1979. A critical evaluation of intrinsic and extrinsic factors responsible for niche restriction in parasites. Am Nat, 114: 648–671.

Rohde, K. 1982. Ecology of Marine Parasites. Univ Queensland Press, St. Lucia, p 245.

Rohde, K. 1985. Increased viviparity of marine parasites at high latitudes. Hydrobiologia, 127: 197–201.

Rohde, K. 1987. Different populations of Scomber australasicus in New Zealand and southeastern Australia, demonstrated by a simple method using monogenean sclerites. J Fish Biol, 30: 651–657.

Rohde, K. 1993. Ecology of Marine Parasites: an Introduction to Marine Parasitology. Cab International, Willingford, UK, p 298.

Rohde, K. 1997. The origins of parasitism in the Platyhelminthes: a summary interpreted on the basis of recent literature. Int J Parasitol, 27: 739–746.

Rohde, K. and Heap, M. 1998. Latitudinal differences in species and community richness and in community structure of metazoan endo- and ecto-parasites of marine teleost fish. Int J Parasitol, 28: 461–474.

Rohde, K. 2001. The Aspidogastrea: an archaic group of Platyhelminthes. In: Interrelationships of Platyhelminthes. (eds) Littlewood, D.T.J. and Bray, R.A., Taylor and Francis, London, pp 159–167.

Romero, R. 1987. Analisi cel.lular quantitative del creixement I de la reproduccio a diferents especies de planaries. Ph.D. Thesis, University of Barcelona.

Rose, C. and Shostak, S. 1968. The transformation of gastrodermal cells to neoblasts in regenerating Phagocata gracilis (Leidy). Exp Cell Res, 50: 553–561.

274 *Reproduction and Development in Platyhelminthes*

Rosen, R. and Dick, T.A. 1983. Development and infectivity of the procercoid of *Triaenophorus crassus* Forel and mortality of the first intermediate host. Can J Zool, 61: 2120–2128.

Roubal, F.R. 1994. Observations on the eggs and fecundity of dactylogyrid and diplectanid monogeneans from the Australian marine sparid fish, *Acanthopagrus australis*. Folia Parasitol, 41: 220–222.

Rychel, A.L. and Swalla, B. 2009. Regeneration in hemichordates and echinoderms. In: *Stem Cells in Marine Organisms* (eds) Rinkevich, B. and Matranga, V., Springer Verlag, Dordrecht, pp 245–256.

Sakaguchi, Y. 1980. Karyotype and gametogenesis of the common liver fluke, *Fasciola* sp. Jap J Parasitol, 29: 507–513.

Sakamoto, T. 1982. Histochemistry and histoenzymology of *Echinococcus*. I. Histochemical observation on general structure of *Echinococcus multilocularis*. Mem Faculty Agri, Kagoshima Univ, 18: 127–139.

Sakanari, J. and Moser, M. 1985. Salinity and temperature effects on the eggs, coracidia, and procercoids of *Lacistorhynchus tenuis* (Cestoda: Trypanorhyncha) and induced mortality in a first intermediate host. J Parasitol, 71: 583–587.

Sakanari, J.A. and Moser, M. 1989. Complete life cycle of the elamobranch cestode, *Lacistorhynchus dollfusi* Beveridge and Sakanari, 1987 (Trypanorhyncha). J Parasitol, 75: 806–808.

Sakurai, T. 1981. Sexual induction by feeding in an asexual strain of the freshwater planarian, *Dugesia japonica japonica*. Annot Zool Jap, 54103–54112.

Saladin, K.S. 1979. Behavioral parasitology and perspectives on miracidial host-finding. Z Parasitenkd, 60: 197–210.

Salo, E. and Baguna, J. 1984a. Regeneration and pattern formation in planarians. I. The pattern of mitosis in anterior and posterior regeneration in *Dugesia* (G) *tigrina*, and a new proposal of blastema formation. J Embryol Exp Morph, 83: 63–80.

Salo, E. and Baguna, J. 1984b. Determinative events and size-independence of pattern formation during planarian regeneration. J Embryol Exp Morph, 82: 178.

Salo, E. and Baguna, J. 1985a. Cell movement in intact and regenerating planarians. Quantitation using chromosomal, nuclear and cytoplasmic markers. J Emryol Exp Morph, 89: 57–70.

Salo, E. and Baguna, J. 1985b. El control de la prolifeacio cel.lular a planaries: accio de nauropeptids mitogens (Substancia P) i factors dependents de densitat cellular. In: *Biologia del Desencolupament*. Societat Catalana de Biologia, Barcelona, 3: 1–18.

Salo, E. and Baguna, J. 1986. Stimulation of cellular proliferation and differentiation in the intact and regenerating planarian *Dugesia* (G) *tigrina* by the neuropeptide substance P. J Exp Zool, 237: 129–135.

Sandeman, I.M. and Burt, M.D.B. 1972. Biology of *Bothrimonus* (= *Diplocotyle*) (Pseudophyllidea: Cestoda): ecology, life cycle, and evolution; a review and synthesis. J Fish Res Bd Can, 29: 1381–1395.

Sasal, P., Trouve, S., Muller-Graft, C. and Morand, S. 1999. Specificity and host predictability: a comparative analysis among monogenean parasites of fish. J Anim Ecol, 68: 437–444.

Sato, K., Sugita, T., Kobayashi, K. et al. 2001. Localization of mitochondrial ribosomal RNA on the chromatoid bodies of marine planarian polyclad embryos. Dev Growth Differ, 43: 107–114.

Schallig, H.D.F.H., Sassen, M.J.M., Hordijk, P.L. and de Jong Brink, M. 1991. *Trichobilharzia ocellata*: influence of infection and cercarial induction of the release of schistosomin, a snail neuropeptide antagonizing female gonodotropic hormones. Parasitology, 102: 85–91.

Scharer, L. and Wedekind, C. 1999. Lifetime reproductive output in a hermaphrodite cestode when reproducing alone or in pairs: A time cost of pairing. Evol Ecol, 13: 381–394.

Scharer, L. and Wedekind, C. 2001. Social situation, sperm competition and sex allocation in a simultaneous hermaphrodite parasite, the cestode *Schistocephalus solidus*. J Evol Biol, 14: 942–953.

Scharer, L., Karlsson, L.M., Christen, M. and Wedekind, C. 2001. Size-dependent sex allocation in a simultaneous hermaphrodite parasite. J Evol Biol, 14: 55–67.

Scharer, L., Joss, G. and Saundner, P. 2004. Mating behaviour of the marine turbellarians *Macrostomum* sp: these worms suck. Mar Biol, 145: 373–380.

*References* 275

Schelkle, B., Faria, P.J., Johnson, M.B. et al. 2012. Mixed infections and hybridization in monogenean parasites. PLoS ONE, 7: e39506.

Schjorring, S. and Jager, I. 2007. Incestuous mate preference by a simultaneous hermaphrodite with strong inbreeding depression. Evolution, 61: 423–430.

Schmidt, G.D. and Robertes, L.S. 1985. *Foundations of Parasitology*. McGraw-Hill, New York, p 670.

Schmidt, G.D. 1986. *CRC Handbook of Tapeworm Identification*. CRC Press, Boca Raton, Florida, p 675.

Schockaert, E.R. 1996. Turbellarians. In: *Methods for the Examination of Organismal Diversity in Soils and Sediments*. (ed) Hall, G.S. CAB International, Willingford, UK, pp 211–225.

Schockaert, E.R., Hooge, M., Sluys, R. et al. 2008. Global diversity of free living flatworms (Platyhelminthes, "Turbellaria") in freshwater. Hydrobiologia, 595: 41–48.

Scholz, T. 1991a. Studies on the development of the cestode *Proteocephalus neglectus* La Rue, 1911 (Cestoda: Proteocephalidae) under experimental conditions. Folia Parasitol, 38: 30–55.

Scholz, T. 1991b. Early development of *Khawia sinensis* Hsu, 1935 (Cestoda: Caryophyllidea), a carp parasite. Folia Parasitol, 38: 133–142.

Scholz, T. 1993. Development of *Proteocephalus torulosus* in the intermediate host under experimental conditions. J Helminthol, 67: 316–324.

Scholz, T. 1997. Life-cycle of *Bothriocephalus claviceps*, a specific parasite of eels. J Helminthol, 71: 241–248.

Scholz, T. 1999. Life cycles of species of *Proteocephalus*, parasites of fishes in the Palearctic Region: a review. J Helminthol, 73: 1–19.

Scholz, T. and Kuchta, R. 2012. *Bothriocephalus acheilognathi* Yamaguti, 1934. In: *Fish Parasites Pathobiology and Protection*. (eds) Woo, P.T.K. and Buchmann, K., CAB International, Willingford, UK, pp 282–297.

Scholz, T., Garcia, H.H., Kuchta, R. and Wicht, B. 2009. Update on the human broad tapeworm (Genus *Diphyllobothrium*), including clinical relevance. Clinic Microbiol Rev, 22: 146–160.

Schotthoefer, A.M., Cole, R.A. and Beasley, V.R. 2003. Relationship of tadpole stage to location of echinostome cercariae encystment and the consequences for tadpole survival. J Parasitol, 89: 475–482.

Schurmann, W., Betz, S. and Peter, R. 1998. Separation and subtyping of planarian neoblasts by density–gradient centrifugation and staining. Hydrobiologia, 383: 117–124.

Schweizer, G., Braun, U., Deplazes, P. and Torgerson, P.R. 2005. Estimating the financial losses due to bovine fasciolosis in Switzerland. Vet Res, 157: 188–193.

Schweizer, G., Meli, M.L., Torgerson, P.R. et al. 2007. Prevalence of *Fasciola hepatica* in the intermediate host *Lymnaea truncatula* detected by real time TaqMan PCR in populations from 70 Swiss farms with cattle husbandry. Vet Parasitol, 150: 164–169.

Scott, M.E. 1982. Reproductive potential of *Gyrodactylus bullatarudis* (Monogenea) on guppies (*Poecilia reticulata*). Parasitology, 85: 217–236.

Scott, M.E. and Nokes, D.J. 1984. Temperature-dependent reproduction and survival of *Gyrodactylus bullatarudis* (Monogenea) on guppies (*Poecilia reticulata*). Parasitology, 89: 221–227.

Sekera, E. 1906. Uber die Verbreitung Selbstfruchtung bei den Rhabdocoeliden. Zool Anz, 30: 142–144.

Sewell, R.B.S. 1922. *Cercariae indicae*. Ind J Med Res, 10: 1–370.

Sheiman, I.M. Sedel'nikov, Z.V., Shkutin, M.F. and Kreshchenko, N.D. 2006. Asexual reproduction of planarians: Metric studies. Rus J Dev Biol, 37: 102–107.

Shibata, N., Umesono, Y., Orii, H. et al. 1999. Expressiono of *vasa* (*vas*)-related genes in germline cells and totipotent somatic stem cell of planarians. Dev Biol, 206: 73–87.

Shibata, N., Hayashi, T., Fukumura, R. et al. 2012. Comprehensive gene expression analyses in pluripotent stem cells of a planarian, *Dugesia japonica*. Int J Dev Biol, 56: 93–102.

Shine, R., Cogger, H.G., Reed, R.R. et al. 2003. Aquatic and terrestrial locomotor speeds of amphibious sea–snakes (Serpentes, Laticaudidae). J Zool Lond, 259: 261–268.

Shinn, A., Pratoomyot, J., Bron, J. et al. 2015. Economic impacts of aquatic parasites on global finfish production. Global Aquacult Advocate, Septembet/October 82–84.

276  *Reproduction and Development in Platyhelminthes*

Shinn, G.L. 1985. Reproduction of *Anoplodium hymanae*, a turbellarian flatworm (Neorhabdocoela, Umagillidae) inhabiting the coelom of sea cucumbers: production of egg capsules, and escape of infective stages without evisceration of the host. Biol Bull, 169: 182–198.

Shoop, W.L. and Corkum, K.C. 1983. Transmammary infection of paratenic and definitive hosts with *Alaria marcinae* (Trematoda) mesocercariae. J Parasitol, 69: 731–735.

Shoop, W.L. 1988. Tramatode transmission patterns. J Parasitol, 74: 46–59.

Short, R.B. and Menzel, M.Y. 1959. Chromosomes in parthenogenetic miracidia and embryonic cercariae of *Schistosomatium douthitti*. Exp Parasitol, 8: 249–264.

Sigh, J. and Buchmann, K. 2000. Association between epidermal thionin-positive cells and skin parasitic infections in brown trout *Salmo trutta*. Dis Aquat Org, 41: 135–139.

Sikes, J.M. and Newmark, P.A. 2013. Restoration of anterior regeneration in a planarian with limited regenerative ability. Nature, 500: 77–80.

Silan, P., Euzet, I., Marlland, C. and Cabral, P. 1987. Le Biotope des ectoparasites branchiaux de poissons: facteurs de variations dans le modele Bars-Monogenea. Bull Ecol, 18: 383–391.

Sillman, E.I. 1962. The life history of *Azygia longa* (Leidy 1851) (Trematoda: Digenea), and notes on *A. acuminata* Goldberger 1911. Trans Am Microsc Soc, 81: 43–65.

Simkova, A., Desdevises, Y., Gelnar, M. and Morand, S. 2000. Co-existence of nine gill ectoparasites (*Dactylogyrus*: Monogenea) parasitizing the roach (*Rutilus rutilus* L.): history and present ecology. Int J Parasitol, 30: 1077–1088.

Simkova, A., Verneau, O., Gelnar, M. and Morand, S. 2006. Specificity and specialization of congeneric monogeneans parasiting cyprinid. Evolution, 60: 1023–1037.

Sinnappah, N.D., Lim, L.-H.S., Rohde, K. et al. 2001. A paedomorphic parasite associated with a neotenic amphibian host: Phylogenetic evidence suggests a revised systematic position for Sphyranuridae within anuran and turtle polystomatoineans. Mol Phyl Evol, 18: 189–201.

Sire, C., Durand, P., Pointier, J.P. and Theron, A. 1999. Genetic diversity and recruitment pattern of *Schistosoma mansoni* in a *Biomphalaria glabrata* snail population: a field study using randon-amplified polymorphic DNA markers. J Parasitol, 85: 436–441.

Sirgel, W.F., Artigas, P., Bargues, M.D. and Mas-Coma, S. 2012. Life cycle of *Renylaima capensis*, a brachylaimid trematode of shrews and slugs in South Africa: two-host and three-host transmission modalities suggested by epizootiology and DNA sequencing. Parasites Vect, 5: 169.

Sluiters, J.F., Brussaard–Wust, C.M. and Meuleman, E.A. 1980. The relationship between miracidial dose, production of cercariae, and reproductive activity of the host in the combination *Trichobilharzia ocellata* and *Lymnaea stagnalis*. Z Parasitenkd, 63: 13–26.

Smith, H.W. 1931. The absorption and excretion of water and salts by the elasmobranch fishes. I. Fresh water elasmobranchs. Am J Physiol, 98: 279–295.

Smyth, J.D. 1946. Studies on tapeworm physiology. 1. Cultivation of *Schistocephalus solidus in vitro*. J Exp Biol, 23: 47–73.

Smyth, J.D. 1949. Studies on tapeworm physiology. 4. Further observations on the development of *Ligula intestinalis in vitro*. J Exp Biol, 26: 1–14.

Smyth, J.D. 1954. Studies on tapeworm physiology. 7. Fertilizatin of *Schistocephalus solidus in vitro*. Exp Parasitol, 3: 64–67.

Smyth, J.D. 1964. Observations on the scolex of *Echinococcus granulosus*, with special reference to the occurrence and cytochemistry of secretory cells in the rostellum. Parasitology, 54: 515–526.

Smyth, J.D. and Smyth, M.M. 1969. Self insemination in *Echinococcus granulosus in vivo*. J Helminthol, 43: 363–367.

Smyth, J.D. and Halton, D.W. 1983. *The Physiology of Trematodes*. Cambridge University Press, Cambridge, p 468.

Smythe, A.B. and Font, W.F. 2001. Phylogenetic analysis of *Alloglossidium* (Digenea: Macroderoididae) and related genera: Life-cycle evolution and taxonomic revision. J Parasitol, 87: 386–391.

Sofi, T.A., Ahmad, F., Sheikh, B.A. et al. 2015. Chromosomes and cytogenetics of helminthes (Turbellaria, Trematoda, Cestoda, Nematoda and Acanthocephala). Neotrop Helminthol, 9: 113–162.

## References    277

Sokolova, I.M. 1995. Influence of trematodes on the demography of *Littorina saxatilis* (Gastropoda: Prosobranchia: Littorinidae) in the White Sea. Dis Aquat Org, 21: 91–101.

Soleng, A., Poleo, A.B.S., Alstad, N.E.W. and Bakke, T.A. 1999. Aqueous aluminium eliminates *Gyrodactylus salaris* (Platyhelminthes, Monogenea) infections in Atlantic salmon. Parasitology, 119: 19–25.

Soleng, A. and Bakke, T.A. 2001. The susceptibility of grayling (*Thymallus thymallus*) to experimental infections with the monogenean *Gyrodactylus salaries*. Int J Parasitol, 31: 793–797.

Solomatova, V.P. and Luzin, A.V. 1988. Gyrodactylosis of carps in fish tanks located on dischards waters of the Kostromsk electric power plant and some problem of the biology of *Gyrodactylus katharineri*. In: *Investigations of Monogeneans in the USSR*. (ed) Skarlato, O.A., Oxonian Press, New Delhi, pp 143–151.

Sorensen, R.E. and Minchella, D.J. 1998. Parasite influences on host life history: *Echinostoma revolutum* parasitism of *Lymnaea elodes* snails. Oecologia, 115: 188–195.

Sorensen, R.E. and Minchella, D.J. 2001. Snail-trematode life history interactions: past trends and future directions. Parasitology, 123S: 3–18.

Sousa, W.P. 1983. Host life history and the effect of parasitic castration on growth: a field study of *Cerithidea californica* Haldeman (Gastropoda: Prosobranchia) and its trematode parasites. J Exp Mar Biol Ecol, 73: 273–296.

Spakulova, M. and Casanova, J.C. 2004. Current knowledge on B chromosomes in natural populations of helminth parasites: a review. Cytogent Genome Res, 106: 222–229.

Stocchino, G.A. and Manconi, R. 2013. Overview of life cycles in model species of the genus *Dugesia* (Platyhelminthes: Tricladida). Ital J Zool, 80: 319–328.

Storhas, M., Weinzierl, R.P. and Michiels, N.K. 2000. Paternal sex in parthenogenetic planarians: A tool to investigate the accumulation of deleterious mutations. J Evol Biol, 13: 1–8.

Stunkard, H.W. 2005. The life history of *Cryptocotyle lingua* (Creplin), with notes on the physiology of the metacercariae. J Morph Physiol, 50: 143–191.

Sturrock, B.M. 1966. The influence of infection with *Schistosoma mansoni* on the growth rate and reproduction of *Biomphalarai pfeifferi*. Ann Trop Med Parasitol, 60: 187–197.

Sunyer, J.O., Tort, L. and Lambris, J.D. 1997. Diversity of the third form of complement C3 in fish: Functional characterization of five forms of C3 in the diploid fish *Sparus aurata*. Biochem J, 326: 877–881.

Takahashi, M.K. and Parris, M.J. 2008. Life cycle polyphenism as a factor affecting ecological divergence within *Notophthalmus viridescens*. Oecologia, 158: 23–34.

Takeda, H., Nishimura, K. and Agata, K. 2009. Planarians maintain a constant ratio of different cell types during changes in body size by using the stem cell system. Zool Sci, 26: 805–813.

Tamura, S., Oki, I. and Kawakutsu, M. 1991. Karyological and taxonomic studies of *Dugesia japonica* from the Southwest Islands of Japan II. Hydrobiologia, 227: 157–162.

Tan, T.C.J., Rahman, R., Jaber-Hijazi, F. et al. 2012. Telomere maintenance and telomerase activity are differentially regulated in asexual and sexual worms. Proc Natl Acad Sci USA, 109: 4209–4214.

Tapeworm. 2009. New World Encyclopedia

Tasaka, K., Yokoyama, N., Nodono, H. et al. 2013. Innate sexuality determines the mechanisms of telomere maintenance. Int J Dev Biol, 57: 69–72.

Taskinen, J. 1998. Influence of trematode parasitism on the growth of a bivalve host in the field. Int J Parasitol, 28: 599–602.

Tavares-Dias, M. and Martins, M.L. 2017. An overall estimation of losses caused by diseases in the Brazilian fish farms. J Parasitol Dis, 41: 913–918.

Taylor, M.G., Amin, M.B.A. and Nelson, G.S. 1969. "Parthenogenesis" in *Schistosoma mattheei*. J Helminthol, 43: 197–206.

Tekin-Ozan, S., Kir, I. and Barlas, M. 2008. Helminth parasites of common carp (*Cyprinus carpio* L., 1758) in Beysehir Lake and population dynamics related to month and host size. Turk J Fish Aquat Sci, 8: 201–205.

Tennant, D.H. 1906. A study on the life history of *Bucephalus haimeanus*; a parasite of the oyster. Q J Microsc Sci, 49: 635–690.

278  *Reproduction and Development in Platyhelminthes*

Terasaki, K. 1980. Comparative studies on the karyotypes of *Paragonimus westermani* (s.str.) and *P. pulmonalis*. Jap J Parasitol, 29: 239–243.

Theron, A. and Gerard, C. 1994. Development of accessory sexual organs in *Biomphalaria glabrata* (Planorbidae) in relation to timing of infection of *Schistosoma mansoni*: consequences for energy utilization patterns by the parasite. J Moll Stud, 60: 25–31.

Theunissen, M., Tiedt, L. and Du Preez, L.H. 2014. The morphology and attachment of *Protopolystoma xenopodis* (Monogenea: Polystomatidae) infecting the African clawed frog *Xenopus laevis*. Parasite, 21: 20.

Thomas, A.P. 1883. The life history of liver-fluke (*Fasciola hepatica*). Q J Microsc Sci, 23: 99–133.

Thompson, D.P. and Geary, T.G. 1995. The structure and function of helminth surfaces. In: *Biochemisry and Molecular Biology of Parasites*. (ed) Marr, J.J. and Muller, M., Academic Press, New York, pp 203–232.

Thoney, D.A. 1986a. Post–larval growth of *Microcotyle sebastis* (Platyhelminthes: Monogenea), a gill parasite of the black rockfish. Trans Am Microsc Soc, 105: 170–181.

Thoney, D.A. 1986b. The development and ecology of the oncomiracidium of *Microcotyle sebastis* (Platyhelminthes: Monogenea), a gill parasite of the black rockfish. Trans Am Microsc Soc, 105: 38–50.

Thoney, D.A. and Burreson, E.M. 1988. Lack of specific humoral antibody response in *Leistomus xanthurus* (Pisces: Serranidae) to parasitic copepods and monogeneans. J Parasitol, 74: 191–194.

Thornhill, J.A., Jones, J.T. and Kusel, J.R. 1986. Increased ovoposition and growth in immature *Biomphalaria glabrata* after exposure to *Schistosoma mansoni*. Parasitology, 93: 443–450.

Threadgold, T. 1984. Parasitic platyhelminths. In: *Biology of the Integument*. (eds) Berieter-Hahn, J., Matolsky, A.G. and Richards, K.S., Springer-Verlag, Berlin, 1: 132–191.

Tielens, A. and Horemans, A.M.C. 1992. The facultative anaerobic energy metabolism of *Schistosoma mansoni* sporocysts. Mol Biochem Parasitol, 56: 49–57.

Tielens, A.G., Horemans, A.M., Dunnewijk, R. et al. 1992. The facultative anaerobic energy metabolism of *Schistosoma mansoni* sporocysts. Mol Biochem Parasitol, 56: 49–57.

Tierney, J.F. 1994. Effects of *Schistocephalus solidus* (Cestoda) on the food intake and diet of the three-spined stickleback, *Gasterosteus aculeatus*. J Fish Biol, 44: 731–735.

Tierney, J.F., Huntingford, F.A. and Crompton, D.W.T. 1996. Body condition and reproductive status in sticklebacks exposed to a single wave of *Schistocephalus solidus* infection. J Fish Biol, 49: 483–493.

Timi, J.T. and Mackenzie, K. 2015. Parasites in fisheries and mariculture. Parasitology, 142: 1–4.

Tinsley, R.C. and Owen, R.W. 1975. Studies on the biology of *Protopolystoma xenopodis* (Monogenoidea): the oncomiracidium and life-cycle. Parasitology, 71: 445–463.

Tinsley, R.C. 1978. Oviposition, hatching and the oncomiracidium of *Eupolystoma anterorchis* (Monogenoidea). Parasitology, 77: 121–132.

Tinsley, R.C. 1983. Ovoviviparity in platyhelminth life cycles. Parasitology, 86S: 161–196.

Tinsley, R.C. and Earle, C. 1983. Invasion of vertebrate lungs by the polystomatid monogeneans *Pseudodiplorchis americanus* and *Neodisplorchis scaphipodis*. Parasitology, 86: 501–517.

Tinsley, R.C. and Jackson, J.A. 2002. Host factors limiting monogenean infections: a case study. Int J Parasitol, 32: 353–365.

Tocque, K. and Tinsley, R.C. 1994. The relationship between *Pseudodiplorchis americanus* (Monogenea) density and host resources under controlled environmental conditions. Parasitology, 108: 175–183.

Toledo, A., Cruz, C., Fragoso, G. et al. 1997. *In vitro* culture of *Taenia crassipes* larval cells and cyst regeneration after injection into mice. J Parasitol, 83: 181–193.

Toledo, R., Carpena, I., Espert, A. et al. 2006. A quantitative approach to the experimental transmission success of *Echinostoma friedi* (Trematoda: Echinostomatidae) in rats. J Parasitol, 92: 16–20.

Travares-Dias, M. and Martinus, M.L. 2017. An overall estimation of losses caused by diseases in the Brazilian fish farms. J Parasitol Dis, 41: 913–918.

Trouve, S. and Morand, S. 1998. Evolution of parasites' fecundity. Int J Parasitol, 28: 1817–1819.

Trouve, S., Sasal, P., Jourdane, J. et al. 1998. The evolution of life history traits in parasitic and free–living platyhelminthes: a new perspective. Oecologia, 115: 370–378.

Trubiroha, A., Wuertz, S., Frank, S.N. et al. 2009. Expression of gonadotropin subunits in roach (*Rutilus rutilus*, Cyprinidae) infected with plerocercoids of the tapeworm *Ligula intestinalis* (Cestoda). Int J Parasitol, 39: 1465–1473.

Trubiroha, A., Kroupova, H., Wuertz, S. et al. 2010. Naturally-induced endocrine disruption by the parasite *Ligula intestinalis* (Cestoda) in roach (*Rutilus rutilus*). Gen Comp Endocrinol, 166: 234–240.

Trujillo-Gonzalez, A., Constantinoiu, C.C., Rowe, R. and Hutson, K.S. 2015. Tracking transparent monogenean parasites on fish from infection to maturity. Int J Parasitol, 4: 316–322.

Tsutsumi, N., Mushiake, K., Mori, K. et al. 2002. Effects of temperature on the egg-laying of the monogenean *Neoheterobothrium hirame*. Fish Pathol, 37: 41–43.

Tsutsumi, N., Yoshinaga, T., Kamaishi, T. et al. 2003. Effects of temperature on the development and longevity of the monogenean *Neoheterobothrium hirame* on Japanese flounder *Paralichthys olivaceus*. Fish Pathol, 38: 41–47.

Tubbs, L.A., Poortenaar, C.W., Sewell, M.A. and Diggles, B.K. 2005. Effects of temperature on fecundity *in vitro*, egg hatching and reproductive development of *Benedenia seriolae* and *Zeuxapta seriolae* (Monogenea) parasitic on yellowtail kingfish *Seriola lalandi*. Int J Parasitol, 35: 315–327.

Tyler II, G.A. 2006. In: *A Monograph on the Diphyllidea*. (ed) Ratcliffe, B.C., Bull Univ Nebraska State Mus, p 142.

Tyler, S., Artois, T., Schilling, S. et al. 2018. World list of turbellarian worms: http://www.marinesecies.org/turbellarians on 2018-07-21.

Ubelaker, J.E. and Olsen, O.W. 1972. Life cycle of *Phyllodistomum bufonis* (Digenea: Gorgoderidae) from the boreal toad, *Bufo boreas*. Proc Helminth Soc, Washington, 39: 94–100.

Ubels, J.L., DeJong, R.J., Hoolsema, B. et al. 2018. Impairment of retinal function in yellow perch (*Perca flavescens*) by *Diplostomum baeri* metacercaria. Int J Parasitol Parasite Wildl, 6-7: 171–179.

Ulmer, M.J. and James, H.A. 1976. Studies on the helminth fauna of Iowa II. Cestodes of amphibians. Proc Helminth Soc Washington, 43: 191–200.

Umesono, Y., Tasaki, J., Nishimura, Y. et al. 2013. The molecular logic for planarian regeneration along the antero-posterior axis. Nature, 500: 73–76.

Veit, P., Bilger, B., Schad, V. et al. 1995. Influence of environmental factors on the infectivity of *Echinococcus multilocularis* eggs. Parasitology, 110: 79–86.

Vicoso, B. and Bachtrog, D. 2011. Lack of global dosage compensation in *Schistosoma mansoni*, a female-heterogametic parasite. Genome Biol Evol, 3: 230–235.

Vignoles, P., Rondelaud, D. and Dreyfuss, G. 2003. A first infection of *Galba truncatula* with *Fasciola hepatica* modified the prevalence of a subsequent infection and cercarial production in the F1 generation. Parasitol Res, 91: 349–352.

Vladimirov, V.L. 1971. The immunity of fishes in the case of dactylogyrusis (in Russian). Parasitologiya, 5: 51–58. English translation: Parasitology (Riverdale), 1: 56–58.

Vowinckel, C. 1970. The role of illumination and temperature in the control of sexual reproduction in the planarian *Dugesia tigrina* (Girard). Biol Bull, 138: 77–87.

Vreys, C. and Michiels, N.K. 1998. Sperm trading by volume in a hermaphroditic flatworm with mutual penis intromission. Anim Behav, 56: 777–785.

Wagner, D.E., Wang, I.E. and Reddien, P.W. 2011. Clonogenic neoblasts are pluripotent adult stem cells that underlie planarian regeneration. Science, 13: 811–816.

Wang, Bo., Collins, J.J. and Newmark, P.A. 2013. Functional genomic characterization of neoblast-like stem cells in larval *Schistosoma mansoni*. eLIFE, 2: e00768. DOI: 10.7557/eLife.00768.

Wang, G., Kim, J.-H., Sameshima, M. and Ogawa, K. 1997. Detection of antibodies against the monogenean *Heterobothrium okamotoi* in Tiger Puffer by ELISA. Fish Pathol, 32: 179–180.

Wang, Y., Zayas, R.M., Guo, T. and Newmark, P.A. 2007. *nanos* function is essential for development and regeneration of planarian germ cells. Proc Natl Acad Sci USA, 104: 5901–5906.

Webster, J.P. and Woolhouse, M.E.J. 1999. Cost of resistance: relationship between reduced fertility and increased resistance in a snail-schistosome host-parasite system. Proc R Soc, 266B: 391–396.

280 *Reproduction and Development in Platyhelminthes*

Wedekind, C. 1997. The infectivity, growth and virulence of the cestode *Schistocephalus solidus* in the first intermediate host, the copepod *Macrocyclops albidus*. Parasitology, 115: 317–324.

Wedekind, C., Strahm, D. and Scharer, L. 1998. Evidence for strategic egg production in a hermaphroditic cestode. Parasitology, 117: 373–382.

Weinzierl, R.P., Berthold, K., Beukeboom, L.W. and Michiels, N.K. 1998. Reduced male allocation in the parthenogenetic hermaphrodite *Dugesia polychroa*. Evolution, 52: 109–115.

Weinzierl, R.P., Schmidt, P. and Michiels, N.K. 1999. High fecundity and low fertility in parthenogenetic planarians. Invert Biol, 118: 87–94.

Wenomoser, D. and Reddien, P.W. 2010. Planarian regeneration involves distinct stem cell responses to wounds and tissue absence. Dev Biol, 15: 979–991.

West, A.J. and Roubal, F.R. 1998. Experiments on the longevity, fecundity and migration of *Anoplodiscus cirrusspiralis* (Monogenea) on the marine fish *Pagrus auratus* (Bloch & Schneider) (Sparidae). J Fish Dis, 21: 299–303.

Whitfield, P.J. and Evans, N.A. 1983. Parthenogenesis and asexual multiplication among parasitic platyhelminths. Parasitology, 86: 121–160.

Whitfield, P.J., Anderson, R.M. and Bundy, D.A.P. 1986. Host specific components of the reproductive success of *Transversotrema patialense* (Digenea: Transversotrematidae). Parasitology, 92: 683–698.

Whittington, I.D. 1987. Hatching in two monogenean parasites from the common dogfish (*Schyliorhinus canicula*): The polyopisthocotylean gill parasite, *Hexabothrium appendiculatum* and the microbothriid skin parasite, *Leptocotyle minor*. J Mar Biol Ass UK, 67: 729–756.

Whittington, I.D. 1990. The egg bundles of the monogenean *Dionchus remorae* and their attachment to the gills of the remora, *Echeneis naucrates*. Int J Parasitol, 20: 45–49.

Whittington, I.D. 1997. Reproduction and host-location among the parasitic platyhelminthes. Int J Parasitol, 27: 705–714.

Whittington, I.D. 1998. Diversity down under: monogeneans in the antipodes Australia with a prediction of monogenean biodiversity worldwide. Int J Parasitol, 28: 1481–1493.

Whittington, I.D. 2004. The Capsalidae (Monogenea: Monopisthocotylea): a review of diversity, classification and phylogeny with a note about species complexes. Folia Parasitol, 51: 109–122.

Whittington, I.D. and Kearn, G.C. 2011. Hatching strategies in monogenean (platyhelminth) parasites that facilitate host infection. Integ Comp Biol, 51: 91–99.

Widmer, E.A. and Olsen, O.W. 1967. The life history of *Oochoristica osheroffi* Meggitt, 1934 (Cyclophyllidea: Anoplocephalidae). J Parasitol, 53: 343–349.

Williams, H.H. and McVicar, A. 1968. Sperm transfer in Tetraphyllidea (Platyhelminthes: Cestoda). Nytt Mag Zool, 16: 61–71.

Wilson, R.A. and Denison, J. 1956. Studies on the activity of the miracidium of the common live fluke, *Fasciola hepatica*. Comp Biochem Physiol, 32: 301–313.

Wilson, R.A. and Denison, J. 1980. The parasitic castration and gigantism of *Lymnaea truncatula* infected with larval stages of *Fasciola hepatica*. Z Parasitenkd, 61: 109–119.

Windsor, D.A. 1998. Most of the species on earth are parasites. Int J Parasitol, 28: 1939–1942.

Wolff, E. and Dubois, F. 1947. Sur une method d'irradiation localizsee permettant de metre en evidence la migration des cellules de regeneration chez les planarians. Cr Seanc Soc Biol, 141: 903–906.

Wolff, E. and Dubois, F. 1948. Sur la migration des cellules de regeneration chez les planarians. Rev Suisse Zool, 55: 218–227.

Woo, P.T.K. 1996. Protective immune response of fish to parasitic flagellates. Annu Rev Fish Dis, 6: 121–131.

Wood, C.L., Byers, J.E. and Cottingham, K.L. 2007. Parasites alter community structure. Proc Natl Acad Sci USA, 104: 9335–9339.

Woodhead, A.S. and Calow, P. 1979. Energy–partioning strategies during egg production in semelparous and iteroparous triclads. J Anim Ecol, 48: 491–499.

Yamabata, N., Yoshinaga, T. and Ogawa, K. 2004. Effects of water temperature on egg production and egg viability of the monogenean *Heterobothrium okamotoi* infecting tiger puffer *Takifugu rubripes*. Fish Pathol, 39: 215–217.

## References 281

Yamaguti, S. 1963. *Systema Helminthum Monogenea and Aspidocotylea*. Interscience, London, Vol 4, p 602.

Yoshizawa, Y., Wakabayashi, K. and Shinozawa, T. 1991. Inhibition of planarian regeneration by melatonin. Hydrobiologia, 227: 31–40.

Young, P.C. 1970. The species of Monogenoides recorded from Australian fishes and notes on their zoogeography. Anal Inst Biol Universid Naci Auto Mexico, 4, Serie Zool, 1: 163–176.

Zhokhov, A.E. 1991. The structure of communities of trematodes in populations of the mollusc *Pisidium amnicum*. Parazitologiya, 25: 426–434.

Zietara, M.S. and Lumme, J. 2002. Speciation by host switch and adaptive radiation in a fish parasite genus *Gyrodactylus* (Monogenea,Gyrodactylidae). Evolution, 56: 2445–2458.

# Author Index

## A

Abebe, R., 4
Aboobaker, A.A., 1, 69, 79
Abrous, A., 162, 165
Adams, J.R. see Lester, R.J.G.
Adell, T., 78
Adou, Y.E., 101-103, 105, 109-110,
Aguirre-Macedo, M.L., 11-12, 22-23, 25
Akesson, B., 43, 81, 83-84, 89
Akoll, P., 101, 110
Allen, J.D., 17, 33, 63-64
Alvarado, A.S. see Reddien, P.W.
Alvarez-Presas, M., 48
Alvite, G., 39
Anderson, G.A., 167
Anderson, R.M., 161
Anderson, R.M. see Keymer, A.E. Madhavi, R.
Antonelli, L., 105
Arme, C., 232
Armstrong, J.C., 157
Auld, S.K.J.R., 14, 171
Avery, R.A. see Ghazal, A.M.
Ax, P., 84
Ayalneh, B., 4

## B

Bachtrog, D. see Vicoso, B.
Badets, M., 93, 104-105, 114-116, 118, 120, 124
Bagge, A.M., 111
Baguna, J., 40-42, 66, 68, 71-78, 80, 230, 235
Baguna, J. see Salo, E.
Baker, J.A. see Heins, D.C.
Bakhraibah, A.O., 105
Bakke, T.A., 95, 129-135
Bakke, T.A. see Jansen, P.A. Soleng, A.
Ball, I.R. see Benazzi, M.
Ballabeni, P., 178-179
Barber, I., 217, 224-225

Bardhan, A., 4
Barger, M.A., 164
Barkar, S.C., 141-142, 159
Barnes, R., 6
Bartel, M.H., 212, 215
Barton, G.D. see Pratt, I.
Basch, N. see Basch, P.F.
Basch, P.F., 157
Beck, M.A., 165
Beckerdite, F.W. see Corkum, K.C.
Bednarz, S., 156
Behensky, C., 44
Beisner, B.E., 55-56
Benazzi, M., 65, 71, 81, 84-85, 88-89
Benazzi, M. see Grasso, M.
Benazzi–Lentati, G., 65-66
Benesh, D.P., 29, 220
Berger, J., 37
Best, J.B., 89
Best, J.B. see Morita, M.
Beukeboom, L.W., 65-66
Birstein, V.J., 227
Blackburn, T.M., see Gaston, K.J.
Blackshaw, R.P., 74
Blackwelder, R.E., 166
Blahoua, G.K., 102-103, 105, 108-110, 112
Blasco-Costa, I., 8-9, 11
Blouin, M.S. see Criscione, C.D.
Blythe, M.J., 79
Boag, B., 74
Boeger, W.A., 133
Bondad-Reantaso, M.G., 113, 118-119, 122, 136
Boray, J.C., 4
Boufana, B., 195
Bourns, T.K.R. see McClelland, G.
Brant, S.V., 228
Braun, F., 131
Brazil-Sato, M.C. see Monteiro, C.M.
Bray, R.A. see Gibson, D.I., Littlewood, D.T.J.

284  *Reproduction and Development in Platyhelminthes*

Brehm, K., 40, 44, 222
Bresciani, J. see Buchmann, K.
Brindley, P.J. see Jurberg, A.D.
Brondsted, A., 3, 71-72, 81, 87-88
Bronsted, H.V. see Brondstead, A.
Brooks, D.R. see Moore, J.
Brumpt, E., 157
Bryant, C., 159
Buchmann, K., 94, 104, 106, 134-136
Buchmann, K. see Lindenstrom, T. Sigh, J.
Bullard, S.A. see Jensen, K.
Bullock, T.H., 46
Burreson, E.M. see Thoney, D.A.
Burt, M.D.B. see Sandeman, I.M.
Bush, A.O., 21, 23-25

**C**

Cable, J., 131-132
Cable, R.M. see Khalil, G.M.
Caira, J.N., 2, 6, 9, 195
Calentine, R.L., 213
Calow, P., 58-59, 61, 179
Calow, P. see Jennings, J.B., Woodhead, A.S.
Campbell, R.A., 195
Cannon, L.R.G., 47, 54
Cardona, A., 80
Carter, V., 232
Casanova, J.C. see Spakulova, M.
Cebria, F., 40, 45-46, 78-79
Cecchini, S., 113, 122, 125
Cerda, J.R., 214-215
Chambers, C.B., 113, 123
Chan, B., 93
Chandebois, R., 71, 76, 78
Chapman, A.D., 6
Chapuis, E., 190-192
Charbagi-Barbirou, K., 227, 231
Charnov, E. L., 60
Cheng, L. see Kerr, C.L.
Cheng, Y. see Tang, G.W.
Child, C.M., 71, 78, 87
Ching, H.L., 151
Chintala, M.M., 26, 62
Choisy, M., 171
Chong, T., 229
Christen, M., 219
Christensen, N. O., 47, 65-68, 160, 163, 165
Chu, K.Y., 168
Ciordia-Davila, H., 156, 159
Clark, W.C., 156
Clausen, K.T., 177
Collins, J.J. III., 7-8, 14, 29, 42-44, 82, 230
Comai, L., 66

Combes, C., 170
Cone, D.K., 134
Conn, D.B., 220
Corkum, K.C., 173
Corkum, K.C. see Shoop, W.L.
Cowles, M.W., 79
Crews, A.E., 173-175, 187
Cribb, T.H., 6-7, 9-10, 139-141, 146-148, 151-
    152, 166, 236, 239
Cribb, T.H. see Barkar, S.C., Downie, A.J.,
    Poulin, R.
Criscione, C.D., 154
Cruz-Lacierda, E. R. see Erazo-Pagador, G.
Cunningham, E., 225
Curtis, W.C., 13, 71, 81

**D**

Davison, J., 89
Dawood, I.K., see Chu, K.Y.
De Jong-Brink, M., 144, 180, 248
De Montaudouin, X., 163
De Mulder, K., 40-43, 77, 81
De Robertis, E.M., 78
Deblock, S., 171
Delogu, V., 66
Denison, J. see Wilson, R.A.
Deplazes, P. see Eckert, J.
Deri, P. see Benazzi-Lantati, G.
Desdevises, Y., 135
Dhakal, S., 207
Diaz Briz, L.M., 146
Dick, T.A. see Rosen, R.
Diggles, B.K., 119-120, 122-124
Dmitrieva, E.V., 129, 132
Dobrovolskij, A.A. see Galaktionov, K.V.
Domenici, L., 55-56
Donges, J., 18, 143, 158, 163
Dorchies, P.H., 4
Dorovskikh, G.N., 132
Downie, A.J., 9
Dronen, N.O, Jr., 17, 34, 169-170
Du Preez, L.H. see Kok, D.J.
Dubois, F., 2, 40, 73
Dubois, F. see Wolff, E.
Dubois, S., 181
Dumont, H.J., 48, 50, 55-57
Dzik, J.M., 36-38

**E**

Earle, C. see Tinsley, R.C.
Eckert, J., 214
Egger, B., 6, 69-71, 77, 80-82

## Author Index 285

Ehlers, U., 5
Eisenhoffer, G.T., 79
Erazo-Pagador, G., 15, 119, 122, 124
Ernst, I., 94, 122
Ernst, I. see Chambers, C.B.
Esch, G.W., 6, 18, 138, 142, 144, 153, 155, 159, 169, 172-174
Esch, G.W. see Crews, A.E., Fernandez, J., Keas, B.E.,
Esteves, A. see Alvite, G.
Evans, N.A., 164
Evans, N.A. see Whitfield, P.J.

**F**

Faliex, E., 101
Faltynkova, A., 9, 150
FAO, 3
Fernandez, J., 181, 184-185
Ferrari-Hoeinghaus, A.P., 105
Ferreira, S.M., 186
Fiore, L., 57
Fluchter, J. see Pandian, T.J.
Font, W.F. see Smythe, A.B.
Franklin, C.E. see Pillans, R.D.
Fredensborg, B.L., 174-175, 177-178, 190-192
Fried, B., 145
Fried, B. see Pechenik, J.A.
Friedlander, M.R., 79
Fripp, P.J. see Mason, P.R.

**G**

Galaktionov, K.V., 18, 141-143, 145, 147, 155-156, 168
Galaktionov, K.V. see Gonchar, A.
Galletti, M.C. see Delogu, V.
Gamble, H.R., 232
Gannicott, A.M.T., 119
Gardner, S.L., 6
Gaston, K.J., 21
Geary, T.G. see Thompson, D.P.,
Gerard, C., 180-181, 183
Gerard, C. see Theron, A.
Gercken, J., 135
Gharbi, S.E. see Lambert, A.
Ghazal, A.M., 223-224
Gibson, D.I., 20, 22, 151, 244
Giribet, G. see Laumer, C.E.
Girstmair, J., 42
Gonchar, A., 147, 166
Gonzales-Lanza, C., 93, 101, 105
Gonzalez-Moreno, O., 142
Goodchild, C.G., 189

Gorbushin, A.M., 184-185
Gould, S.J., 237
Grabda-Kazubska, B., 171-172
Gracenea, M. see Gonzalez-Moreno, O.
Granovitch, A.I., 34, 175, 190-192
Grasso, M., 81, 88
Grasso, M. see Benazzi, M.
Gremigni, V., 76
Gremigni, V. see Domenici, L.
Grobler, N.J., 114
Grossman, A.I., 228
Guegan, J.F., 104, 112, 124
Guilford, H.G., 157
Guneydag, S., 214-216
Gupta, S.C., 4
Gurley, K.A., 78
Gururajan, R., 41
Gustafsson, M.K.S., 230
Gutierrez, P.A., 112

**H**

Haas, W., 142, 187
Hafer, N. see Benesh, D.P.
Hager, A., 216
Haight, M., 156
Hall, S.R., 188
Halton, D.W., 34-37, 39, 120
Halton, D.W. see Smyth, J.D.
Handberg-Thorsager, M.,
Hanna, R.E.B. see Henderson, D.J.
Hansen, H., 10, 12, 95
Hanson, E.D., 89
Harkema, R., 214
Harrington, W.C., 12
Harris, A.H., 155
Harris, B.G. see Komuniecki, R.
Harris, P.D., 20, 93, 127, 129-131, 136-137
Harris, P.D. see Cable, J.
Hartmann, M., 81, 87
Hartvigsen, R. see Halvorsen, O.
Hase, S., 81, 88
Hass, W. see Loy, C.
Hass, W., 162
Hassanine, R.M.EL-S. 9, 163, 181
Hauser, J., 81
Heap, M. see Rohde, K.
Heins, D.C., 29, 216, 225
Heitkamp, U., 55-56
Henderson, D.J., 37
Herrmann, H.K., 173, 239, 242
Hirazawa, N., 93, 102
Hoai, T.D., 118-120, 123
Hoberg, E.P., 20, 23

286 *Reproduction and Development in Platyhelminthes*

Hodasi, J.K.M., 178, 184, 188
Hopkins, C.A., see McCaig, M.L.O.
Horemans, A.M.C. see Tielens, A.
Hori, I., 40-41, 230
Horridge, G.A. see Bullock, T.H.
Horsfall, M.W., 214
Hoshi, M., 89
Hug, L.A. see Roger, A.J.
Hugueny, B. see Guegan, J.F.
Huston, D.C., 10, 147, 159
Hutson, K.S. see Hoai, T.D.
Huxham, M., 178, 191
Huyse, T., 157
Hyman, L.H., 2, 5-6, 13, 17-18, 27-29, 47, 52-
        53, 59-60, 82-83, 93-94, 138, 141, 144, 155,
        158, 163, 165-168, 170, 179, 189, 194-201,
        203-204, 208-214, 221, 228, 245

**I**

Iglesias, M., 78
Imbert-Establet, D., 157
Irwin, S.W.B., 145, 159
Ishii, Y., 155
Ishikawa, K., 89
Ishizuka, H., 89
Iyaji, F.O., 105

**J**

Jackson, J.A. see Tinsley, R.C.
Jager, I. see Schjorring, S.
Jagersten, G., 5
James, B.L., 159
James, H.A. see Ulmer, M.J.
Janardanan, K.P., 242
Jansen, P.A., 135
Jarecka, L., 200
Jarrol, E.L.Jr., 17, 34, 223
Jennings, J.B., 15, 29, 33, 47, 54, 58, 67
Jensen, K.T. see Mouritsen, K.M.
Jensen, K., 9, 10, 195
Jensen, K. see Caira, J.N.
Jensen, T., 176
Jhansilakshmibai, K. see Madhavi, R.
Jianying, Z., 94-96
Jocelyn, M.C., 151
Joffe, B.I., 46, 91
Johansson, L.C., 24
Johnson, M.B., 132
Johnson, T.J., 186
Johnstone, T.H., 143
Jokela, J., 177, 179, 181
Jokela, J. see Navarro, B.S.

Jokiel, P.L., 54
Jourdane, J., 43, 144, 156
Jurberg, A.D., 3

**K**

Kanneworff, B. see Christensen, A.M.
Kavana, N.J., 205-206
Kearn, G.C., 15, 27-28, 34, 93-95, 106, 113-114,
        117, 119-121, 123-124, 135
Kearn, G.C. see Whittington, I.D.
Keas, B.E., 169, 187, 189
Keeney, D.B., 141, 154, 158-159
Kenk, R., 81-83, 87, 89
Kennedy, V.S. see Chintala, M.M.
Kerr, C.L., 229
Keymer, A.E., 215-216
Khalil, G.M., 156-157
Khidr, A.A., 102, 110
Knakievicz, T., 227
Kobayashi, C., 78, 88
Kobayashi, K., 15, 55, 84, 86, 89-90
Kocur, R., 168
Koie, M., 141, 147
Kok, D.J., 107-108, 118
Komuniecki, R., 39
Koziol, U., 40, 43, 235
Kreshchenko, N. see Reuter, M.
Kreshchenko, N.D., 230
Krichinskaya, E.B., 81
Krist, A.C., 180, 183
Kuales, G., 229
Kube, J. see Probst, S.,
Kuchta, R., 203
Kuchta, R. see Scholz, T.

**L**

Lackenby, J.A., 125, 227, 231
Ladurner, P., 76
Ladurner P. see Pfister, D.
Lagrue, C., 9, 172-173
Lambert, A., 15, 126
Lange, C.S., 73-75
Laumer, C.E., 5, 7
Lawson, J.R., 164
Lazaro, E.M., 81, 85
Lee, B., 103
Lefebvre, F., 10, 237-239, 242
Lefevre, T., 169
Lender, T.H., 71
Lester, R.J.G., 137
Lester, R.J.G. see Cannon, L.R.G.
Lim, J.H., 4

Lim, L.H.S., 95
Lindenstrom, T., 137
Lindenstrom, T. see Buchmann, K.
Littlewood, D.T.J., 2, 6, 84, 94, 139-140, 147, 151-152, 195-196, 209, 236, 242-245
Littlewood, D.T.J. see Caira, J.N.
Liu, S.Y., 78
Lively, C. see Jokela, J.
Lively, C.M. see Krist, A.C.
Llewellyn, J., 101-102, 107-108, 110
Llewellyn, J. see MacDonald, S.
Lo, C.M., 20, 101, 109, 173, 176
Loker, E.S., 143-144, 163
Loot, G., 201, 208, 211, 214-216
Lorch, S., 227
Love, S., 4
Loverde, P. see Minchella, D.J.
Lowenberger, C.A., 160, 164, 175-177
Loy, C., 187
Lumme, J. see Zietara, M.S.
Luscher, A., 217, 219
Luzin, A.V. see Solomatova, V.P.
Lynch, J.E., 215

**M**

MacArthur, R.H., 33
MacDonald, S., 119, 121
Maciel, P.O., 122
MacInnis, A.J., 162
Mackenzie, K. see Timi, J.T.
Mackiewicz, J.S., 27-28, 194, 199, 201, 204, 206, 211-214, 216
Mackiewicz, J.S. see Orosova, M.
Madanire-Moyo, G.N., 102-103, 108
Madhavi, R., 8, 134, 166-167, 171, 174
Mair, G.R., 230
Makanga, B., 179, 188
Malinowski, P.T., 71
Malmberg, G., 131
Malone, J.B., 4
Manconi, R. see Stocchino, G.A.
Mann, B.Q. see Lee, B.,
Marchiondo, A.A., 194
Marco, A., 229
Marcogliese, D.J., 169-170
Marie-Orleach, L., 60
Martins, M.L. see Tavares-Dias, M.
Martorelli, S.R. see Gutierrez, P.A.
Mason, P.R., 160-161
Mathavan, S., 33
Matrokhina, S.N. see Dorovskikh, G.N.
Matsumoto, M. see Nodono, H.

Mazeri, S., 4
Mazzanti, C., 136
McCaig, M.L.O., 29, 208
McClelland, G., 178, 183, 187-188
McCusker, P., 3-4, 43
McVicar, A. see Williams, H.H.
Mehlhorn, H., 141, 160, 163
Meier, M., 183
Meier-Brook, C. see Meier, M.
Meinkoth, N. A., 210
Melvin, D.M., 206
Menzel, M.Y. see Short, R.B.
Mercer, J.G., 232
Mettrick, D.F. see Berger, J.
Meyer, M.C., 206, 208
Meyrowitsch, D., 164
Miceli, C. see Gremigni, V.
Michiels, N., 60
Michiels, N.K. see Vreys, C.
Milinski, M., 217, 219
Milinski, M. see Christen, M.
Minchella, D.J., 154, 188
Minchella, D.J. see Sorensen, R.E.
Mo, T.A., 106, 109
Mohandas, A., 143
Molina, M.D., 78-79
Mondal, M., 37
Monteiro, C.M., 173, 175
Mooney, A.J., 93, 113, 118-119
Moore, J., 221
Moore, M.M., 101
Moraczewski, J., 81
Morag, L., 212
Morand, S. see Guegan, J.F.
Morand, S. see Poulin, R.
Morand, S. see Trouve, S.
Morgan, T.H., 15, 69, 78
Morita, M., 230
Morley, N.J., 164
Moser, M. see Sakanari, J.A.
Mouritsen, K.M., 177, 181
Mouton, S., 68, 89
Mukhin, V.A., 205
Muller, R., 2-3

**N**

Nagwa, E.A., 4
Nakagawa, H., 71, 89
Nansen, P. see Christensen, N.O.
Navarette-Perea, J., 37
Navarro, B.S., 47, 66
Negovetich, N.J., 175-176

## 288 Reproduction and Development in Platyhelminthes

Newman, L. see Michiels, N.
Newmark, P.A., 41-43, 53, 68, 71, 76, 78-79, 84-85, 92
Newmark, P.A. see Alvarado, A.S. Cebria, F. Collins, J.J. III. Sikes, J.M.
Niewiadomska, K., 142, 170
Nimeth, K.T., 68
Nodono, H., 90
Nokes, D.J. see Scott, M.E.
Nollen, P.M., 217
Norberg, U.M.L. see Johansson, L.C.
Norena, C., 47, 52
Nuttycombe, J.W., 87

## O

Ogawa, K., 93
Oglesby, L.C., 93, 115, 117
Ohashi, H., 43-44
Okano, D., 43
Okino, T., 204
Olsen, O.W. see Ubelaker, J.E., Widmer, E.A.
Opuni, E.K., 231
Orii, H., 41, 43
Owen, R.W. see Tinsley, R.C.
Ozer, A., 105, 107-108, 110
Ozturk, T. see Ozer, A.

## P

Paling, J. E., 103-105
Pallas, P.S., 68
Palmberg, I., 76, 81
Pampoulie, C., 173
Pandian, T.J., 2-4, 8-9, 20, 26, 31-33, 35, 47, 54-55, 57, 65-66, 69-72, 80, 83, 104, 112, 136, 141, 179-180, 199, 205, 208, 232, 245
Pandian, T.J. see Mathavan, S.
Paperna, I., 113
Pappas, P.W., 37
Park, J.K., 6
Parker, G.A., 170
Parris, M.J. see Takahashi, M.K.
Paterson, A.M., 19-20
Petersen, C.P., 78
Payne, A., 4
Pearre, S., 169
Pearson, J.C., 143-145, 156
Pechenik, J.A., 145, 164, 175
Pennycuick, L., 175, 186, 225-226
Peoples, R.C., 151, 175
Perich, M.J. see Rivera, V.R.
Peter, R., 40, 42-43, 81
Peters, A., 60

Peters, L.E. see LaBeau, M.R.
Pfister, D., 80
Phalee, A., 141, 159
Phares, K., 231
Pickering, A.D., 137
Pillans, R.D., 246
Planetary Biodiversity Inventory, 210
Pojmanska, T. see Niewiadomska, K.
Pongratz, N., 66, 85, 227
Poppe, T., 94
Pottinger, T.G. see Pickering, A.D.
Poulin, R., 7-8, 10-12, 15, 18, 20, 25, 27-28, 94-95, 103-104, 130, 138, 171-173, 175, 237-238
Poulin, R. see Blasco-Costa, I., Fredensborg, B.L., Herrmann, H.K., Lagrue, C.
Premvati, G., 142
Presswell, B., 9
Presswell, B. see Poulin, R.
Presswell, R., 9
Price, P.W., 29, 33-34
Probst, S., 180-183
Pulkkinen, K., 224-225

## R

Radha, E., 102-105, 110, 112
Ramm, S.A., 5, 7, 228
Randolph, H., 40, 72
Rantanen, J.T., 174, 176, 179
Rau, M.E. see Lowenberger, C.A.
Rauch, G., 154
Rawlinson, K.A., 5-7, 15, 26, 33, 47-49, 63-64
Read, C.P. see Pappas, P.W.
Reddien, P.W., 75, 79, 82
Reddien, P.W. see Patersen, C.P., Wenomoser, D.
Reddy, A., 144
Redmond, M.D., 169, 175, 182
Reed, P., 6, 93-94
Rees, G., 143, 155, 159
Renwrantz, L. see Gercken, J.
Reuter, M., 25, 40, 45, 84, 230
Reuter, M. see Joffe, B.I.
Reynoldson, T.B., 48, 61-63
Rieger, R., 43
Rietzler, A.C., 56
Rinaldi, G., 156
Rink, J.C., 1, 69, 75, 79
Riutort, M. see Alvarez-Presas, M.
Rivera, V.R., 81
Robertes, L.S. see Schmidt, G.D.
Roberts, T., 4
Roger, A.J., 90

## Author Index  289

Rohde, K., 18, 22, 25, 95, 103-104, 107-112, 121, 125-127, 138
Romero, R., 68, 71-74, 235
Romero, R. see Baguna, J.
Rose, C., 71
Rosen, R., 201, 206-207, 215
Roubal, F.R., 118-121, 125
Roubal, F.R. see West, A.J.
Ruxton, G.D. see Barber, I.
Rychel, A.L., 80

### S

Sakaguchi, Y., 157
Sakamoto, T., 40
Sakanari, J.A., 199, 209-210, 215, 224
Sakurai, T., 81
Saladin, K.S., 160
Salo, E., 41, 68, 76-78
Salo, E. see Handberg-Thorsager, M.
Sandeman, I.M., 212
Sasal, P., 27-28, 127
Sato, K., 71
Schallig, H.D.F.H., 180
Scharer, L., 59, 217-219
Schelkle, B., 132
Schjorring, S., 219
Schmidt, G.D., 6, 95, 195
Schockaert, E.R., 6, 47-51
Scholz, T., 198-201, 204-209, 211-212, 214-215, 224
Schulz, E. see Ax, P.
Schotthoefer, A.M., 165
Scott, M.E., 120, 122, 133
Sergievsky, S.O. see Granovitch, A.I.
Sewell, R.B.S., 142
Sheiman, I.M., 81
Sheiman, I.M. see Kreshchenko, N.D.
Shephered, B.A. see Blackwelder, R.E.
Shibata, N., 41, 81
Shine, R., 21
Shinn, A., 4
Shinn, G.L., 17
Shoop, W.L., 8, 18, 25, 141, 146, 154-155, 158-159, 244
Short, R.B., 157
Shostak, S. see Rose, C.
Sigh, J., 135
Sikes, J.M., 78
Silan, P., 101
Sillman, E.I., 162
Simkova, A., 127, 134-135
Simpson, R.E. see Johnstone, T.H.
Singh, B.P. see Gupta, S.C.

Sinnappah, N.D., 115
Sire, C., 154
Sirgel, W.F., 142, 146
Sluiters, J.F., 162, 165, 169
Smith, H.W., 246
Smyth, J.D., 2, 29, 36, 204-205, 217
Smyth, M.M. see Smythe, J.D.
Smythe, A.B., 171
Sofi, T.A., 227
Sokolova, I.M., 177, 190, 193
Soleng, A., 137
Solomatova, V.P., 94
Sorensen, R.E., 176, 178, 182-184, 186-188
Sousa, W.P., 176, 184-185
Spakulova, M., 227
Stocchino, G.A., 81, 86
Storhas, M., 66
Stunkard, H.W., 2, 168
Sturrock, B.M., 169, 182, 188-189
Sunyer, J.O., 136
Svensson, P.A. see Barber, I.
Swalla, B. see Rychel, A.L.
Swarnakumari, V.G.M. see Madhavi, R.

### T

Takahashi, M.K., 224
Takeda, H., 75
Tamura, S., 90
Tan, T.C.J., 81, 91-92
Tasaka, K., 90-91
Taskinen, J., 182
Taylor, M.G., 157
Tekaya, S. see Carbagi-Barbirou, K.
Tekin-Ozan, S., 102
Tennant, D.H., 156
Terasaki, K., 157
Theron, A., 183
Theron, A. see Gerard, C.
Theron, A. see Jourdane, J.
Theunissen, M., 124
Thomas, A.P., 141
Thompson, D.P., 37
Thoney, D.A., 15, 120, 124, 136
Thornhill, J.A., 183, 189
Threadgold, T., 36
Tielens, A.G., 179
Tierney, J.F., 224, 226
Timi, J.T., 11
Tinsley, M.C. see Auld, S.K.J.R.
Tinsley, R.C., 93, 114-119, 121-124
Tinsley, R.C. see Gannicott, A.M.
Tinsley, R.C. see Harris, P.D.
Tinsley, R.C. see Tocque, K.

## 290  *Reproduction and Development in Platyhelminthes*

Tocque, K., 93, 116
Toledo, A., 44, 222
Toledo, R., 145, 159
Townsley, S.J. see Jokiel, P.L.
Travares-Dias, M., 4
Trouve, S., 6, 27-31, 33-34
Trubiroha, A., 232
Trujillo-Gonzalez, A., 123-124
Tsutsumi, N., 119-120, 125
Tubbs, L.A., 118-122, 125
Tyler II, G.A., 210
Tyler, S., 6

### U

Ubelaker, J.E., 166
Ubels, J.L., 141, 170, 186
Ulmer, M.J., 204
Umesono, Y., 78

### V

Valtonen, E.T. see Pulkkinen, K.
Valtonen, E.T. see Bagge, A.M.
Veit, P., 206
Verneau, O. see Badets, M.
Vicoso, B., 228
Vignoles, P., 191
Vik, R. see Meyer, M.C.
Vladimirov, V.L., 136
Vreys, C., 52, 60

### W

Wagner, D.E., 68
Wang, Bo., 40, 43-44, 156
Wang, G., 136

Waters, A.J. see Nuttycombe, J.W.
Webster, J.P., 173, 175
Wedekind, C., 26, 217-219
Wedekind, C. see Luscher, A., Scharer, L.
Weinzierl, R.P., 63, 65-66
Wenomoser, D., 76
West, A.J., 119-120
Whitfield, P.J., 27, 34, 143, 158, 169, 220-221
Whittington, I.D., 3, 6, 15, 29, 95, 114-115, 121, 123, 126, 131
Widmer, E.A., 206
Williams, H.H., 159, 216
Williams, J.P.G. see Bryant, C.
Wilson, E.O. see MacArthur, R.H.
Wilson, R.A. see Lawson, J.R.
Wilson, R.A., 160, 164, 178, 181-183, 187-188
Windsor, D.A., 7
Wolff, E., 40
Woo, P.T.K., 135
Wood, C.L., 177
Woodhead, A.S., 57-58
Woodhouse, M.E.J. see Webster, J.P.
Wu, B. see Chan, B.,

### Y

Yamabata, N., 119
Yamaguti, S., 15, 19, 112
Yamasu, T., see Ishikawa, K.
Yoshizawa, Y., 230
Young, P.C., 95, 247

### Z

Zbikowski, J. see Zbikowska, E.
Zhokhov, A.E., 174
Zietara, M.S., 135

# Species Index

## A

*Abarenicola affinis*, 175
*Abothrium gadi*, 6
*Abudefduf*, 18
*Acanthobothrium coronatum*, 201-202
*Acanthocotyle*, 13-14
*A. greeni*, 119
*A. labianchi*, 121
*Acanthocyclops vernalis*, 211
*Acanthodiaptomus*, 211
*Acantholochus unisagittatus*, 105
*Acanthopagrus australis*, 119
*Acanthurus*, 15
*Accacoelium*, 112
*Acipenser*, 201
*Actinocleidus fergusoni*, 110
*Adenopea canata*, 81
*Aedes aegypti*, 163, 176
*Aequorea* spp, 146
*Alaria*, 141, 144-145, 159, 240
*A. marcianae*, 155, 158
*Alaurina*, 48, 81
*A. aethiopica*, 81
*Alburnus alburnus*, 211, 214-215
*Aledo atthis*, 217
*Allobilharzia visceralis*, 228
*Allocreadium fasciatusi*, 8
*Allodiscocotyle* sp, 110-111
*Alloglossidium*, 171, 173, 241
*A. corti*, 171
*A. greeri*, 171
*A. hamrumi*, 171
*A. hirudicola*, 171
*A. macrobdellensis*, 171, 173, 242
*A. renale*, 171
*A. schmidti*, 171
*A. turnbulli*, 171
*Alloglossoides caridicola*, 171
*A. dolandi*, 171
*Allolobophora* sp, 221
*Allomurraytrema robustum*, 118-119, 121

*Alopex lagopus*, 203
*Alutera*, 18
*Alveococcus*, 212
*Amanses*, 15
*Amnicola limosa*, 168
*A. travancorica*, 8
*Ampbitbecium* sp, 105
*Ampelisca macrocephala*, 15, 67-68
*Amphiliana foliacea*, 6, 200-201
*Amphiscolops langerhansi*, 70-71, 81, 89
*Anarhichas lupus*, 35, 38
*Anas boschas*, 208
*Ancylostoma caninum*, 155
*Anguilla* spp, 198
*A. anguilla*, 135-136
*A. dieffenbachia*, 173
*Anodonta piscinalis*, 179, 181-182, 184
*Anoplodiscus cirrusspiralis*, 118-120, 126
*Anoplodium hymanae*, 17
*Antalis entails*, 141
*Anteropora*, 195
*Anthocotyle merluccii*, 107
*Apatemom gracilis*, 27
*Aphalloides codomicola*, 173, 178
*Aplocheilus melastigma*, 8
*Aplodinotus grunniens*, 103
*Aponurus* sp, 173, 176
*Archigetes*, 201, 212
*A. iowensis*, 213, 238
*A. limnodrili*, 213, 238
*A. sieboldi*, 213, 238
*Archilopsis*, 71
*Arctodiaptomus*, 211
*Ardea cinerea*, 8
*A. conera*, 217
*Argeia pugattensis*, 32, 34
*Ariostralis nebulosa*, 142
*Arothron hispidus*, 9
*Artacana proboscidae*, 141
*Arthurdendyus triangulatus*, 230
*Artioposthia triangulata*, 45, 74

## 292 *Reproduction and Development in Platyhelminthes*

*Asellus,* 57
*Asterina miniata,* 65
*Astyanax altiparanae,* 104-105
*Ataenius cognatus,* 221
*Athya innocuous,* 163
*Atractylytocestus huronensis,* 204, 228
*Australapatemon* sp, 151
*Austrobilharzia terrigalensis,* 228
*A. vargilandis,* 175
*Austropeplea tomentosa,* 150
*Axine belone,* 107
*Azygia bucci,* 150
*A. longa,* 162, 167-168

### B

*Baicalarctia gulo,* 6
*Barbus graellis,* 15
*B. guiraonis,* 15
*B. haasi,* 15
*B. issenensis,* 104
*B. nasus,* 104
*B. sacratus,* 104
*Bdellocephala brunnea,* 88, 89
*B. punctata,* 59, 68, 71
*Bdelloura candida,* 13, 47, 69, 71
*Belone beloni,* 107
*Benedenia,* 126
*Benedenia* sp, 101
*B. hawaiiensis,* 15, 126
*B. seriolae,* 113, 118-123, 125-126, 226, 230
*Biacetabulum,* 212
*Bilharzia,* 141
*Bilharziella polonica,* 157, 228
*Biomphalaria glabrata,* 142, 144, 160-163, 169,
  175-176, 179-181, 183, 186, 188-189
*B. pfeifferi,* 182-184, 188-189
*Bipalium kewense,* 2, 27
*Bithynia tentaculata,* 163, 168
*Bivitellobilharzia nairi,* 228
*Boeckella,* 211
*Bothridium pythonis,* 13-14
*Bothriocephalus,* 212, 238
*B. claviceps,* 198-199, 204, 206
*B. rarus,* 17, 34, 223
*Bothriomolus,* 71
*Bothrioplana,* 13-14, 71
*Botrhionomus,* 212
*Brachthemis contaminata,* 33
*Brachydenio rerio,* 176, 188-189
*Brachylaima ilobregatensis,* 142
*Brachyphallus crenatus,* 170
*Brachyurus,* 212

*Branchotenthes octohamatus,* 121
*Bufo bufo,* 165
*B. marinus,* 163
*B. pardalis,* 114, 116-117
*B. regularis,* 117
*Bulinus natalensis,* 183
*Bunocotyle progenetica,* 182, 186
*Bunodera luciopercae,* 170, 176, 179
*Burnellus trichofurcatus,* 144

### C

*Caligus* sp, 111
*C. epidemicus,* 114
*Caranx kalla,* 102-105, 110-112
*C. melampygus,* 110-111
*Carassius auratus gibelio,* 65
*C. a. langsdorfi,* 65
*Carcharhinus leucas,* 246
*Caryophyllaeus,* 200, 212
*Catatropis,* 183
*Catenula,* 71, 81
*Cemocotylella,* 110
*Cemocotylella* sp, 111
*Centropomus anthenocolops,* 105
*C. ergensis,* 105
*C. undeclimatis,* 105
*C. vexus,* 105
*Cephalopholis argus,* 101, 109
*Cerastoderma edule,* 163, 181
*Cercaria loossi,* 175
*C. margaretensis,* 145
*Cerithidea californica,* 71, 174, 176, 183-185
*Cestoplana,* 71
*Chaetodon,* 18
*Chaetogaster limnaei,* 163
*Cheirodon* sp, 169, 176, 188-189
*Chorinemus tol,* 110-111
*Chromis,* 18
*Cichlidogyrus,* 103, 108
*Cichlidogyrus* spp, 101, 102, 107-110
*C. aegypticus,* 103
*C. digitatus,* 103, 105, 109, 112
*C. kouassii,* 101
*C. vexus,* 101, 103, 109, 112
*Cittotaenia,* 196
*Cladorchis,* 13-14
*Clarias gariepinus,* 101, 110
*Clavella devastatrix,* 111
*Clinostomum,* 172, 174
*C. funduloides,* 172
*Clio pyrimidata,* 71
*Coelogynopora,* 71

## Species Index    293

Coitocaecum parvum, 172-173, 175
Compeloma, 65
Concinnocotyle, 115
C. australensis, 93, 115
Convoluta convoluta, 47
Convolutriloba longifissura, 43, 70-71, 81, 83-84, 89, 236
C. okinawa, 81
C. retrogemma, 81, 89
C. clupeformis, 202
Coregonus lavaretus, 133, 225
Corethara sp, 163
Corodoras palaetus, 133
Corophium spp, 176
C. arenarium, 176
C. volutator, 176
Corydoras ehrhardti, 132-133
C. paleatus, 132-133
C. schwartzi, 133
Coryphaenoides mediterraneus, 215-216
Cotugni, 196
Cotylurus flabelliformis, 167-168
Crangon franciscorium, 34
Crassostrea cucullata, 9
C. gigas, 181
C. madrasensis, 181
C. virginica, 181
Crenobia alpina, 227
Crepidostomum cooperi, 141
Cryptocelis alba, 70-71
Cryptocotyle sp, 182, 184-185
C. arenarium, 176
C. lingua, 27, 168, 175, 177-179
Ctenocephalus canis, 204
Cucumaria frondosa, 57
Culex pipens, 163
Cura foremanii, 81
Cyathocephalus, 212
Cyathura carinata, 186
Cyclocoelum microstomum, 142
C. mutabile, 150
Cyclocotyla chrysophryi, 107
Cyclops, 8, 19, 203, 211-213
Cyclops sp, 163, 211
C. abyssorum, 195, 211
C. agilis, 211
C. biscupidatus thomasi, 201-202, 205-208, 215
C. furcifer, 211
C. kolensis, 211
C. lacustis, 211
C. scutifer, 211
C. strenuus, 201, 205-207, 210-211, 215
C. vicinus, 211

Cylindrotaenia, 200, 213
C. americana, 205
Cyprinus carpio, 102

## D

Dactylogyrus, 111-113, 135, 139
Dactylogyrus spp, 109
D. cornu, 105, 107-108, 110
D. crucifer, 111-112
D. extensus, 113
D. homoion, 111, 121
D. microcanthus, 112
D. nanus, 111-112
D. onchoratus, 113
D. skrjabini, 103, 109
D. solidus, 109
D. suecicus, 111-112
D. vastator, 27, 113, 136
Diaptomus ursi, 203
Dalyellia, 52
Daphnia pulex, 163
Dascyllus, 15
D. aruanus, 101
Dasyatis kuhli, 245
D. uarnak, 246
Dasyprocta punctata, 221
Demidospermus spp, 110
Dendritobilharzia pulverulenta, 228
Dendrocoelopsis ezensis, 69, 71
D. lactea, 71
Dendrocoelum dentricticum, 170
D. lacteum, 57-59, 61-62, 68-69, 71, 78, 163
Dentalium spp, 141
D. entale, 141
Derogenes varicus, 170
Devainea proglottina, 13
Diadema antillarum, 65
Diaptomus, 203
D. castor, 211
D. stemmacephalum, 203
Dibothriorhynchus, 196
Dicentrarchus labrax, 93, 101, 105, 113, 125
Diclidophora denticulata, 127
D. derjavani, 134
D. merlangi, 34-36, 38-39, 107
Diclybothrium armatum, 103
Dicrocoelum dentrictum, 25, 141, 165
Didymogaster sylvatica, 221
Dikerogammarus, 201
Dilepis caninum, 204
Dioecocestus asper, 228
Dionchus spp, 114

294 *Reproduction and Development in Platyhelminthes*

*D. remorae*, 114
*Diopatra neapolitana*, 174-175
*Diorchis*, 200
Diorchis sp, 199
*D. myrocae*, 199
*Diphyllobothrium*, 19, 196, 211
*Diphyllobothrium* spp, 204
*D. dentrictum*, 25, 203
*D. latum*, 27, 194, 200-201, 203
*D. sebago*, 206
*Diplectanum aequans*, 93, 101, 105, 113, 121-122, 125-126
*D. laubieri*, 101, 105
*Diplogonoporus (Diphyllobothrium)*, 196
*D. balaenopterae*, 203
*D. grandis*, 203
*Diploproctodaeum arothroni*, 9, 162
*Diplorchis*, 116
*Diplostomulum*,
*D. scheuringi*, 174
*Diplostomum*, 2, 141, 144, 154, 159
*D. baeri*, 170, 186
*D. flexicaudum*, 168
*D. gasterostei*, 175, 186
*D. phoxini*, 178-179
*D. pseudopathaceum*, 154, 167-168
*D. spathaceum*, 163-164, 168
*Diplozoon paradoxum*, 107
*D. homoion gracile*, 121
*Dipylidium*, 197
*D. caninum*, 208
*Discocotyle sagittata*, 103-105, 118-119
*Drepanotrema surinamensis*, 163
*Dugesia* spp, 85, 87
*D. aethiopica*, 81, 85, 86
*D. afromontana*, 81, 85, 87
*D. benazzii*, 55, 65-66, 81, 87, 89
*D. dendriticum*, 230
*D. derotocephala*, 27, 82-83, 86, 88, 89
*D. entrusca*, 81, 87
*D. fissipara*, 81, 84
*D. gonocephala*, 52, 60, 71, 88, 236
*D. hepta*, 85, 86
*D. japonica*, 40, 42, 43, 71, 75, 81, 85, 236
*D. japonica japonica*, 81
*D. lugubris*, 48, 59, 61-62, 65, 71, 73-75, 81, 236
*D. maghrebiana*, 81, 85, 87
*D. mediterranea*, 88
*D. paramensis*, 81, 84
*D. polychroa*, 60
*D. ryukyuensis*, 17, 47, 55, 71, 81, 85, 88-92
*D. sanchezi*, 81, 89

*D. sicula*, 71, 81, 236
*D. subtentaculata*, 71, 76, 81
*D. tahitiensis*, 81
*D. temnocephala*, 2
*D. tigrina*, 2, 6, 40, 71-72, 81, 88, 89, 163, 230
*Dunaliella tertiolecta*, 64

## E

*Echeneis naucrates*, 115
*Echinobothrium*, 13
*E. affine*, 221
*E. benedeni*, 6
*Echinococcus*, 27, 33, 194, 206, 212,
*E. granulosus*, 2, 203, 205, 214-215, 217, 221-222, 232
*E. multilocularis*, 43-44, 206, 222, 235
*E. oligarthrus*, 221
*E. vogeli*, 221
*Echinolittorina austrotrochoides*, 10, 147
*Echinoparyphium*, 183
*E. aconiatum*, 143
*E. recurvatum*, 168
*Echinostoma*, 144, 156, 159, 165, 240
*E. caproni*, 144, 164-165
*E. ilocanum*, 160
*E. liei*, 163-164
*E. revolutum*, 141, 178, 180, 183, 187, 189
*E. trivolvis*, 144-145, 164-165, 184
*Echinostomum*, 184
*E. margarinatum*, 144
*Ectocotyle paguri*, 2
*Elimia symmetrica*, 172
*Enchytraeus variatus*, 32
*Enoplocotyle kidakoi*, 121
*Entobdella hippoglossi*, 27, 121
*E. soleae*, 15, 34, 93, 106, 117-121, 124, 135
*Epinephelus coioides*, 119
*Epischura baicalensis*, 211
*Erpobdella octoculata*, 151
*Esox lucius*, 202-203, 211
*Eubothrium*, 199
*Eucyclops serrulatus*, 211
*Eudiaptomus*, 211
*E. gracilis*, 202, 207, 211, 215
*E. graciloides*, 211
*E. zachariasi*, 211
*Euphausia similis*, 170
*Euplanaria tigrina*, 81, 89
*Eupolystoma*, 116
*E. alluaudi*, 117
*E. anterorchis*, 114, 116, 118, 122, 123
*Eurytemora*, 211

## Species Index    295

## F

Fasciola, 38
Fasciola sp, 38, 146
F. gigantica, 141, 143, 159, 162-163
F. hepatica, 13-14, 27, 33, 35, 38-39, 43, 138, 141,
    143, 145, 150, 156-157, 160-163, 165, 167-
    168, 178, 183-184, 187-193
Fasciolopsis buski, 180
Felis silvestris, 221
Fellodistomum fellis, 35, 36, 38-39
Femcampia erythrocephala, 67
Fonticola morgani, 81

## G

Gadus luscus, 101
Gadus merlangus, 101, 107, 111, 127
Gadus minutus, 127
Galaxias sp, 173
Galba truncatula, 190-191, 193
Gallus gallus, 221
Gambusia affinis, 209
Gammarus, 201, 212-213
Gasterosteus aculeatus, 8, 105, 131, 135, 154,
    175, 186, 215-217, 220, 224-225
Gasterosteus turnbulli, 132, 171
Gastrocotyle indica, 102-104, 109, 111-112
G. trachuri, 101-103, 110
Geocentrophora sphyrocephala, 71
Gigantobilharzia huronensis, 228
G. ocotyla, 27
Girardia tigrina, 45, 74, 81, 85
Glanduloderma myzostomatis, 67
Glaridacris catostomi, 204
G. luruei, 204
Glomeris, 221
Gnathonemus petersii, 201
Gobiomorphus, 172-173
G. cotidianus, 172-173
Gorgocephalus yaaji, 10, 147, 159
Gotocotyle bivaginalis, 124
G. sacunda, 124
Graffilla buccinicola, 54
Gulo gulo, 221
Gymnophalus choledochus, 171, 175
Gyrocotyle, 2
G. fimbriata, 215
G. nybelini, 6
G. rugosa, 201, 215
Gyrodactylus, 44-45, 80, 95, 129-130, 139
Gyrodactylus sp, 112
G. alvica, 129
G. anisopharynx, 132-133

G. arcuatus, 105
G. bullatarudis, 27, 118, 120-121, 122, 132-134
G. colemansis, 134
G. derjavini, 104, 136
G. ehrhardti, 132
G. gasterostei, 130-131
G. indica, 102-103, 105, 109, 130
G. pleuronecti, 130
G. salaris, 95, 106, 129, 132-137
G. schwarzi, 132
G. sphinx, 132
G. stellatus, 101
G. thymalli, 95
G. turnbulli, 132

## H

Haematoloechus coloradensis, 17, 34, 169-170, 224
Haementeria ghilianii, 32
Haliotrema spariensis, 118-119
H. eccentricus, 157
H. occidualis, 169, 174-176, 178, 182-184, 187,
    189
Halitrema sp, 101
Haplobothrium globuliforme, 6, 201, 209
Haplometra cylindracea, 172
Haploops sp, 67-68
Hatschekia sp, 109
Heliotrema, 139
Helisoma anceps, 138, 169, 173-176, 178, 182-
    185, 187, 189
Helix aspersa, 142
Helobdella stagnalis, 151
Helocentrus, 18
Hemiurus sp, 169-170
Heronimus chelydrae, 144, 146
Heterapta sp, 111
Heteraxine heterocerca, 119
Heteraxinoides xanthophilis, 136
Heterobilharzia americana, 157, 228
Heterobothrium okamotoi, 118-119, 136
Heterocope appendiculata, 211
Heteromastus filiformis, 146, 175
Heterophyes, 144, 159
Hexabothrium appendiculatum, 114
Himasthala continua, 177
H. elongata, 177
H. quissetensis, 163, 168
Hippoglossus hippoglossus, 27
Hirudicolotrema richardsoni, 171
Hofstenia giselae, 70-71
Hoploops tubicola, 15, 67
Hydatigera, 212
Hydrobia, 65

296  *Reproduction and Development in Platyhelminthes*

*Hydrobia* spp, 182
*H. ulvae*, 177, 181-182, 184-185
*H. ventrosa*, 57, 181-185
*Hydroides dianthus*, 174-175
*Hydrolagus collici*, 215
*Hyla meridionalis*, 104-105, 114, 115
*H. regilla*, 186
*Hymenolepis*, 196
*H. diminuta*, 29, 36-39, 206-207, 214-217, 232
*H. erinacei*, 208
*H. exigua*, 200, 204
*H. furcata*, 212
*H. microstoma*, 37
*H. nana*, 37, 203-204, 208, 214, 222-223
*Hypophthalmichthys*, 103

**I**

*Ichthyophaga* sp, 47, 54
*Ichthyophonus*, 105
*Ichthyoxenus fushanensis*, 32, 34
*Ilyanassa obsoleta*, 175-176, 184
*Imogine zebra*, 6, 63
*Inemicasifer*, 200
*Isancistrum*, 129
*Isodiametra pulchra*, 40-43, 77, 81
*Isoglaridacris bulbocirrus*, 204
*Isthmiophora melis*, 143
*Itaspiella*, 71

**J**

*Jassa fulcata*, 32, 34

**K**

*Kahawaia truttae*, 107-108
*Kassina senengalensis*, 107-108
*Kenkia rhynchida*, 13-14
*Khawia sinensis*, 197, 199, 205-207, 214-215
*Kronborgia amphipodicola*, 15, 47, 66-68
*Kyphosus cineraseans*, 10

**L**

*Labeo coubie*, 110
*Lacistorhynchus dollfusi*, 209
*L. tenuis*, 199, 215, 224,
*Lamellodiscus acanthopagri*, 118
*L. major*, 119
*Larus argentatus*, 208
*L. pipixum*, 203
*Lates calcarifer*, 119, 123
*Latridopsis ciliaris*, 108
*Lebistes reticulatus*, 163

*Lecithophyllum botryophorum*, 141
*Leiostomus xanthurus*, 136
*Lenkoranoides*, 15
*Lepidopus caudatus*, 215-216
*Leptocotyle minor*, 95, 114, 123, 245
*Leptoplana alcioni*, 70-71
*L. littoralis*, 70-71
*L. saxicola*, 71
*L. tremallaris*, 70
*L. velutinus*, 70-71
*Leucochloridium*, 166, 170, 241
*Levinseniella*, 186
*Ligula*, 194, 212
*L. intestinalis*, 6, 201, 203, 205, 208, 210-211,
    214-215, 231, 233
*Limnodrilus*, 201, 214-215, 238
*Limulus polyphemus*, 47
*Lithobium aenigmaticum*, 2
*Litoria gracilenta*, 116
*Littorina* spp, 147
*L. littorea*, 27, 168, 175-181, 183-184
*L. saxatilis*, 175, 177, 180, 190-193
*Liza haematocheila*, 211
*Lota lota*, 211
*Lumbriculus variegatus*, 72
*Lymnaea acuminata*, 189
*L. catascopium*, 143-144, 166
*L. cubensis*, 150
*L. elodes*, 177-178, 180, 182-184, 187-189
*L. natalensis*, 143, 183
*L. palustris*, 162-163
*L. peregra*, 163, 168, 178-179
*L. stagnalis*, 9, 143, 150, 154, 163, 165, 168-169,
    178-179, 181-183, 187-188
*L. stagnicola*, 180
*L. truncatula*, 138, 143, 150, 165, 168, 178, 181-
    184, 187-189
*L. viatrix*, 150
*Lynx rufus*, 204

**M**

*Macrobdella ditetra*, 173
*Macrobilharzia macrobilharzia*, 228
*Macrocyclops*, 8
*M. albidus*, 211, 217
*M. ater*, 223
*M. distructus*, 8
*Macrophthalmus hirtipes*, 154
*Macrostomum*, 45, 76
*Macrostomum* spp, 82
*M. grande*, 71
*M. hystricinum*, 42-43, 70-71
*M. lignano*, 68, 77, 89

## Species Index 297

M. lineare, 46, 71, 76
M. marinum, 70-71
M. pusillum, 70-71
M. tuba, 71
Margarites helicinus, 145
Maritigrella crozieri, 64
Maritrema novaezealandensis, 141, 154, 158, 177-179, 190-193
M. subdolum, 186
Mediavagina latridis, 107-108
Melanoides, 65
M. tuberculata, 142, 168
Merlangius merlangus, 34
Mesocestoides, 198, 200, 204-205
M. corti, 44, 220, 232
Mesocyclops leuckarti, 8, 210-211, 215
M. oithonoides, 170, 211
M. viridis, 211
Mesostaphanus haliasturis, 141-142, 159
Mesostoma spp, 55
M. ehrenbergii, 52-53, 55-57
M. lingua, 56, 70-71
M. nigrirostrum, 56
M. productum, 56, 71
M. rhynchotum, 56
Mesostomium sp, 48
M. arctica, 48
Microcotyle sebastis, 15, 120, 124
M. spinicirrus, 103
Microcotyloides, 109
Microcotyloides sp, 101, 109-110
Microcyclops varicans, 8
Micromesistus poutassou, 126
Microphallus, 171, 177, 180-181, 183, 239, 241
M. claviformis, 176, 182, 186
M. pirum, 182
M. pseudopygmaeus, 183
Microstomum, 2, 48, 86
Microstomum sp, 42-43, 59
M. lignano, 59, 70
M. linae, 48
M. lineare, 46, 71, 76
M. marinum, 70-71
M. thermale, 48
Moniezia, 196-197
M. expansa, 45, 230
Monobothrioides congolensis, 110
Monocelis sp, 42
M. fusca, 71
M. lineata, 71
Monoecocestus sigmodontis, 206
Mulloidichthys, 15
Multiceps, 212

M. coenuri, 216
Myxolepis collaris, 210-211
Myzophyllobothrium, 13-14

## N

Naso, 18
Nassarius reticulatus, 163
Necturus maculosus, 115
Nematoparataenia, 6
Nematotaenia, 204
Nemertoderma bathycola, 6
Neobenedenia sp, 27, 118-121, 123-124
N. girellae, 44, 93, 102-103, 113, 117, 122, 136
N. melleni, 118, 126
Neoceratodus forsteri, 115
Neodiplorchis, 116
Neogobius melanostomus, 214
Neoheterobothrium hirame, 118-120, 125-126
Neopolystoma, 116
N. palpebrae, 115
Nippotaenia, 6
Nitzschia sturionis, 114
Notocotyle sp, 184
Notocotylus attenuatus, 168
Notophthalmus viridescens, 223
Notoplana australis, 64
N. humilis, 43, 70-71
Notozotbecium sp, 105

## O

Octobothrium merlangi, 111
Octodactylus minor, 126
Oculotrema hippopotami, 115
Oegyrodactylus farlowellae, 131
Oekiocolax plagiostomorum, 47
Oncorhynchus gorbuscha, 211
O. keta, 211
O. masour, 211
O. mykiss, 119, 133, 135-136
O. nerka, 211
Oncosphere lycophore, 198-200
Onoba aculeus, 183
Onthophagus ater, 212
Onthophagus viduus, 212
Oochoristica osheroffi, 206
Ophidaster granifer, 65
Ophryotrocha adherens, 32
O. puerilis puerilis, 32
Opisthoglyphe, 171, 240
O. ranae, 172
Orchestia, 204
Oreochromis mossambicus, 101-103, 105, 108

298    *Reproduction and Development in Platyhelminthes*

*O. niloticus*, 108, 110
*Ornithobilharzia canaliculata*, 228
*O. turkestanicum*, 228
*Orthotrotrema monostomum*, 171
*Ostrea lutaria*, 180
*Otomesostoma auditivum*, 70-71

**P**

*Pagellus centrodontus*, 107
*Paguras auratus*, 119
*Palaemon serratus*, 67
*Palaeorchis crassus*, 174
*Paracalliope fluviatilis*, 172-173, 175
*Paragonimus*, 38
*P. westermani*, 157
*Paralichthys olivaceus*, 119-120, 125-126, 136
*Paramecynostomum diversicolor*, 71
*Paramphistomum*, 144, 159
*P. daubneyi*, 165
*Parapolystoma*, 116
*P. bulliense*, 116
*Paratimonia gobii*, 169
*Paravortex scrubiculariae*, 54
*Paricterotaenia paradoxa*, 221
*Paronatrema* sp, 169-170
*Paronia*, 196-197
*Parorchis acanthus*, 143, 155, 156, 159
*Parupeneus*, 15
*Parvatrema margaritense*, 18, 141, 145, 147, 158, 247
*Pelates quadrilineatus*, 54
*Perca flavescens*, 170, 186
*P. fluviatilis*, 135, 172-173, 175, 203, 211
*Phaenocora typhlops*, 27
*Phagocata gracilis*, 71
*P. velata*, 69, 81, 85, 86
*P. vitta*, 71, 81, 85, 86, 227
*P. vivida*, 81, 88
*Phalacrocorax carbo*, 217
*Phanerothecium caballeroi*, 131
*Pharyngostomoids procyonis*, 155
*Pheidole* sp, 212, 214-215
*Pheidole* spp, 214
*P. bicarinata*, 215
*P. cernae*, 210
*P. dentate*, 212
*P. fervida*, 212
*P. vinelandica*, 204, 212
*Philophthalmus megalurus*, 156
*Phoxinus phoxinus*, 129, 135
*Phyllobothrium dohrnii*, 6, 13
*Phyllodistomum*, 139, 240

*P. elongatum*, 174
*Phyllomedus* sp, 163
*P. marmorata*, 163
*P. virgata*, 34, 169-170, 224
*Pimelodus maculatus*, 110
*Pinctata radiata*, 181
*Pisidium amnicum*, 174, 176, 179
*Plagioporus sinitsini*, 145, 172
*Plagiorchis elegans*, 163-165, 175-177
*Plagiostomum*, 47
*P. girardi*, 71
*Planaria* sp, 163
*P. derotocephala*, 27, 71, 82-83, 86-87, 89
*P. maculata*, 42, 71, 81
*P. torva*, 59, 71
*Planocera californica*, 71
*P. reticulata*, 64, 132
*Planorbis planorbis*, 9, 150
*Platichthys flesus*, 135
*Platycephalus fuscus*, 54
*Platymonas convoluta*, 47
*Pleioplana atomata*, 26, 27, 33, 63
*Pleuronectes vetulus*, 101
*Pneumobites*, 141
*Podiceps cristatus*, 203, 211
*Podocotyle stenotometra*, 169
*Poecilancistrium caryophyllum*, 201, 203
*Poecilia formosa*, 65
*P. recitulata*, 133
*Pollachius virens*, 127
*Polycelis* spp, 61
*P. auriculata*, 69, 71
*P. cornuta*, 81
*P. felina*, 227
*P. nigra*, 61-62, 71, 88
*P. tenuis*, 48, 57-59, 61-62, 71, 91
*Polychoerus caudatus*, 71
*Polycotylus validus*, 13
*Polylabroides multispinosus*, 118-120, 122-124
*Polystoma australis*, 107-108, 118, 124, 163
*P. gallieni*, 104-105, 114-116, 118, 120, 124, 127, 238
*P. integerrimum*, 27, 115-117, 238
*Polystomoidella oblonga*, 116-117, 238
*Polystomoides*, 2, 116
*P. asiaticus*, 115
*Pomacea australis*, 163
*P. glaucus*, 163
*Pomatoschistus microps*, 173
*Postharmostomum helicis*, 163-164
*Posthodiplostomum cuticola*, 163
*Potamopyrgus*, 65
*P. antipodarum*, 172-173, 177, 180, 183

## Species Index  299

Praesagittifera naikaiensis, 71
Priacanthus, 15
Pricea multae, 124
Pristis microdon, 246
Proboscidactyla mutabilis, 146
Prochilodus argenteus, 175, 177
Prosorhynchus, 13
Prosthiostomum sp, 54
Proteocephalus, 6, 200-201, 209, 211
Proteocephalus spp, 199
P. cornuae, 199
P. filicollis, 211
P. longicolis, 199, 210
P. macrocephalus, 199, 210-211
P. neglectus, 205-207, 214-215
P. osculatus, 199, 201, 210-211
P. percae, 199, 211
P. torulosus, 199, 201, 206-207, 210-211, 215
Protogynella, 27
Protomicrocotyle sp, 111
Protopolystoma, 115
P. xenopodis, 115-119, 121, 124
Pseudaxine trachuri, 110
Pseudoceros bifurcatus, 60
P. canadensis, 64
Pseudocoelus japonicus, 170
Pseudodactylogyrus anguillae, 107
P. bini, 93, 136
Pseudodiorchis pistillum, 221
Pseudodiplorchis, 116
P. americanus, 93, 114, 115-117, 119, 121, 124
Pseudograffilla arenicola, 54
Pseudohaplogionaria macnaei, 81
P. unichaeta, 81
P. sutcliffei, 71
Pseudopolystoma, 115
P. dendriticum, 115
Pseudorhabdosynochus lantanensis, 15, 119, 122, 124
Pseudostylochus intermedius, 71
Pseudothoracocotyle gigantica, 103, 124
Ptychogoniomus megastoma, 141
Pucelis litoricola, 71
Pungitus pungitus, 135

## Q

Quadrivisio bengalensis, 32, 34

## R

Radix ovata, 9, 150
R. peregra, 9, 150

Raillietina echinobothrida, 37, 203-204, 212, 215, 224
Raja spp, 121
R. clevata, 119
Rajonchocotyle, 2, 123
R. emarginata, 114, 123
Rana clamitans, 138
R. rugosa, 115
R. temporaria, 115, 117
Renicola reoscovita, 177
Renylaima capensis, 142, 146, 223-224
Rhabomys pumilo, 41, 221
Rhamdia quelen, 104-105
Rhina ancylostoma, 245
Rhipidocotyle campanula, 182, 184
R. fennica, 179, 181-182, 184
Rhodomonas lens, 64
Rhynchoscolex simplex, 47
Ribeiroia, 186
R. ondatrae, 169, 175, 182
Rutilus rutilus, 111-112, 135, 203, 208, 211, 216, 232-233

## S

Sabussowia diocia, 66, 228
S. hastata, 228
S. macrostoma, 228
S. papillosa, 228
S. ronaldi, 228
S. verrucosa, 228
Saccocoelioides nanii, 175
Sagitta elegans, 170
S. setosa, 170
Salmon, 105
Salmo salar, 106, 109, 129, 133-134, 206, 208, 211
S. trutta, 104-105, 133-135
Salvelinus alpinus, 129, 133-135
S. fontinalis, 129, 133-135
S. namaycush, 133-134
S. umbla, 195
Sander canadiensis, 211
S. vitreus, 211
Sanguinicola, 141
S. inermis, 3, 141
Sardinops melanosticta, 211
Sargassum, 48
Scaphiopus couchii, 93, 114, 116-117, 119
Scarus, 18
Scarus sp, 105
S. rivulatus, 47, 54
Schistocephalus, 208, 212
S. solidus, 8, 26-27, 29 38, 212, 215-219, 224-226

## 300  Reproduction and Development in Platyhelminthes

*Schistosoma*, 29, 35, 38-39, 141, 161, 227
*Schistosoma* sp, 146
*Schistosoma* spp, 2, 29, 33, 38
*S. haematobium*, 2, 6, 157, 162, 228
*S. hippotami*, 228
*S. incognitum*, 228
*S. indicum*, 228
*S. intercalatum*, 228
*S. japonicum*, 29, 38, 142, 157, 162, 228
*S. malayensis*, 228
*S. mansoni*, 35-36, 38, 40, 42-44, 144, 156-157,
 160-165, 169, 175, 179-180, 182-183, 188-
 189, 227-229
*S. margrebowei*, 183
*S. mattheei*, 157
*S. mekongi*, 228
*Schistosomatium douthitti*, 143-144, 157, 162-
 163, 166, 228
*Schmidtea mediterranae*, 68, 71-75, 77-79, 81, 85,
 90, 92, 229, 235
*S. polychroa*, 47, 65-66, 92
*Scleroductus* sp, 105
*Scolex pleuronectis*, 214-216
*Scolopax rusticola*, 221
*Scomberomorus commersoni*, 103, 107-108, 110,
 124-125
*Scutogyrus longicornis*, 103
*Sebastes melanops*, 120
*Segmentina trochoides*, 180
*Seriola* spp, 113, 123
*S. dumerili*, 93, 102-103, 113, 119
*S. lalandi*, 93, 113, 119-121, 125-126
*S. quinqueradiata*, 93, 119, 120
*Solea solea*, 15, 34, 118-120, 123, 135
*Somateria*, 147
*S. mollissima*, 145
*Sparganum proliferum*, 220
*Spariocotyle chrysophrii*, 105
*Spelotrema*, 13-14
*Sphyranura*, 115
*S. oligorchis*, 115
*Spirometra*, 38, 200, 203-204
*Spirometra* sp, 205-206
*S. erinaceieuropaei*, 204
*S. mansoni*, 38
*S. mansonoides*, 194, 231
*S. theileri*, 231
*Spirorchis*, 141
*Spirorchis* sp, 174
*S. scripta*, 189
*Stagnicola elodes*, 177
*S. emerginata*, 168
*S. exilis*, 168

*Staphylepis cantaniana*, 221
*Stegastes nigricans*, 176
*Stegodexamiene anguillae*, 173, 239, 242
*Stenostomum*, 59
*S. grande*, 71, 81, 86
*S. leucops*, 81, 86
*S. tenuicauda*, 29, 81, 86
*S. unicolor*, 81, 86
*Stenotherus odoratus*, 93, 115
*Stephanostomum*, 139
*Sterna paradisea*, 203
*Stichocotyle nephrops*, 6
*Stichorchis subtriquetrus*, 6, 143-144, 159
*Sticopus californicus*, 17
*Strigea*, 141
*S. subquetrus*, 158
*Strongyloides* spp, 155
*Styliorhinus canalicula*, 95, 245
*Stylochus ellipticus*, 15, 26-27, 33, 61-64
*S. frontalis*, 47
*S. pilidium*, 15
*S. uniporus*, 64
*Stylostomum sanjuania*, 64
*Succinea*, 170, 223
*Syncoelium*, 112
*Syncoelium* sp, 170
*Syndesimus antillarum*, 47, 54
*S. dendrostomum*, 54
*S. echinorum*, 54
*S. franciscana*, 54
*Synodus*, 18

## T

*Taenia*, 2, 33, 212
*Taenia* spp, 216
*T. crassiceps*, 220-221
*T. endothoracicus*, 220-221
*T. multiceps*, 220-222
*T. parva*, 220-221
*T. pisiformis*, 220-221
*T. saginata*, 2, 6, 27
*T. selousi*, 220-221
*T. serialis*, 220-221
*T. solium*, 13, 27, 37, 194, 203
*T. twitchelli*, 220-221
*Takifugu rubripes*, 136
*Talpa*, 221
*Tenebrio molitor*, 206
*Terapon puta*, 101, 109-110
*Tetramorium caespitum*, 204, 212
*Thymallus thymallus*, 129, 133
*Thysanozoon brocchii*, 70-71

## Species Index    301

*Tigriopus californicus*, 17, 209, 224
*Tilapia guineensis*, 101-103, 109-110
*T. zillii*, 102-103, 109-110, 112
*Toxacara canis*, 155
*Trachurus trachurus*, 101-103, 110, 214
*Transversotrema patialense*, 27, 168-169, 174, 176, 188-189
*Triaenophorus*, 201
*T. crassus*, 195, 201-202, 205-207, 225-226
*T. nodosus*, 195
*Triakis scyllium*, 245
*T. semifasciata*, 209
*Tribolium confusum*, 214-216
*Trichobilharzia franki*, 228
*T. ocellata*, 156, 162, 168-169, 178-180, 182-183, 187-188
*T. regent*, 228
*T. szidati*, 228
*Trichodina carassii*, 113
*Tristoma*, 13-14
*Tristriata anatis*, 147
*Tubifex*, 65, 201, 213, 214-215
*Turtonia minuta*, 141, 145
*Tylocephalum*, 6
*Typhloplana*, 47
*Typlocoelum cybium*, 158
*Tyturus*, 24

## U

*Udonella caligorum*, 114
*Uncinania lucasi*, 155
*Urocleidoides mastigatus*, 105
*Urocleidus ferox*, 110
*Utricularia* sp, 163

## V

*Vallisia* sp, 111
*Vimba vimba tenelli*, 108, 110

## W

*Wedlia retrovitalis*, 228
*W. submaxillinis*, 228

## X

*Xanthichthys*, 18
*Xenopus laevis*, 116-119

## Z

*Zeacumantus subcarinatus*, 141, 154, 174-179, 190-193
*Zeuxapta seriolae*, 93, 113, 118-123, 125-126
*Zygocotyle lunata*, 174

# Subject Index

## A

Acidic climate, 205
Addition, 76, 159, 180-182, 212, 235
Aestivation, 116, 239
Amputation, 42, 68-69, 73, 77, 80-82, 84
Antagonism, 111
Architomy, 81, 83-84, 89
Attractive strategy, 169
Autotomy, 83

## B

B chromosomes, 66, 85, 227
Blastema, 41-42, 45, 68-71, 75-80, 230
Branching sporocysts, 181

## C

Castration, 177-182
Cell-cell signaling, 229
Chemical cue, 15, 18, 123, 172
Chromatoid body, 41, 79
Cladogram, 5, 7, 48-49
Clonal mixing, 152, 154, 246
Competitive exclusion, 184
Condition factor, 101, 103, 127, 177, 186
Counting mechanism, 75
Crowding effect, 56-57, 106, 215

## D

Decoy mechanism, 163
Diseases,
    Fasciolosis, 3-4
    Schistosomiasis, 2-4, 12
    Sparganosis, 204
Dwarf male, 67

## E

Egg,
    Dormant, 47, 55-57
    Subitaneous, 47, 55-57

Ectolecithol, 1, 6
Endolecithol, 6
Endemism, 50
Endomitosis, 65
Epimorphosis, 68-69, 72, 78
Eutelic, 130

## F

Fecundity,
    Batch, 26, 53, 56, 61-62
    Lifetime, 26, 120
    Relative, 26-27, 62-63, 132, 135, 219
Feeding platform, 54
Fissiparous genes, 84

## G

Generation time, 29-30, 56, 61, 129, 131-132
Gigantism, 178-181, 183-184, 193
Grafting, 41, 81-83

## H

Host specificity,
    Ecological, 126
    Euryxenic, 211
    Geographical, 126
    Oioxenic, 130, 246
    Phylogenetic, 126
    Stenoxenic, 15
Host switching, 130, 135, 147

## I

Intracellular digestion, 37-38, 205
Incubation, 63-64, 67, 113, 121-124, 160, 162-163

## L

Lables,
    BrdU pulse chase, 40
    $_3$H thyrimidine, 76, 217

Fluorescent, 59, 91
Green Fluorescent Protein gene, 59
Latitude, 25, 101, 104, 109, 227
Life span, 27-30, 61, 75, 90-92, 103, 117, 123-124, 129, 132-133, 139, 142-143, 154, 159, 163-164, 169, 178-179, 183, 186, 189, 191-193, 199, 213, 217

## M

Markers,
Cytoplasmic, 41
Microsatellite, 132
Mitochondrial, 195
Nuclear, 195
Microtriches, 37
Morphallaxis, 68-71
Muller's ratchet, 80

## N

Na+ mediated transport, 37
Neoteny, 212, 237-238
Neuropeptides, 40, 45-46, 72, 180, 229-230
Neurotransmitters, 72

## O

Ontogenetic pathways, 198, 202, 213
Ovoviviparity, 116
Viviparity, 131-132, 238
Hyperviviparity, 131-132, 238
Oyster leech, 47

## P

Parasitic,
sink, 212
tag, 11-12
Paratenics, 154, 209
Paratomy, 81, 83-84
Parthenogensis,
Apomictic, 64, 143, 157-158, 204
Automictic, 65

Peptide hormones, 229, 231
Penis fencing, 60
Polarity, 78, 81-83
Polyembryony, 155, 158
Poikilothermy, 25, 234
Premeiotic doubling, 65
Progenesis, 131-132, 171-173, 212-213, 234, 237-240

## R

Ramets, 80-84
Response,
Proximate, 190, 226
Ultimate, 190, 193, 226
Respiratory pigments, 39, 55
Recruitment, 26, 33-34, 55, 63-64, 177, 190
Reproductive life span, 27, 29, 61

## S

Senescence, 1, 3, 74, 90-92
Sperm,
Allosperm, 60, 65
Autosperm, 60-61
Spermatophore, 60
Susceptibility, 67, 101, 104, 110, 133-134, 136, 165, 169, 171, 173, 176
Substitution, 143, 180-181
Stunting, 180-181, 183-184, 193

## T

Transgenic line, 59
Trophic transmission, 16, 138, 167, 195, 247
Tegument, 35-37, 136, 138, 142
Transmission efficiency, 17, 169

## V

Venereal infection, 117

# Author's Biography

Recipient of the S.S. Bhatnagar Prize, the highest Indian award for scientists, one of the ten National Professorships, T.J. Pandian has served as editor/member of editorial boards of many international journals. His books on Animal Energetics (Academic Press) identify him as a prolific but precise writer. His five volumes on Sexuality, Sex Determination and Differentiation in Fishes, published by CRC Press, are ranked with five stars. He is presently authoring a multi-volume series on Reproduction and Development of Aquatic Invertebrates, of which the volumes on Crustacea, Mollusca, Echinodermata and Prochordata, and Annelida have already been published. The next one, Platyhelminthes, is in your hands and Minor Phyla is being prepared.